D1201748

The Surface Chemistry of Natural Particles

The Surface Chemistry of Natural Particles

Garrison Sposito

2004

OXFORD
UNIVERSITY PRESS

Oxford New York
Auckland Bangkok Buenos Aires Cape Town Chennai
Dar es Salaam Delhi Hong Kong Istanbul Karachi Kolkata
Kuala Lumpur Madrid Melbourne Mexico City Mumbai Nairobi
São Paulo Shanghai Taipei Tokyo Toronto

Published by Oxford University Press, Inc.
198 Madison Avenue, New York, New York, 10016

www.oup.com

Library of Congress Cataloging-in-Publication Data
Sposito, Garrison, 1939–
The surface chemistry of natural particles / Garrison Sposito.
p. cm.
Includes bibliographical references.
ISBN 0-19-511780-8
1. Sediments—Surfaces. 2. Adsorption. 3. Solid-liquid interfaces.
4. Surface chemistry. I. Title.
QE471.2 .S697 2003
541.3′3—dc21 2003000331

9 8 7 6 5 4 3 2 1

Printed in the United States of America
on acid-free paper

For Doug, Dina, Frank,
Jennifer, Sara, and,
of course, Cristina

ὅ τι καλὸν φίλον ἀεί

—Euripedes, *Bacchæ*

Preface

Twenty years have passed since the ink began to dry on the last manuscript pages of *The Surface Chemistry of Soils*, a monograph which developed its basic leitmotivs from classic paradigms in analytical and physical chemistry, those insouciant disciplinary muses of the modern progenitors of natural particle surface chemistry, Paul Schindler and Werner Stumm. The modus operandi in the monograph was revisionist history: a reinterpretation of well known macroscopic surface chemical phenomena—ion adsorption, electrokinetic motion, and colloidal flocculation—from the heuristic point of view offered by concepts in coordination chemistry. On this approach, one hoped, the milieu of understanding could be transformed, from the noisy antediluvian marketplace of competing (albeit inadequate) theories, each clamoring for validation within an ambiguous phenomenology, to an organized modern workshop hosting a pluralistic methodology guided by a unique molecular perspective. (One thinks here of the wry comparison made between England and France by Voltaire.)

This book is intended to give an account of the extraordinary development of experiment and theory in natural particle surface chemistry over the two decades following publication of *The Surface Chemistry of Soils*. Features of this multifaceted development have been traced from time to time in symposium publications and in critical reviews, but not in an advanced textbook. The purpose of *The Surface Chemistry of Natural Particles* is to try to fill this lacuna with a broadly based discussion of molecular spectroscopy, kinetics, and equilibria as they apply to natural particle surface reactions in aqueous media, with emphasis on insights gained over the past few years. This discussion, divided among five chapters, is complemented by lengthy annotations, augmented by propitious reading suggestions, and punctuated by substantial end-of-chapter problem sets that require a critical reading of important technical journal articles. The five problem sets, each listed under the epigram Research Matters, involve careful analysis and close attention to detail, the distinguishing characteristics of any good scientific investigation.

Chapter 1 is a review of qualitative notions about ion adsorption by natural particles in contact with aqueous electrolyte solutions. It begins with an operational definition of adsorption and a brief description of its phenomenology, then moves to the molecular picture underlying adsorbate/adsorbent structure as illuminated by the Pauling rules, all of this being a prelude to understanding the origins of particle surface charge and its dependence on pH. The chapter ends with a discussion of the useful macroscopic synopsis of ion adsorption behavior provided by the Schindler diagram.

Chapter 2 gives an introduction to the strategies and scientific underpinnings of four spectroscopic approaches that have proved to be especially effective for detecting adsorbate species on natural particles: X-ray and infrared absorption spectroscopy, electron spin resonance spectroscopy, and nuclear magnetic resonance spectroscopy. Examples are presented that illustrate the degree to which each approach meets the criteria of selectivity, sensitivity, and noninvasiveness that are essential to the study of adsorbate molecular structure in the presence of water. Chapters 3 and 4 involve detailed consideration of current models of surface chemical kinetics and equilibria as applied to natural particles. The purview of these chapters is necessarily broad, ranging from ion adsorption to surface oxidation-reduction to mineral dissolution, and from specimen minerals to heterogeneous organic adsorbents. Chapter 5 describes natural particle colloidal phenomena as revealed by photon scattering spectroscopy and elucidated by mass-fractal concepts. This chapter is intended as an introduction to an emerging subdiscipline of natural particle surface chemistry whose importance in environmental contexts is expanding rapidly.

I thank Professor James J. Morgan for continual encouragement, wisdom, and support in ways too numerous and subtle to recount here. I am grateful also to Professor Domenico Grasso and to Rebecca Sutton for their comments on part of the manuscript for this book while in an early draft form. I owe much to Angela Zabel for her careful preparation of the typescript and to Cynthia Borcena for her equally exquisite rendering of the figures that adorn the chapters of this book. Responsibility for any errors is, of course, my own.

Contents

1 Ions at the Particle–Aqueous Solution Interface, 3

1.1 Adsorption 3
1.2 Adsorbate Structure 11
1.3 Surface Charge 20
1.4 Points of Zero Charge 25
1.5 Ion Adsorption Trends 33
Notes 37
For Further Reading 40
Research Matters 40

2 The Spectroscopic Detection of Surface Species, 43

2.1 Strategy 43
2.2 X-ray Absorption Spectroscopy 49
2.3 Infrared Spectroscopy 55
2.4 Electron Spin Resonance Spectroscopy 60
2.5 Nuclear Magnetic Resonance Spectroscopy 65
Notes 71
For Further Reading 76
Research Matters 77

3 Surface Chemical Kinetics, 79

3.1 Phenomenology 79
3.2 Specific Adsorption Reactions 85

3.3 Surface Oxidation-Reduction Reactions 94
3.4 Proton-Promoted Mineral Dissolution Reactions 101
3.5 Ligand-Promoted Mineral Dissolution Reactions 106
Notes 115
For Further Reading 123
Research Matters 123

4 Modeling Ion Adsorption, 126

4.1 Modeling the Diffuse Ion Swarm 126
4.2 Modeling Surface Complexes 140
4.3 Modeling Temperature Effects 148
4.4 Modeling Affinity 157
4.5 Modeling Natural Particle Adsorption Reactions 163
Notes 170
For Further Reading 180
Research Matters 180

5 Colloidal Phenomena, 182

5.1 Probing Particle Structure 182
5.2 Particle Formation Kinetics 193
5.3 The Stability Ratio 202
5.4 The von Smoluchowski Rate Law 211
5.5 Solving the von Smoluchowski Equation 219
Notes 226
For Further Reading 235
Research Matters 235

Index 239

The Surface Chemistry of Natural Particles

1

Ions at the Particle–Aqueous Solution Interface

1.1 Adsorption

Adsorption is the process through which a chemical substance accumulates at the common boundary of two contiguous phases. If the reaction produces enrichment of the substance in an interfacial layer, the process is termed *positive adsorption*. If instead a depletion of the substance is produced, the process is termed *negative adsorption*. If one of the contiguous phases involved is solid and the other fluid, the solid phase is termed the *adsorbent* and the matter which accumulates at its surface is an *adsorbate*. A chemical species in the fluid phase that potentially can be adsorbed is termed an *adsorptive*. If the adsorbate is immobilized on the adsorbent surface over a time scale that is long, say when compared to that for diffusive motions of the adsorptive, the adsorbate, together with the site on the adsorbent surface to which it is bound, are termed an *adsorption complex*.[1]

Adsorption experiments typically are performed in a sequence of three essential steps: (1) *reaction* of an adsorptive with an adsorbent contacting a fluid phase of known composition under controlled temperature and applied pressure for a prescribed period of time; (2) *separation* of the adsorbent from the fluid phase after reaction; and (3) *quantitation* of the chemical substance undergoing adsorption, both in the supernatant fluid phase and in the separated adsorbent slurry that includes any entrained fluid phase. The reaction step can be performed in either a closed system (batch reactor) or an open system (flowthrough reactor), and can proceed over a time period that is either quite short (adsorption kinetics) or very long (adsorption equilibration) as compared with the natural timescale for achieving a steady composition in the reacting fluid phase.[2] The separation step is similarly open to choice, with centrifugation, filtration, or gravitational settling being convenient methods to achieve it. The quantitation step, in principle, should be designed not only to determine the moles of adsorbate and unreacted adsorptive, but also to verify whether unwanted side-reactions, such as precipitation of the adsorptive or dissolution of the adsorbent, have influenced the adsorption experiment.[3]

The quantitative description of adsorption reactions is based on the concept of *surface excess*,[1] originally developed by Willard Gibbs.[4] In the study of adsorption by natural particles, a convenient mathematical formulation of this concept is

$$n_i^{(j)} = n_i - \left(n_j x_i / x_j\right) \tag{1.1}$$

where $n_i^{(j)}$ is the specific surface excess of substance i *relative to substance* j; n_i or n_j is the number of moles of substance i or j in the separated adsorbent slurry per unit mass of dry adsorbent; and x_i or x_j is the mole fraction of substance i or j in the supernatant fluid phase. (The *mole fraction* of a substance in any mixture is the ratio of moles of the substance in the mixture to the total moles of all substances in the mixture.[5]) This definition of *relative* surface excess assumes no entry of substance i or j into the bulk interior of the adsorbent.[1] Formally, $n_i^{(j)}$ is the surface excess of substance i, per unit mass of adsorbent, relative to the partitioning of some reference substance j between the separated slurry and the supernatant fluid phase.[1] The role of the reference substance j becomes evident on noting that, by eq. 1.1,

$$n_j^{(j)} = n_j - \left(n_j x_j / x_j\right) \equiv 0 \tag{1.2}$$

By definition, there is no relative surface excess of j. Thus $n_i^{(j)}$ is a surface excess of i referred to an interface at which no net accumulation of j occurs.[1] There is no adsorption of the substance i when the equipartitioning condition,

$$\frac{n_i}{n_j} = \frac{x_i}{x_j} \tag{1.3}$$

is met. Otherwise, the value found for $n_i^{(j)}$ can be either positive or negative, because eq. 1.1 describes a net interfacial accumulation of the substance i relative to that of substance j.

The conventional choice for the reference substance j in eq. 1.1 is the solvent in the fluid phase which contacts a solid adsorbent. Therefore, in the case of a natural particle adsorbing a solute from an aqueous solution, eq. 1.1 can be reduced to the expression:

$$n_i^{(w)} = n_i - \theta_w m_i \tag{1.4}$$

where θ_w is the mass of water in the adsorbent slurry per unit mass of adsorbent particles (the gravimetric water content of the slurry) and m_i is the molality[5] of the adsorbing substance i in the aqueous solution isolated after reaction. The molality m_i often can be represented accurately by c_i, the molar concentration[5] of i in the supernatant aqueous solution phase. Equation 1.4 represents the specific surface excess of substance i relative to a reference aqueous solution that contains θ_w kilograms of water plus substance i at the molality m_i. This excess is thus assigned to a surface at which there is no net accumulation of water. If water in the interstitial space of the slurry is not adsorbed, then this surface can be taken to be congruent with the geometric boundaries of the adsorbing particles.[4] If some of the interstitial water is adsorbed, say, within a region of width 1.0 nm at the boundary of a particle,[6] then the surface of zero net accumulation of water would differ slightly from the geometric particle surface.

If the reactant fluid phase is air, the reference substance could be nitrogen gas. In the case of a soil adsorbing a vapor from air, it is expected that the number of moles of nitrogen gas in the interstitial space will be negligible because of the very low density of air. For example, a cubic centimeter of soil with a porosity of 0.5 contains only about 1.5 μmol nitrogen gas, but could contain 10^4 times as many moles of adsorbed water vapor. Thus n_j in eq. 1.1 can be safely neglected, and the relative surface excess can be expressed accurately by the equation:

$$n_i^{(N_2)} = n_i \tag{1.5}$$

for any gaseous substance i adsorbed strongly from air by natural particles. Even if nitrogen gas itself is adsorbed by a soil (e.g., to determine its specific surface area), then only that gas should be present in the soil interstitial space and eq. 1.5 is still the correct expression with which to calculate the amount adsorbed, since the number of moles of nonadsorbed gas in the soil interstitial space will again be negligibly small.

As a numerical example of eq. 1.4,[7] consider a montmorillonitic soil containing calcite ($CaCO_3$) and gypsum ($CaSO_4 \cdot 2H_2O$) that is first equilibrated with a mixed $NaCl/CaCl_2$ solution, followed by centrifugation to separate a supernatant solution from a soil slurry whose gravimetric water content is 0.562 kg kg^{-1}. Quantitation of the electrolyte composition in the slurry and the supernatant solution yields the data set:

$n_{Na} = 13.20$ mmol kg^{-1}	$c_{Na} = 12.67$ mmol L^{-1}
$n_{Ca} = 79.25$ mmol kg^{-1}	$c_{Ca} = 7.28$ mmol L^{-1}
$n_{Cl} = 13.90$ mmol kg^{-1}	$c_{Cl} = 25.03$ mmol L^{-1}
$n_{HCO_3} = 25.10$ mmol kg^{-1}	$c_{HCO_3} = 0.27$ mmol L^{-1}
$n_{SO_4} = 5.00$ mmol kg^{-1}	$c_{SO_4} = 0.98$ mmol L^{-1}

It is likely that the presence of bicarbonate and sulfate, and, therefore, a portion of the calcium present, can be attributed wholly to adsorbent dissolution. Charge-balance considerations then would reduce n_{Ca} and c_{Ca} accordingly to the expressions:

$$n'_{Ca} \equiv n_{Ca} - \tfrac{1}{2} n_{HCO_3} - n_{SO_4} = 61.70 \text{ mmol kg}^{-1}$$
$$c'_{Ca} \equiv c_{Ca} - \tfrac{1}{2} c_{HCO_3} - c_{SO_4} = 6.17 \text{ mmol L}^{-1}$$

By eq. 1.4, ignoring the minute difference between molar and molal concentrations, the specific surface excesses of Na, Ca, and Cl are:

$$n_{\rm Na}^{(w)} = 13.20 - (0.562)\ 12.67 = +6.07 \text{ mmol kg}^{-1}$$
$$n_{\rm Ca}^{(w)} = 61.70 - (0.562)\ 6.17 = +58.23 \text{ mmol kg}^{-1}$$
$$n_{\rm Cl}^{(w)} = 13.90 - (0.562)\ 25.03 = -0.17 \text{ mmol kg}^{-1}$$

Both cations are positively adsorbed by the soil under the conditions of measurement, while chloride is negatively adsorbed, consistent with the montmorillonitic character of the soil.[7]

Adsorption data of the type just illustrated, when obtained at fixed temperature and applied pressure, often are investigated in response to imposed changes in pH or in the concentration of the adsorptive in the supernatant aqueous solution. A graph of $n_i^{(w)}$ versus pH which displays an S-curve shape is termed an *adsorption edge* (fig. 1.1), whereas a graph that displays the mirror image of an S-curve is termed an *adsorption envelope*,[8] although significant deviations from these two generic groups often are observed (fig. 1.2). A graph of $n_i^{(w)}$ versus c_i (or m_i) is termed a *surface excess isotherm* or, less formally, an *adsorption isotherm*.[1] Isotherm data sets also can be classified into generic groups based on the distinctive shapes of their graphs (fig. 1.3).[8,9] A variety of model-based equations to represent these generic shapes is available and can be applied to parameterize adsorption data sets in a way that facilitates comparison among different adsorbents or the discovery of trends with respect to varying composition properties (pH, background electrolyte, competing adsorptives, etc.).[10]

Studies of the pH-dependence of $n_i^{(w)}$ usually are performed under a condition of fixed total moles of substance i, with the dependent variable then more conveniently taken as the mole fraction of i found in the adsorbate form. A model-based equation to represent generic adsorption edges or envelopes in parametric

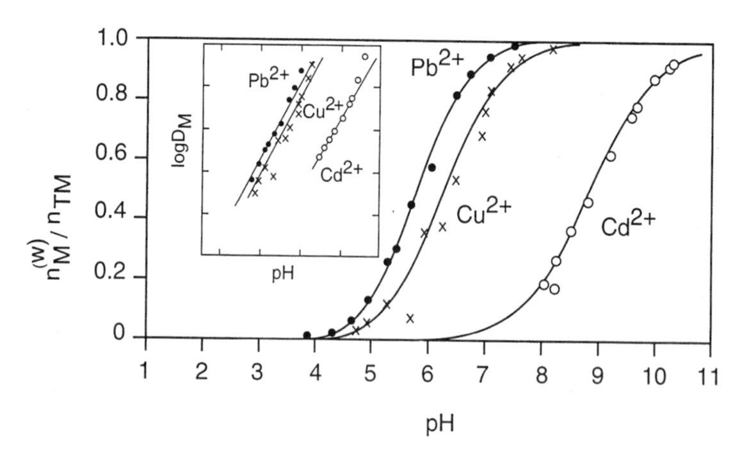

Fig. 1.1 Adsorption edges for three trace metal cations (M = Pb^{2+}, Cu^{2+}, and Cd^{2+}) reacted with amorphous $Fe(OH)_3$ (n_{TM} = 0.33 mol kg^{-1}, c_s = 0.06 kg m^{-3}). The inset graph shows Kurbatov plots of these data, with log D replacing ln D in eq. 1.7. Data from Z.-J. Wang and W. Stumm, *Neth. J. Agric. Sci.* **35**:231 (1987).

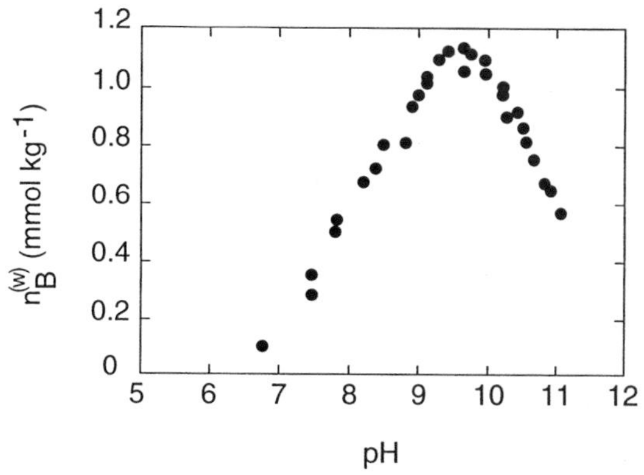

Fig. 1.2 Adsorption envelope for borate reacted with an Entisol soil ($n_{TB} = 2.31$ mmol kg^{-1}, $c_s = 200$ kg m^{-3}). The portion of the graph for pH < 9.5 resembles the adsorption edge described by eq. 1.12 for "metal-like" adsorption, whereas for pH > 9.5 it resembles the adsorption envelope described by eq. 1.12 for "ligand-like" adsorption. Data from S. Goldberg and R. A. Glaubig, *Soil Sci. Soc. Am. J.* **50**:1173 (1986).

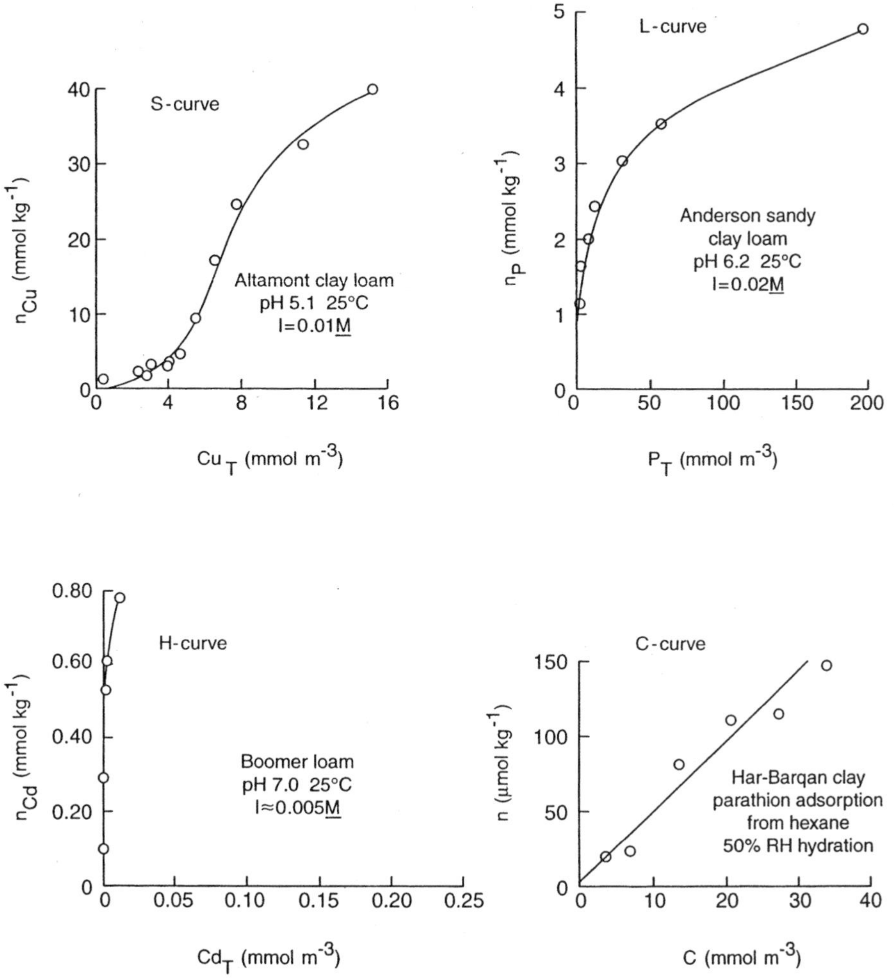

Fig. 1.3 The four generic groups of adsorption isotherm, based on the classification scheme proposed by Giles and coworkers.[8,9] After figure 4.1 in G. Sposito, *The Surface Chemistry of Soils*, Oxford University Press, New York, 1984.

form can be applied if a graph of the logarithm of the ratio of adsorbate-to-supernatant-solution mole fraction of i,

$$D_i \equiv \frac{x_{\text{iads}}}{x_{\text{isoln}}} = \frac{n_i^{(w)} c_s}{m_i} = c_s\, K_{di} \tag{1.6}$$

versus pH is linear over a sufficiently broad range of pH values:

$$\ln D_i = a + b \text{ pH} \tag{1.7}$$

where a, b are empirical parameters, c_s is the adsorbent solids concentration in the reacting suspension (kg m^{-3}, equal to $1000/\theta_w$), and K_{di} (m^3 kg^{-1}) is the *distribution coefficient* for substance i,[8]

$$K_{di} \equiv \frac{n_i^{(w)}}{m_i} \tag{1.8}$$

The parameter D_i is the *distribution ratio* for substance i, and a graph of ln D against pH according to eq. 1.7 is a *Kurbatov plot*,[11,12] as illustrated in fig. 1.1 (inset).

A geometric interpretation of the empirical parameters a, b in eq. 1.7 can be made as follows. The pH value at which half of the moles of substance i added are in an adsorbed chemical form is designated pH_{50}. Since $D_i = 1$ at this pH value according to eq. 1.6, it follows from eq. 1.7 that

$$\text{pH}_{50} = \left|\frac{a}{b}\right| \tag{1.9}$$

Moreover, after substituting $x_{\text{iads}}/(1 - x_{\text{iads}})$ for D_i in eq. 1.7, one can derive the result:

$$\left(\frac{dx_{\text{iads}}}{d\text{pH}}\right)_{\text{pH}=\text{pH}_{50}} = \frac{b}{4} \tag{1.10}$$

Therefore, eq. 1.7 can be rewritten in the form:

$$\ln D_i = b(\text{pH} - \text{pH}_{50}) \tag{1.11}$$

or, as a model equation for $n_i^{(w)}$,

$$n_i^{(w)} = n_{\text{Ti}}\{1 + \exp[-b(\text{pH} - \text{pH}_{50})]\}^{-1} \tag{1.12}$$

where n_{Ti} is the (fixed) total moles of i in the system per unit mass of adsorbent and b is expressed in eq. 1.10.[11]

The slope parameter b is thus seen to provide a quantitative measure of the steepness of an adsorption edge or envelope at pH_{50}. These characteristics of a plot of $n_i^{(w)}$ against pH usually cannot be attributed to any one feature of the adsorption reaction, but the sign of b can be used to classify the plot as either "metal-like" ($b > 0$) or "ligand-like" ($b < 0$), in the spirit of figs. 1.1 and 1.2.[13] Under the condition that n_{Ti} is very small when compared to the maximum possible surface excess for substance i, the absolute value of b can sometimes be related to the stoichiometry of the adsorption reaction.[12] The values of pH_{50} for a series of substances reacted with the same adsorbent solid under identical chemical conditions provide a relative measure of the adsorptive selectivity of the solid phase, with small pH_{50} values implying an evident high selectivity in the case of "metal-like" pH-dependence, and the inverse for "ligand-like" pH-dependence.[11–13]

Adsorption isotherm data also can be rationalized empirically through the behavior of the distribution coefficient (K_d, now dropping the subscript i) as the value of the surface excess increases. The C-curve in fig. 1.3, for example, corresponds to a distribution coefficient that is independent of the surface excess, whereas the S-curve corresponds to one that increases initially as a power of the surface excess. The L- and H-curve isotherms, by contrast, correspond to a distribution coefficient that decreases with increasing surface excess, the rate of decrease being perforce larger in the case of the H-curve.[8] More generally, it is observed that a graph of K_d against the surface excess is a decreasing function convex to the x-axis (fig. 1.4). If the value of K_d tends to a finite constant as the surface excess tends to zero, and if K_d extrapolates to zero at some finite value of the surface excess, then the corresponding adsorption isotherm data can *always* be fit to a two-term series of the form:

$$n_i^{(w)} = \frac{b_1 K_1 c_i}{1 + K_1 c_i} + \frac{b_2 K_2 c_i}{1 + K_2 c_i} \tag{1.13}$$

where b_1, b_2, K_1, and K_2 are empirical parameters and c_i is an aqueous-phase concentration.

Equation 1.13 can be derived rigorously by making use of the mathematical properties of Stieltjes transforms,[14] but its correctness as a universal approximation can be appreciated along simpler lines of reasoning after using eq. 1.8 to substitute for c_i (or m_i) in terms of K_{di} and $n_i^{(w)}$ in order to generate the second-degree algebraic equation (again dropping the subscript i and superscript w):

$$K_d^2 + (K_1 + K_2)K_d n + K_1 K_2 n^2 - (b_1 K_1 + b_2 K_2)K_d - bK_1 K_2 n = 0 \tag{1.14}$$

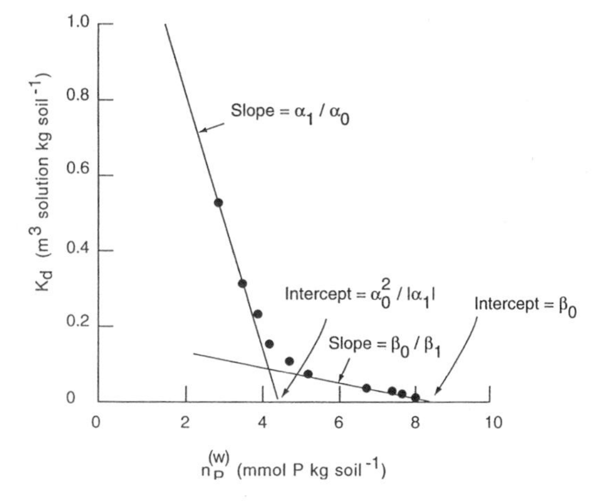

Fig. 1.4 Graph of K_d against the surface excess of phosphate adsorbed by a clay loam soil. The lines through the data points correspond to eqs. 1.16 and 1.19. After fig. 4.2 in G. Sposito, *The Surface Chemistry of Soils*, Oxford University Press, New York, 1984.

where $b \equiv b_1 + b_2$ is not to be confused with b in eq. 1.7! The derivative of K_d with respect to n follows from eq. 1.14:

$$\frac{dK_d}{dn} = -\frac{(K_1 + K_2)K_d + 2K_1K_2n - bK_1K_2}{2K_d + (K_1 + K_2)n - (b_1K_1 + b_2K_2)} \tag{1.15}$$

As n tends to zero, eqs. 1.14 and 1.15 can be combined to show that

$$K_d \approx \alpha_0 + (\alpha_1/\alpha_0)n \qquad (n \downarrow 0) \tag{1.16}$$

where

$$\alpha_0 = b_1K_1 + b_2K_2 \tag{1.17}$$

$$\alpha_1 = -\left(b_1K_1^2 + b_2K_2^2\right) \tag{1.18}$$

Thus, the distribution coefficient is linear in n near the origin, with a slope equal to $\alpha_1/\alpha_0 < 0$. According to eq. 1.16, the x-intercept of the linear expression is $\alpha_0^2/|\alpha_1|$, as indicated in fig. 1.4. On the other hand, when K_d is zero, $n = b$ according to eq. 1.14. Equation 1.15 can then be used to demonstrate that

$$K_d \approx \left(\beta_0^2/|\beta_1|\right) + (\beta_0/\beta_1)n \qquad (n \uparrow b) \tag{1.19}$$

where

$$\beta_0 = b = b_1 + b_2 \tag{1.20}$$

$$\beta_1 = -\frac{b_1}{K_1} - \frac{b_2}{K_2} \tag{1.21}$$

Thus the distribution coefficient again becomes linear in n as it tends to b, which is its maximum value according to eq. 1.13. The slope of the line is $\beta_0/\beta_1 < 0$, and its x-intercept is, of course, b. If adsorption data are plotted as in fig. 1.4, then the limiting slopes and the two x-intercepts can be determined graphically. The values found can be substituted into eqs. 1.17, 1.18, 1.20, and 1.21 to solve *uniquely* for the empirical parameters b_1, K_1, b_2, and K_2[14]. These parameters, however, like those in eq. 1.7, have no particular *chemical* significance in terms of adsorption reactions. They are, on the other hand, convenient for comparing among adsorption data sets and for solute transport modeling in terms of purely mathematical representations of adsorption.[15]

The lack of mechanistic significance in adsorption isotherm equations has long been known,[16] and is nicely illustrated by a data-fitting exercise discussed by Kinniburgh et al.,[17] who compared eq. 1.13 with a very different model adsorption isotherm equation (Tóth isotherm) using their own data on the reaction of Zn^{2+} with the poorly crystalline hydrous iron oxide, ferrihydrite $[(Fe_2O_3 \cdot 2H_2O)_n \quad n \approx 5]$. Both model equations fit the data quite well ($r^2 > 0.998$, with very small standard deviations of the residuals), despite having completely different interpretations in terms of "site affinities" for adsorption (two discrete sites versus a continuum of "site affinities" with strong "negative skew"—see Problem 5 at the end of this chapter).[17]

No unambiguous mechanistic interpretation of these and other adsorption isotherm models can be had on the basis of goodness-of-fit criteria alone, a conclusion that extends even to the issue of determining whether an adsorption reaction has ever occurred, as opposed to a precipitation reaction.[16,18] Not only do the data sets for this latter reaction yield plots similar to that in fig. 1.4 under a broad variety of experimental settings, but they also are often consistent with undersaturation conditions because of coprecipitation phenomena,[18,19] making identification of the reaction mechanism even more problematic. When no molecular-scale data on which to base a decision as to mechanism are available, the prima facie loss of an adsorptive from aqueous solution to the solid phase can be termed *sorption* in order to avoid the implication that either adsorption or precipitation has taken place.[20] As a general rule, a surface precipitation mechanism is favored by high adsorptive concentrations and long reaction times in sorption processes.[21]

1.2 Adsorbate Structure

The reactions between adsorptive ions and natural particles can be portrayed as a web of sorption processes mediated by two parameters, timescale and surface coverage.[20–22] Surface complexes (i.e., adsorption complexes[1]) are the expected products of these reactions when timescales are sufficiently short and surface coverage is sufficiently low, "sufficiently" always being defined operationally in terms of conditions attendant to the sorption process (fig. 1.5). For example, timescales on the order of hours and a surface excess below 5.4 mmol kg^{-1} are "sufficiently small" to allow surface complexation to be the dominant mode of adsorption for Co^{2+} on kaolinite particles suspended in $NaNO_3$ solution at pH 8.[23] For Cd^{2+} reacting at 25°C with goethite particles suspended in $NaClO_4$ solution at pH 7.5,[24] the timescale that is "sufficiently short" is also hours, but the "sufficiently low" surface excess can be as high as 130 mmol kg^{-1}.

As time scales are lengthened and surface coverage increases, or as chemical conditions are altered (e.g., pH changes) for a fixed reaction time, adsorbate "islands" comprising a small number of adsorptive ions bound closely together

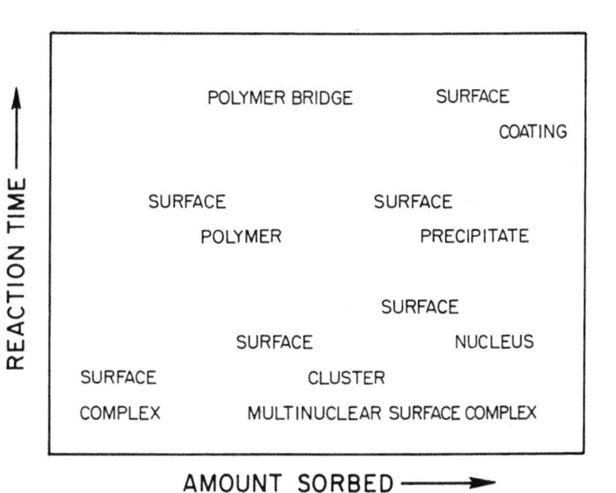

Fig. 1.5 Regions of stability, in the two-dimensional field defined by the quantity of sorbate and the sorption reaction time, for the products of sorption reactions with natural particles.[22]

will form. These reaction products are termed *multinuclear surface complexes* (fig. 1.5) by analogy with their counterpart in aqueous solution chemistry. They are more likely for adsorptive ions that readily form polymeric structures in aqueous solution.[24] Multinuclear surface complexes may in turn grow with time to become colloidal structures that are precursors of surface polymers or, if they are well organized on a three-dimensional lattice, of surface precipitates.[20,25] Figure 1.6 illustrates the growth process from surface complex, to surface nucleus, to surface precipitate for Cr^{3+} sorbed by poorly crystalline goethite at pH 4.[26]

Adsorption complexes can be classified as *outer-sphere surface complexes* or *inner-sphere surface complexes*.[27] By analogy with its counterpart in aqueous solution, an outer-sphere surface complex has at least one water molecule interposed between the adsorbate species and the adsorbent site to which it is bound to form an adsorption complex. Thus, an outer-sphere surface complex contains a solvated adsorbate species. An inner-sphere surface complex, by contrast, has no water molecule interposed between the adsorbate species and the adsorbent site that binds it. Therefore, the formation of the complex involves a desolvated adsorbate species, although the latter may be partially solvated by water molecules that do not intervene in the bond to the adsorbent site.

Figure 1.7 illustrates the structure of the outer-sphere surface complex formed between Na^+ and a surface site on the basal plane of the clay mineral, montmorillonite (see the *Clay Minerals* inset).[28] At about 50% relative humidity, this layer-type aluminosilicate forms a stable hydrate in which there are two monolayers of water molecules. The Na^+ that are counterions for the negative structural charge developed as a result of isomorphic substitutions within the clay mineral layer tend to adsorb as solvated species on the basal plane (a plane of hexagonal rings of oxygen ions known as a siloxane surface) near deficits of negative charge

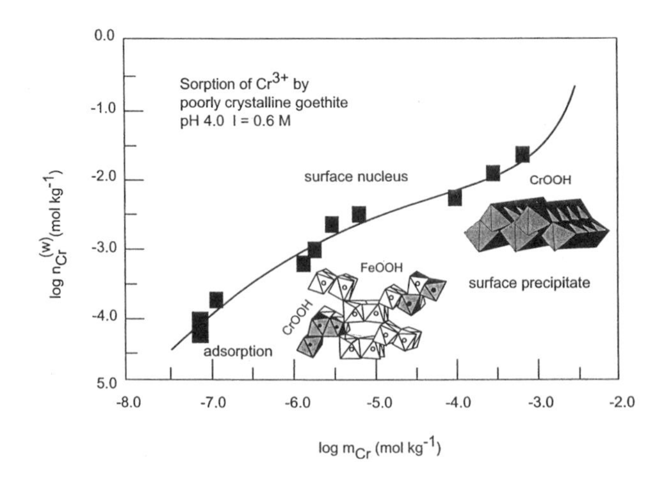

Fig. 1.6 Sorption isotherm for Cr^{3+}reacted with poorly crystalline goethite (c_s = 10 kg m^{-3}) illustrating movement from lower left to upper right in the stability field of fig.1.5. After Manceau et al.[26]

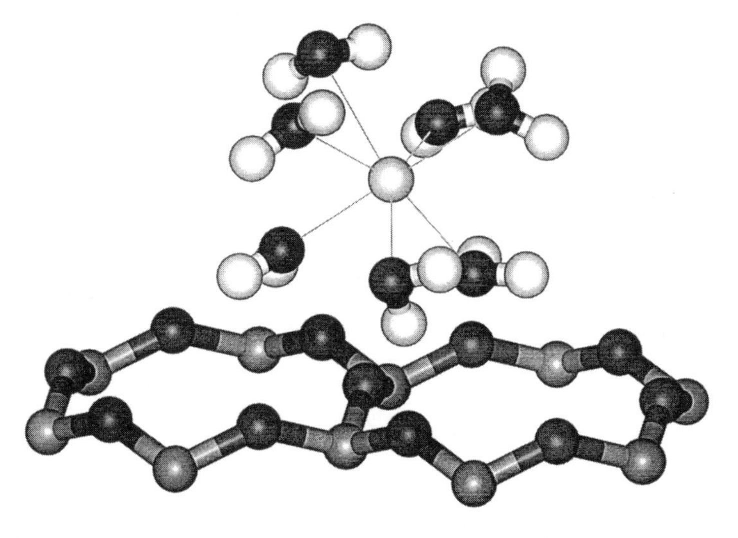

Fig. 1.7 An outer-sphere surface complex formed between Na^+ and an octahedral charge site in montmorillonite (see the *Clay Minerals* inset). The solvation shell comprises seven water molecules coordinated to Na^+ at distances varying from 0.22 to 0.25 nm, with an average distance of 0.23 ± 0.05 nm, just as is observed in concentrated aqueous solutions of $NaNO_3$, for example. (Original figure courtesy of S.-H. Park.)

originating in the octahedral sheet from substitution of a bivalent cation for Al^{3+}.[29] This mode of adsorption occurs as a result of the strong solvating characteristics of Na^+ and the physical impediment to direct contact between Na^+ and the site of negative charge posed by the layer structure itself. The way in which this negative charge is distributed on the siloxane surface is not well known, but if the charge tends to be delocalized there, that would also lend itself to outer-sphere surface complexation. It is pertinent to note that the structure of the solvation complex in fig. 1.7 is very similar to that of solvated Na^+ in concentrated aqueous solutions.[6,29]

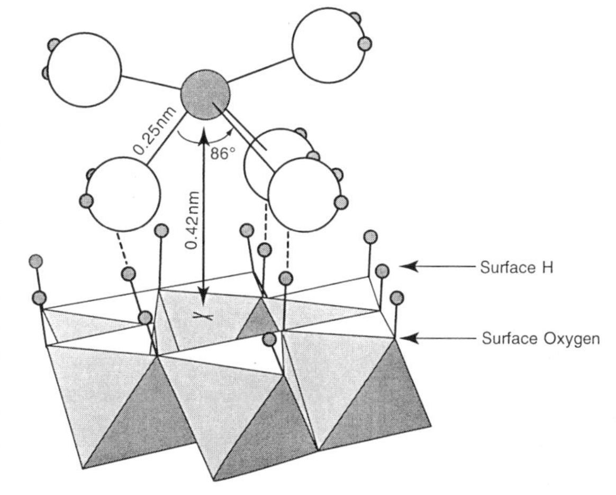

Fig. 1.8 An outer-sphere surface complex formed between Pb^{2+} and surface OH in the (0001) plane in corundum (see the *Metal Oxides* inset). The complex is stabilized by hydrogen bonds (dashed lines) between solvation shell water molecules and the surface OH groups. The cation is coordinated to five water molecules. After Bargar et al.[31]

Clay Minerals

Clay minerals are aluminosilicates that, like the micas, are sandwiches of tetrahedral and octahedral sheet structures. This bonding of the tetrahedral and octahedral sheets occurs through the apical oxygen ions in the former. The clay minerals usually are classified into three *layer types*, distinguished by the number of tetrahedral and octahedral sheets combined, and further into five *groups*, differentiated by the kinds of isomorphic cation substitutions that occur. Layer types are illustrated below, and the groups are described in the table beneath. The 1:1 layer type consists of one tetrahedral and one octahedral sheet. It is represented by the kaolinite group, with the general chemical formula $[Si_4]Al_4O_{10}(OH)_8 \cdot nH_2O$, where the cation enclosed in square brackets is in tetrahedral coordination and n is the number of moles of hydration water. Usually, there is no significant isomorphic substitution in this clay mineral. As is common with clay minerals in natural particles, the octahedral sheet has two-thirds cation site occupancy (dioctahedral sheet). The 2:1 layer type has two tetrahedral sheets that sandwich an octahedral sheet. The three clay mineral groups with this structure are illite, vermiculite, and smectite. If a, b, and c are the stoichiometric coefficients of Si, octahedral Al, and Fe(III), respectively, in the chemical formulas of these groups, then

$$x = (8 - a) + (4 - b - c) = 12 - a - b - c$$

is the *layer charge*, the number of moles of excess electron charge per chemical formula that is produced by isomorphic substitution. As indicated in the table below, the three 2:1 groups differ from one another in two principal ways. The layer charge decreases in the order illite > vermiculite > smectite, and the vermiculite group is further distinguished from the smectite group by a greater extent of isomorphic substitution in the tetrahedral sheet. Among the smectites, those in which the substitution of Al for Si exceeds that of Fe(II) or Mg for Al are called *beidellite*, and those in which the reverse is true are called *montmorillonite*. In any of the 2:1 minerals, the layer charge is balanced by cations that reside on the plane of oxygen ions (siloxane surface) of the tetrahedral sheets. These interlayer cations are represented by M in the chemical formula.

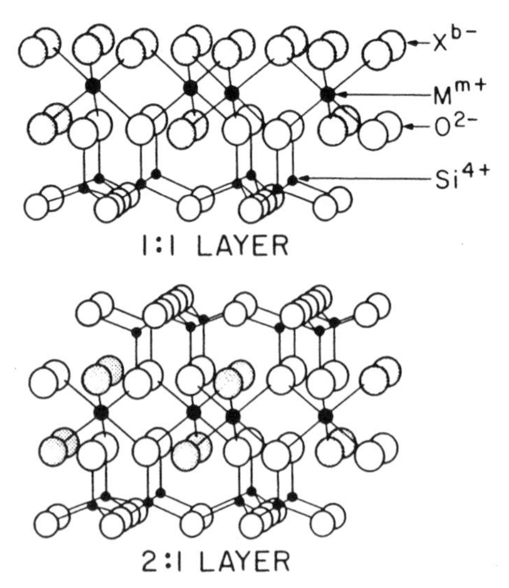

Clay Mineral Groups

Group	Layer Type	Layer Charge (x)	Typical Chemical Formula[a]
Kaolinite	1:1	< 0.01	$[Si_4]Al_4O_{10}(OH)_8 \cdot nH_2O$ ($n = 0$ or 4)
Illite	2:1	1.4–2.0	
Vermiculite	2:1	1.2–1.8	$M_x[Si_aAl_{8-a}]Al_bFe(III)_c(Mg, Fe(II))_{4-b-c}O_{20}(OH)_4$
Smectite	2:1	0.4–1.2	$(0 \leq a \leq 8, 0 \leq b \leq 4, 0 \leq c \leq 4)$[b]

[a] $n = 0$ is kaolinite and $n = 4$ is halloysite; M indicates a monovalent interlayer cation; [ ] indicates a tetrahedral sheet.
[b] For illite, $a = 6.81 \pm 0.04$, $b = 3.06 \pm 0.08$, $c = 0.44 \pm 0.06$; whereas for smectite, $a = 7.67 \pm 0.02$, $b = 2.98 \pm 0.02$, $c = 0.37 \pm 0.02$. See C. E. Weaver and L. D. Pollard, *The Chemistry of Clay Minerals*, Elsevier, Amsterdam, 1973.

Figure 1.8 illustrates the structure of an outer-sphere surface complex formed between Pb^{2+} and a hydrated surface site on an exposed (0001) plane in the Al oxide, corundum (see the *Metal Oxides* inset[30]).[31] This plane comprises triangular rings of six linked oxygen ions in the exposed faces of AlO_6^{3-} octahedra, with each O^{2-} bonded to a pair of neighboring Al^{3+}.[31] Like the basal planes in gibbsite, this surface is protonated to electroneutrality in aqueous media at pH ≈ 7. The outer-sphere surface complex has Pb^{2+} coordinated to three water molecules that are also hydrogen-bonded to surface hydroxyls on the border of an octahedral cavity in the center of one of the triangular rings of protonated oxygen ions.[31,32] Two other water molecules solvate the adsorbed Pb^{2+} to give a total solvation-shell coordination number of five, similar to what is observed for Pb^{2+} in concentrated aqueous solutions.[31]

Figure 1.9 illustrates the structure of an inner-sphere surface complex formed between K^+ and the siloxane surface of montmorillonite when hydrated by the two monolayers of water molecules. The K^+ counterion is bound directly to a negatively charged site on the surface, which has been generated by isomorphic substitution of Al^{3+} for Si^{4+} in one of the SiO_4^{4-} making up a tetrahedral sheet in the layer structure.[28,30] The negative charge is well localized on the triad of

Metal Oxides, Oxyhydroxides, and Hydroxides

Name	Chemical formula[a]	Name	Chemical formula[a]
Anatase	TiO_2	Goethite	α-FeOOH
Birnessite	(Na,K,Mn)Mn(IV,III)O_2	Hematite	α-Fe_2O_3
Boehmite	γ-AlOOH	Ilmenite	$FeTiO_3$
Corundum	α-Al_2O_3	Manganite	γ-MnOOH
Ferrihydrite	$(Fe_2O_3 \cdot 2H_2O)_5$	Maghemite[b]	γ-Fe_2O_3
Gibbsite	γ(-Al(OH)$_3$)	Magnetite[b]	$FeFe_2O_4$

[a] γ denotes cubic close-packing of anions, whereas α denotes hexagonal close-packing.
[b] Some Fe(III) are in tetrahedral coordination.

Metal Oxides

Because of their great abundance in the lithosphere and their low solubility in the normal range of soil pH values, aluminum, iron, and manganese form the most important oxide, oxyhydroxide, and hydroxide minerals. These minerals are listed in the table on page 15, and two representative octahedral structures are shown in the figure beneath. Among the iron oxide compounds, goethite is the one most often found in natural particles. However, under oxic condition and in the presence of iron-complexing ligands that inhibit crystallization (e.g., organic ligands or silicate anions), ferrihydrite $[(Fe_2O_3 \cdot 2H_2O)_n, n \approx 5]$ may precipitate. This poorly crystalline solid comprises sheets of octahedra with Fe(III) coordinated to O, OH, and OH_2 in a defect-sprinkled arrangement similar to that in hematite. The oxygen ions in goethite lie in planes, and the Fe^{3+} cations are coordinated in distorted octahedra that share edges. Some of the octahedral vertex ions are hydroxyl groups that form hydrogen bonds with neighboring oxygen ions. Isomorphic substitution of Al for Fe in goethite is common. Gibbsite is the most important of the aluminum minerals. The dioctahedral sheets included in the structure are bound together by hydrogen bonds between opposed hydroxyl groups. Hydrogen bonding also occurs between hydroxyl groups along the edges of unfilled octahedra within a sheet, thereby producing additional distortion of the aluminum octahedra beyond the distortion caused by the sharing of edges. The most common manganese mineral is birnessite. Birnessite contains sheets of MnO_6^{8-} octahedra, some of which are empty (cation vacancy) or have Mn(III) substituted for Mn(IV), the structural charge being compensated by interlayer cations coordinated to water molecules.

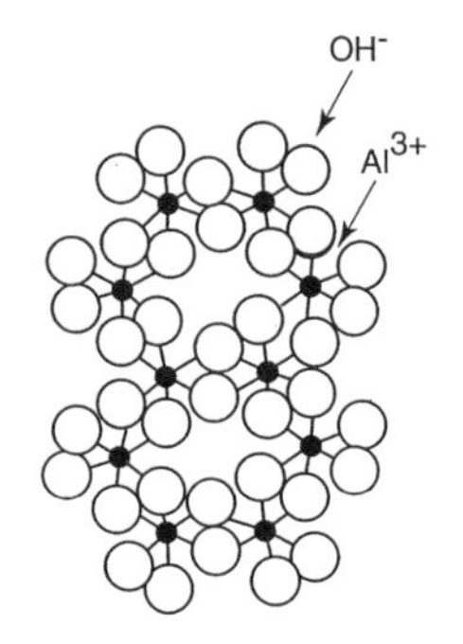

Gibbsite

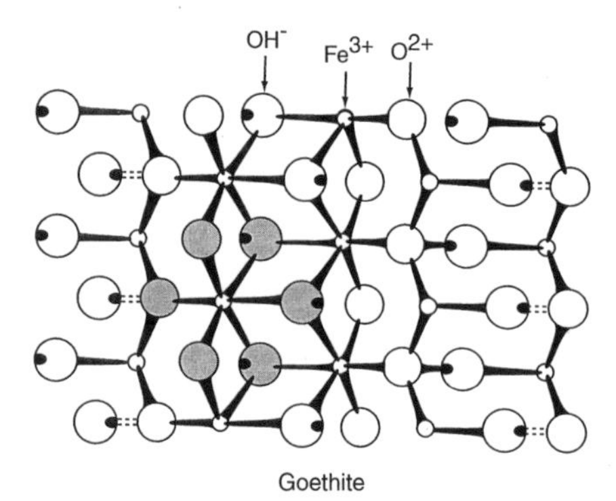

Goethite

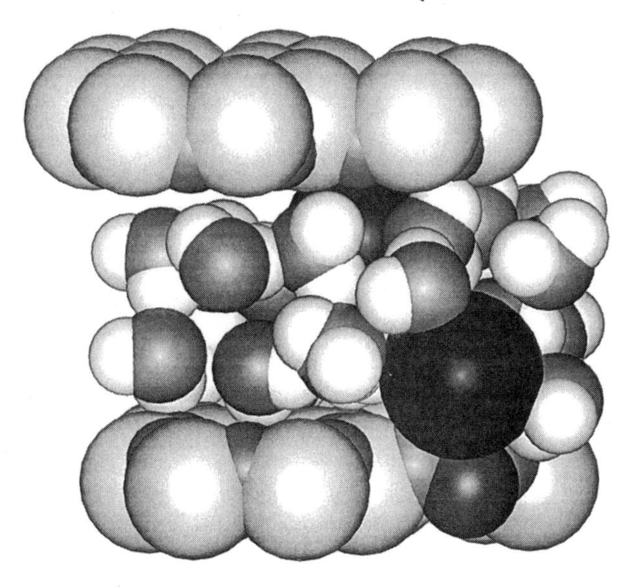

Fig. 1.9 An inner-sphere surface complex formed between K^+ and a tetrahedral charge site in montmorillonite (see the *Clay Minerals* inset). The cation is coordinated to eight oxygen atoms, two of which are in the nearest siloxane surface of the clay mineral. (Original figure courtesy of R. A. Sutton.)

oxygen ions which form the base of the tetrahedron, and this localization, together with the poor solvating tendency of K^+ facilitates inner-sphere complexation. Note, however, that the adsorbed K^+ is partially solvated by neighboring water molecules in the region between the two clay layers. Removal of these water molecules by desiccation of the clay mineral can lead to the K^+ entering the roughly hexagonal cavity bordered by oxygen ions in the siloxane surface so as to reside nearer to the source of negative charge.[8] The nearly perfect match between the diameter of K^+ and that of the cavity (0.26 nm) enhances this possibility.

Figure 1.10 illustrates an inner-sphere surface complex between Cd^{2+} and a hydroxylated surface site on the (100) plane in goethite.[33] This plane features reactive OH groups bound to just a single Fe^{3+}, which allows them to protonate, as well as dissociate, protons.[34] Adsorbed Cd^{2+} can bind to a pair of the OH groups after their protons are dissociated, thereby linking the corners of two adjacent $FeO_3(OH)_3$ octahedra in the mineral structure.[33] The Cd^{2+} complex may also be solvated. Note the similarity between the octahedral structure of the partially solvated adsorbed Cd^{2+} and that of the $FeO_3(OH)_3$ octahedra comprised in the mineral structure. Inner-sphere surface complexation of bivalent metal cations by metal-oxide minerals often may be conceptualized as a *distorted mode of crystal growth*, with the adsorbate playing the role of an added octahedral unit. Similarly, inner-sphere surface complexation of anions (chiefly oxyanions, such as arsenite[35]) by metal oxides may be conceptualized as a distorted mode of crystal growth, with the adsorbate playing the role of the

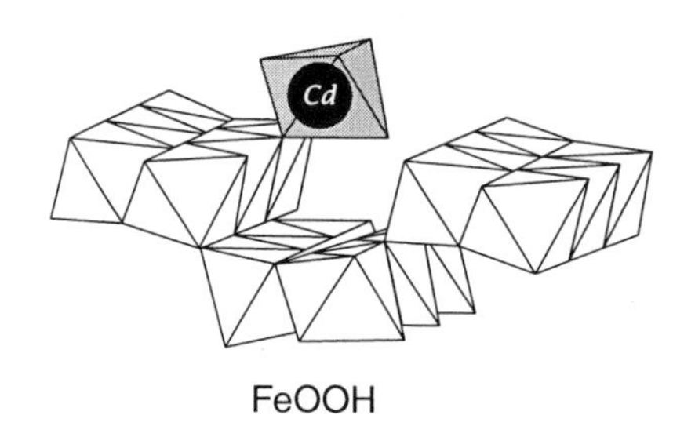

Fig. 1.10 An inner-sphere surface complex formed between Cd^{2+} and a pair of ionized surface OH groups in the (100) plane in goethite (see the *Metal Oxides* inset). The cation is coordinated to six oxygen atoms, two of which are on the goethite surface in neighboring $FeO_3(OH)_3$ octahedra. The two surface O are bonded to just one Fe^{3+}, which gives them a high propensity for cation adsorption. After Spadini et al.[33]

oxygen or hydroxyl ions it replaces in order to bind directly to the metal cation in the structure.[25]

The varieties of adsorbate structures assumed by the surface complexes of ionic adsorptives are usually adduced from spectroscopic data (see Chapter 2) informed by the trends observed in adsorption experiments.[25] This process, in turn, is aided by whatever knowledge is available concerning the adsorbent surface, particularly the stereochemistry and reactivity of its *surface functional groups*,[27] the molecular units bound into an adsorbent at its periphery that contribute to the reactions of its adsorption sites. Given a concept of inner-sphere surface complexation as the attempt of an adsorbent to extend its bulk structure through the acquisition of an adsorbate, it is logical to consider whether the guidelines developed for stable ionic crystals can be applied to surface complexes, as well. These guidelines, known as the *Pauling Rules*,[30] may be summarized as follows:

Rule 1. A polyhedron of anions is formed about each cation. The cation–anion distance is determined by the sum of the respective radii, and the coordination number is determined by the radius ratio of cation to anion.

Minimum Radius Ratio	Coordination Number
1.00	12
0.732	8
0.414	6
0.225	4

Rule 2. In a stable crystal structure, the sum of the strengths of the bonds that reach an anion from adjacent cations is equal to the absolute value of the anion valence.

Rule 3. The cations maintain as large a separation as possible and have anions interspersed between them so as to screen their charges. In geometric terms, this means that anion polyhedra tend not to share edges or especially faces. If edges are shared, they are shortened.

Rule 4. In a structure comprising different kinds of cation, those of high valence and small coordination number tend not to share polyhedron elements with one another.

Rule 5. The number of essentially different kinds of ions in a crystal structure tends to be as small as possible. Thus the number of types of coordination polyhedra in a close-packed array of anions tends to be a minimum.

The Pauling Rules are evident applications of Euclidean geometry and the coulomb law that serve to constrain the varieties of molecular structure than can occur in stable ionic crystals. That they can be employed in the same manner for adsorbates has only recently been appreciated.[32,36] Their utility can be illustrated by an application of Rule 2, which involves the concept of bond strength and local charge balance in a crystal. The *bond strength* s_{ij} from a cation i to an anion j is defined by the equation:

$$s_{ij} \equiv \frac{Z_i}{N_{ij}} \tag{1.22}$$

where Z_i is the cation valence (formal oxidation number) and N_{ij} is its coordination number (number of anions of type j in the polyhedron surrounding a cation of type i). As an example, consider the bond strength from Al^{3+} to O^{2-}, which is equal to 0.5 if an octahedron of O^{2-} (coordination number equal to 6) has formed around Al^{3+} (valence equal to 3). Rule 2 states that

$$\sum_i s_{ij} = |Z_j| \tag{1.23}$$

where Z_j is the valence (formal oxidation number) of an anion j and the sum on the left side runs over all cations bonded to the anion. For example, in the case of corundum (α-Al_2O_3), eq. 1.23 implies that four Al^{3+} (bond strength equal to 0.5) are bonded to each O^{2-} (valence equal to –2) in the structure. Any proposed structure (e.g., three Al^{3+} bonded to O^{2-}) that would not "saturate" the anion valence with four Al^{3+} is considered unstable, according to Rule 2.

The concept of bond strength has been generalized to that of *bond valence* in order to account for the common experimental observation that bond strengths depend on bond length, generally increasing as bond length decreases.[37] Thus s_{ij} in eq. 1.23 is expressed $s_{ij}(R_{ij})$, where R_{ij} is the length of the bond between cation i and anion j. This innovation can be pictured as a way of expressing the incomplete transfer of electron charge between a cation and an anion in a bond, such that the Pauling estimate of the number of electron charges transferred per bond (eq. 1.22) is modified. For example, the bond valence–bond length relation for Al–O bonds is:[32,37]

$$s_{AlO} = \exp\left[(0.1651 - R_{AlO})/0.037\right] \tag{1.24}$$

where R_{AlO} is an Al–O bond length in nanometers. This is an empirical equation for the bond valence based on extensive statistical analyses of crystal structure data. In corundum, the Al–O bond lengths observed are either 0.185 nm or 0.197 nm, leading to bond valences equal to 0.584 or 0.422, respectively.[32] Therefore, the left side of eq. 1.23 yields 2.01, in agreement with the required value, 2.0. The replacement of eq. 1.22 by eq. 1.24 thus permits observed small deviations from the perfect symmetry implied by Rule 1 to be assimilated into the law of charge balance as represented in eq. 1.23.

A surface oxygen ion on the (0001) plane in corundum is bonded to only two Al^{3+}, not four.[31] Therefore, the oxygen ion valence is "undersaturated" unless the O^{2-} bonds to another cation. This could be a proton, whose bond strength when coordinated to one oxygen ion would be +1, according to eq. 1.22. Equation 1.23 is then satisfied by a protonated surface oxygen bound to a pair of structural Al^{3+} or, equivalently, by a hydroxyl surface functional group on an exposed (0001) plane in corundum. The bond valence for H^+ coordinated to O^{2-} can be calculated with the empirical equation:[32]

$$s_{HO} = 0.024(R_{HO} - 0.0677)^{-1} \tag{1.25}$$

where R_{HO} is approximately 0.1 nm, leading to $s_{HO} \approx 0.74$, instead of 1.0 as eq. 1.22 would predict. Either estimate of s_{HO} suggests that a hydroxylated (0001) plane in corundum would be a stable surface, as is observed experimentally.[31] Another possibility for the additional cation bound to a surface oxygen ion is Pb^{2+}, for which the bond valence can be calculated with the empirical equation:[32]

$$s_{PbO} = \exp\left[(0.204 - R_{PbO})/0.037\right] \tag{1.26}$$

A Pb–O bond length of 0.228 nm[37] leads to $s_{PbO} = 0.52$ according to eq. 1.26. This result lies well below s_{HO}, suggesting that a single Pb^{2+} bound to the surface O^{2-} would leave its valence "undersaturated" using the criterion in eq. 1.23. On the other hand, merely adding Pb^{2+} to an already-protonated surface O^{2-} would result in significant "oversaturation". This line of reasoning is consistent with the outer-sphere complex structure in fig. 1.8, which features a stably hydroxylated (0001) plane with adsorbed Pb^{2+} remaining solvated, not in direct contact with the surface functional groups. Hydrogen bonds between three of the solvating water molecules and the surface hydroxyl groups then serve to stabilize the surface complex.[32] Pauling Rule 2 is thus useful in understanding how this structure is preferred over an inner-sphere arrangement for Pb^{2+} adsorption.

1.3 Surface Charge

Natural particles develop surface charge from isomorphic substitutions and structural disorder (including defects) in minerals, and from adsorption reactions with ionic species in aqueous solution for both minerals and organic matter. The specific surface charge,[5] conveniently measured in units of moles of protonic charge per kilogram ($mol_c\ kg^{-1}$), that is created by these two mechanisms is classified into three broad categories:[38] (i) *structural charge*, denoted σ_o, (ii) *net adsorbed proton charge*, denoted σ_H, and (iii) *net adsorbed ion charge*, denoted $\Delta q \equiv q_+ - q_-$, where

$$q_+ \equiv \sum_{\text{cations}} Z_i n_i^{(w)} \qquad q_- \equiv \sum_{\text{anions}} |Z_i| n_i^{(w)} \tag{1.27}$$

defines the *total adsorbed cation* (q_+) and *anion* (q_-) *charge* in terms of the specific surface excess, given by eq. 1.4, and the valence of each type of ion that is adsorbed by a particle. The sum in eq. 1.27 is over all adsorbed cationic or anionic species, *except* for protons or hydroxide ions in surface complexes.[38] These latter species are taken into account uniquely in σ_H, thus respecting the great importance of H and OH as constituents of many kinds of surface-reactive natural particles.[22]

Structural charge occurs in both metal oxides and clay minerals, in the latter of which it is proportional to the layer charge:[39]

$$\sigma_o = -10^3 \frac{x}{M_r} \tag{1.28}$$

where M_r is the relative molecular mass of the unit cell.[5] According to the table provided in the Clay Minerals inset,

$$\begin{aligned} M_r = {} & 28.1a + 27(8 - a + b) + 55.8c \\ & + 24.3(4 - b - c) + 388 \end{aligned} \tag{1.29}$$

for the relative molecular mass, in daltons, of a 2:1 dioctahedral clay mineral, if the octahedral sheet has only Mg substitution. Thus $M_r \approx 730$ Da, given typical values of the stoichiometric coefficients in eq. 1.29, and σ_o ranges from –2.7 mol_c kg^{-1} for high-charge illite to –0.69 mol_c kg^{-1} for low-charge smectite, based on eq. 1.28 and the defined ranges of layer charge for these clay minerals.

Structural charge arising from clay minerals in a soil or sediment sample is quantitated conventionally as Cs-accessible surface charge following a reaction of the sample with 0.05 *m* CsCl at pH 5.5–6.0.[3,40] Briefly, the adsorbent is saturated with Cs by repeated washing in CsCl, with a final supernatant solution ionic strength of 0.05 *m*. Following centrifugation, the supernatant solution is discarded and the remaining entrained CsCl solution is removed by washing with ethanol. The samples are then dried at 65°C for 48 h to enhance formation of inner-sphere Cs surface complexes (see fig. 1.9). Next, the samples are washed in 0.01 *m* LiCl solution to eliminate outer-sphere surface complexes of Cs. The suspension is centrifuged, and the supernatant LiCl solution is removed for analysis, leaving only a slurry containing the adsorbent sample and entrained LiCl solution. Finally, Cs is extracted with 1 *m* ammonium acetate (NH_4OAc) solution, and the LiCl and NH_4OAc solutions are analyzed for Cs. Permanent structural charge, σ_o, is then calculated as minus the difference between moles Cs in the NH_4OAc extract and moles Cs in the entrained LiCl solution, per kg of dry adsorbent, as prescribed by eq. 1.4:

$$-\sigma_o = n_{NH_4OAc} - \theta_w m_{LiCl} \tag{1.30}$$

where θ_w is the mass of entrained solution per unit mass of dry adsorbent. This method is reliable even for highly heterogeneous samples that comprise both crystalline and amorphous minerals, organic matter, and biota. Its sensitivity is such that $|\sigma_o|$ values < 1 $mmol_c$ kg^{-1} are detectable.[40] In general, σ_o determined as Cs-accessible structural charge can be attributed to the presence of 2:1 layer type clay minerals.[41]

Typically, the net adsorbed proton charge, σ_H, is determined for aqueous natural particle suspensions by electrometric titration as a function of pH under chosen conditions of ionic strength, background electrolyte concentration and composition, and particle suspension density, at fixed temperature and pressure.[42] A common approach involves the use of a glass electrode and a double-junction reference electrode in the potentiometric titration cell:

glass electrode	suspension of solid particles in background electrolyte solution	background electrolyte solution	liquid junctions	reference electrode

The emf of the electrode assembly is measured while known volumes of either strong acid or strong base are added to the suspension. These data, in turn, are converted to proton concentrations with the help of a calibration curve prepared from similar titration data obtained without suspended particles in the cell.[42] Values of the *apparent net proton surface excess*, δn_H, measured in moles, are calculated with the equation:

$$\delta n_H \equiv \left(c_A - c_B + [OH^-] - [H^+]\right) V \tag{1.31}$$

where c_A is the concentration of strong acid added, c_B is the concentration of strong base added, $[OH^-]$ and $[H^+]$ are free hydroxide ion and proton concentrations, respectively, and V is the suspension volume. The *apparent net adsorbed proton charge* is then:

$$\delta\sigma_{H,titr} = \frac{\delta n_H}{M} \tag{1.32}$$

where M is the dry mass of particles in the suspension. The experimental parameter $\delta\sigma_{H,titr}$ has units of moles of protonic charge per kilogram.

The definition of $\delta\sigma_{H,titr}$ indicates that it is a net proton specific surface excess (specific surface excess of H^+ less that of OH^-) measured *relative* to the net proton specific surface excess that exists before either strong acid or base is added to a particle suspension. The initial value of the net proton surface excess cannot be determined from titration data alone,[43] so $\delta\sigma_{H,titr}$ must be renormalized by adding to it some datum value from an *independent* experimental determination. For example, at very low pH, δn_H might achieve a maximum positive value, as it does for synthetic polyelectrolyte anions. In practice, unless a well-defined plateau appears in δn_H as pH decreases or the number of moles of protonatable surface functional groups is known, this kind of renormalization procedure is not practicable. Moreover, implicit in the use of δn_H to calculate $\delta\sigma_{H,titr}$ is the critical assumption that the only protons consumed upon addition of strong acid or base are those that have reacted with solid particles to form surface complexes. This assumption cannot be true unless δn_H has been corrected for side-reactions of added protons or hydroxide ions with dissolved chemical species (e.g., CO_2) or with the bulk structure of the solid adsorbent.[41,42,44] To these complications must be added the ambiguities created by the liquid junctions in the titration cell and by adsorbed protons or hydroxide ions that are in the diffuse ion swarm and not immobilized in surface complexes.[38] Thus, the relationship between $\delta\sigma_{H,titr}$ and σ_H is problematic.

Net adsorbed ion charge, Δq, conventionally is measured by reacting an adsorbent with an "index" electrolyte solution at a given pH value and ionic strength, so as to create a homoionic adsorbate (aside from adsorbed H^+ or OH^-), then determining the specific surface excess of the cation and anion in the "index" electrolyte.[42] Equation 1.27 is applied to calculate Δq with the data obtained. The reaction of bivalent "index" cations (e.g., Mg^{2+} or Ba^{2+}) with a negatively charged adsorbent leads to what is termed a *cation exchange capacity* (CEC)[45]:

$$\mathrm{CEC} \equiv \Delta q \quad \text{if } n_-^{(w)} \leq 0 \tag{1.33}$$

where $n_-^{(w)}$ is the specific surface excess of the "index" anion (usually Cl^-). An *anion exchange capacity* (AEC) can be defined analogously, given the results of a measurement of Δq using an "index" electrolyte with a monovalent anion and a positively charged adsorbent[42]:

$$\mathrm{AEC} \equiv -\Delta q \quad \text{if } n_+^{(w)} \leq 0 \tag{1.34}$$

where $n_+^{(w)}$ is the specific surface excess of the "index" cation (usually Na^+ or Li^+). The important conceptual point here is that *negative adsorption of anions (cations) contributes to CEC (AEC) in addition to positive adsorption of cations (anions).*

To see this point in detail, consider the example calculation of surface excess quantities in a montmorillonitic soil[7] presented in section 1.1. In the soil slurry, the condition of overall electroneutrality can be expressed:

$$2n'_{\mathrm{Ca}} + n_{\mathrm{Na}} = n_{\mathrm{Cl}} + \mathrm{CEC} \tag{1.35}$$

which states that the positive charge carried by Ca^{2+} and Na^- must be balanced by the negative charge carried by Cl^- and by the montmorillonite adsorbent, that is, its CEC value. In the supernatant solution, the equivalent condition is:

$$2c'_{\mathrm{Ca}} + c_{\mathrm{Na}} = c_{\mathrm{Cl}} \tag{1.36}$$

Multiplying both sides of eq. 1.36 by the gravimetric water content of the soil slurry, θ_w, and subtracting the result from eq. 1.35, one finds the expression:

$$2\left(n'_{\mathrm{Ca}} - \theta_w c'_{\mathrm{Ca}}\right) + \left(n_{\mathrm{Na}} - \theta_w c_{\mathrm{Na}}\right) = \left(n_{\mathrm{Cl}} - \theta_w c_{\mathrm{Cl}}\right) + \mathrm{CEC}$$

or

$$q_+ - q_- = \mathrm{CEC} \tag{1.37}$$

upon rearranging the anion term and applying the definitions in eq. 1.27. In the present example, $q_+ = 122.53$ mmol$_c$ kg^{-1}, $q_- = -0.17$ mmol$_c$ kg^{-1}, and, therefore, CEC $= 122.7$ mmol$_c$ kg^{-1}. If q_- had been positive, the right side of eq. 1.35

would contain the difference, CEC – AEC, instead of CEC alone, to account for positive charge on the adsorbent as well. Therefore, more generally,

$$q_+ - q_- = \text{CEC} - \text{AEC} \tag{1.38}$$

of which eqs. 1.33 and 1.34 are special cases. Values of CEC for soils range from 0.05 to 1.4 $\text{mol}_c\ \text{kg}^{-1}$, generally increasing significantly with the content of humified organic matter, whose CEC contribution alone ranges from 1–10 $\text{mol}_c\ \text{kg}^{-1}$, far exceeding the surface charge contributed by clay or oxide minerals.[46] Values of AEC for soils range from negligible to about 0.01 $\text{mol}_c\ \text{kg}^{-1}$, generally increasing with the content of iron and aluminum oxide minerals.[46]

Applications of the three categories of surface charge to interpreting the reactions of natural particles often are facilitated by some auxiliary definitions. The sum of structural and net adsorbed proton charge defines the *intrinsic charge*, σ_{in},[38]

$$\sigma_{\text{in}} \equiv \sigma_{\text{o}} + \sigma_{\text{H}} \tag{1.39}$$

which is intended to represent components of surface charge that developed *in toto* from adsorbent structure. The net adsorbed ion charge can be decomposed in molecular-level terms[27]:

$$\Delta q \equiv \sigma_{\text{IS}} + \sigma_{\text{OS}} + \sigma_{\text{d}} \tag{1.40}$$

which refer, respectively, to the net charge of ions adsorbed in inner-sphere surface complexes (IS), in outer-sphere surface complexes (OS), or in the diffuse ion swarm (*d*).[1] The utility of eq. 1.40 depends on the extent to which experimental detection and quantitation of these surface species is possible.

The partitioning of surface complexes into inner sphere and outer sphere is not always possible (or required), however, and eq. 1.40 can alternatively be written in the simpler form:

$$\Delta q \equiv \sigma_{\text{S}} + \sigma_{\text{d}} \tag{1.41}$$

where σ_{S} denotes the *Stern layer charge*[1] representing all adsorbed ions not in the diffuse ion swarm. This latter conceptual distinction, based largely on adsorbed ion mobility, is epitomized by defining the *net total particle charge*, σ_p[1,38]:

$$\sigma_{\text{p}} \equiv \sigma_{\text{in}} + \sigma_{\text{S}} \tag{1.42}$$

which is the specific surface charge contributed by the adsorbent structure and by adsorbed ions that are immobilized into surface complexes, that is, adsorbed ions that do not engage in translational motions relative to the adsorbent that may be likened to the diffusive motions of a free ion in aqueous solution. The adsorbed ions that do engage in more or less free diffusive motions must nonetheless contribute a net charge that balances the net total particle charge[1,27]:

$$\sigma_{\text{p}} + \sigma_{\text{d}} = 0 \tag{1.43}$$

Equation 1.43 is the *condition of surface charge balance* for an aqueous system of natural particles. It states simply that, any electric charge natural particles may bear in an aqueous system is always balanced by a counterion charge in the swarm of electrolyte ions near their surfaces. An alternative form of eq. 1.43 can be written down at once after combining eqs. 1.39, and 1.41–1.43:

$$\sigma_o + \sigma_H + \Delta q = 0 \tag{1.44}$$

Equation 1.44 attests to the necessary overall electroneutrality of any sample of natural particles that has been equilibrated with an aqueous electrolyte solution. Structural charge and the portion of net particle charge attributable to surface-complexed protons or hydroxide ions must be balanced with the net surface charge that is contributed by all other adsorbed ions and by H^+ or OH^- in the diffuse ion swarm.

Equation 1.44 can be used to test experimental surface charge data for self-consistency. A convenient approach is to plot Δq against σ_H over a range of pH values for which these two surface-charge components have been measured.[46] A simple rearrangement of eq. 1.44,

$$\Delta q = -\sigma_H - \sigma_o \tag{1.45}$$

shows that the slope of this "Chorover plot" must be equal to −1, with both its y- and x-intercepts equal to $-\sigma_o$.[43,46] Figure 1.11 illustrates the application of eq. 1.45 to a kaolinitic Oxisol (a highly weathered tropical soil comprising kaolinite, metal oxides, and quartz intermixed with organic matter) that was equilibrated with LiCl solution at three different ionic strengths over the pH range 2–6.[47] The line through the data is based on a linear regression equation,

$$\Delta q = -1.01 \pm 0.07\sigma_H + 12.5 \pm 0.8 \qquad (r^2 = 0.92)$$

with both Δq and σ_H expressed in units of $\text{mmol}_c\ \text{kg}^{-1}$, and with 95% confidence intervals following the values of the slope and intercept. The value of σ_o measured independently is $-12.5 \pm 0.04\ \text{mmol}_c\ \text{kg}^{-1}$, in excellent agreement with the y- and x-intercepts. Chorover plots confirming charge balance necessarily should precede any interpretation of experimental surface charge data in terms of a chemical model of the solid-aqueous solution interface. We note in passing that the combination of eqs. 1.38 and 1.44 exposes immediately the conditional nature of CEC and AEC data: any chemical factors that influence σ_H will also change these two capacity variables.[45,46] Finally, eq. 1.44 shows that a datum value to use in connection with the renormalization of $\delta\sigma_{H,titr}$ in eq. 1.32 can be obtained by determining the pH value at which $\Delta q + \sigma_o = 0$ (i.e., $\sigma_H = 0$).[46–48]

1.4 Points of Zero Charge

Points of zero charge have been defined as pH values for which one of the categories of surface charge is equal to zero, at some ambient temperature, applied pressure, and aqueous solution composition.[1,38,42] Other variables than pH (for example, the negative logarithm of the aqueous solution activity of any

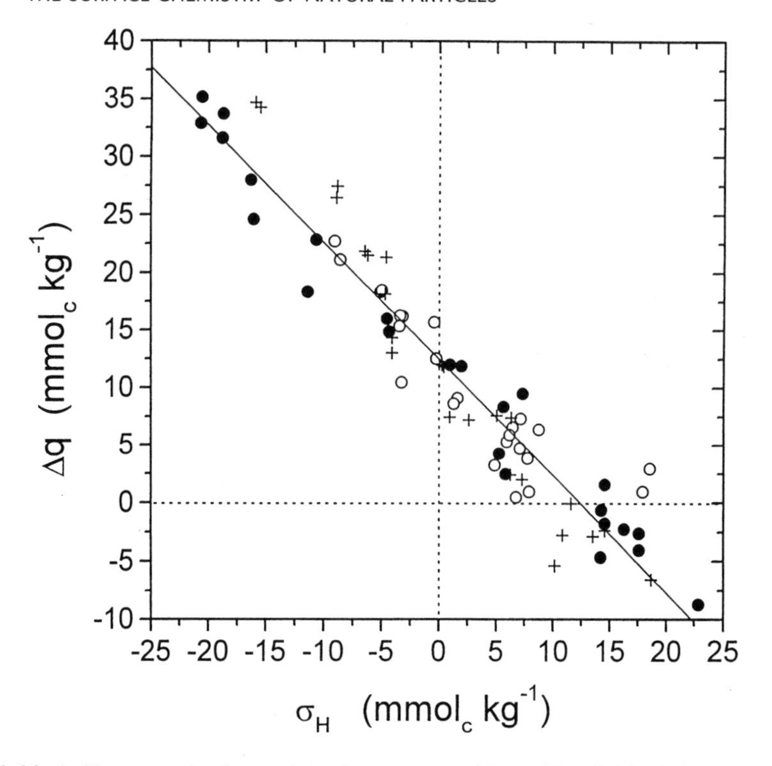

Fig. 1.11 A Chorover plot for surface charge on a cultivated kaolinitic Oxisol suspended in LiCl background electrolyte solution (c_s = 10 kg m^{-3}). The ionic strength of the background electrolyte solution was varied from 1 mol m^{-3} (open circles), to 5 mol m^{-3} (crosses), to 10 mol m^{-3} (filled circles), whereas pH was varied from 2 to 6. The straight full line is a regression line, whereas the dotted lines define the two coordinate axes. Charge balance (eq. 1.44) requires a Chorover plot to be linear, with unit slope and with x- and y-intercepts both equal to the absolute value of the structural charge. (Original figure courtesy of J. D. Chorover.)

ion that adsorbs primarily through inner-sphere surface complexation) can be utilized to define a point of zero charge,[49] but pH remains the exemplary variable. Four standard definitions are listed in the table below[1,38,42]:

IUPAC Symbol	Name	Definition
p.z.n.p.c.	point of zero net proton charge	$\sigma_H = 0$
p.z.n.c.	point of zero net charge	$\sigma_{in} = 0$
p.z.c.	point of zero charge	$\sigma_p = 0$
i.e.p.	isoelectric point	$u = 0$

where u is electrophoretic mobility.[50]

The p.z.n.p.c. is the pH value at which the net adsorbed proton charge is equal to zero. The most straightforward method to determine this pH value would be to measure Δq as a function of pH and then locate the pH value at which $\Delta q = -\sigma_o$, thus taking direct advantage of eq. 1.44 (fig. 1.12), given that a separate measurement of σ_o has been made. Most published reports of p.z.n.p.c. values based on the use of potentiometric measurements to determine σ_H resort to the device of choosing $\sigma_o \equiv 0$ at the crossover point of two $\delta\sigma_{H,titr}$ versus pH curves that have been determined at different ionic strengths.[50] Unfortunately, each such curve is, in principle, offset differently from a true σ_H curve by an *unknown* $\delta\sigma_{H,titr}$ that corresponds to the particular initial state of the titrated system, thus making the crossover point illusory.

Equation 1.44 also imposes a constraint on changes in the net adsorbed proton charge and/or net adsorbed ion charge that may occur in response to imposed

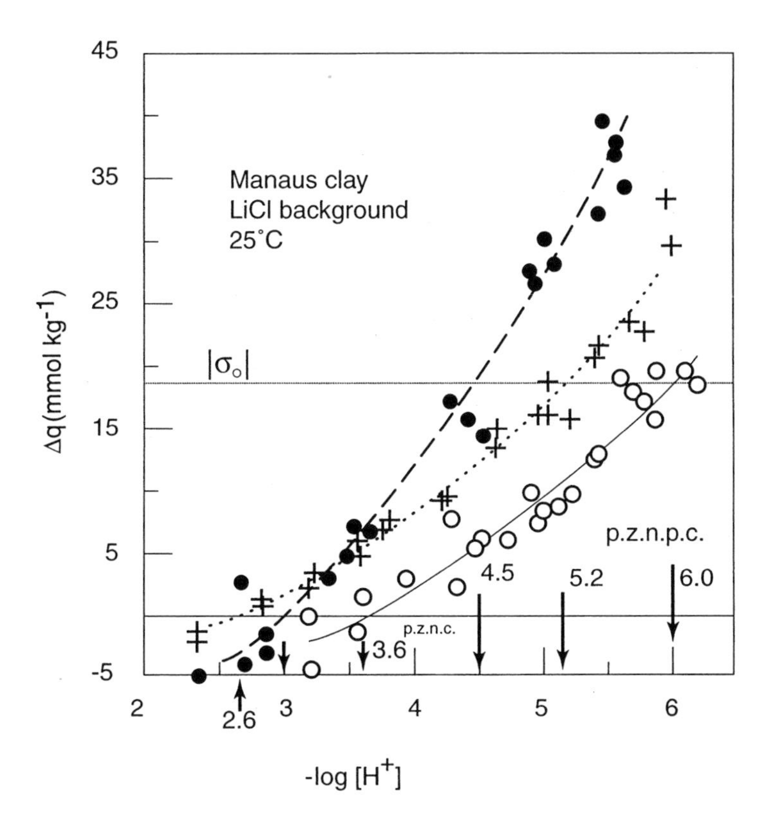

Fig. 1.12 Graph of net adsorbed ion charge versus pH for an uncultivated kaolinitic Oxisol suspended in LiCl background electrolyte solution under the same conditions as in fig. 1.11. The curves through the data points are quadratic regression lines. The horizontal line in the middle of the graph denotes the absolute value of the structural charge, and its intersections with the Δq curves define the p.z.n.p.c. (6.0, 5.2, and 4.5 for ionic strengths 1(○), 5(+), and 10 (•) mol m^{-3}, respectively). The lower horizontal line defines the p.z.n.c. (3.6, 2.6, and 3.0, for ionic strengths 1 (○), 5(+), and 10 mol m^{-3}, respectively). Note the strong ionic strength dependence of these two points of zero charge. Data from Chorover and Sposito.[47]

changes in adsorbate or adsorptive composition at fixed temperature (T) and applied pressure (P):

$$\delta\sigma_H + \delta\Delta q = 0 \tag{1.46}$$

where δ represents an infinitesimal shift caused by any mechanism that does not alter the adsorbent bulk structure or composition (i.e., σ_o is to remain constant during any process causing the shift δ). For example, if the ionic strength (I) of the aqueous solution equilibrated with a solid adsorbent is changed at fixed T and P,

$$\left(\frac{\partial\sigma_H}{\partial I}\right)_{T,P} + \left(\frac{\partial\Delta q}{\partial I}\right)_{T,P} = 0 \tag{1.47}$$

This constraint may be applied to the definition of the *point of zero salt effect* (p.z.s.e.),[42]

$$\left(\frac{\partial\sigma_H}{\partial I}\right)_{T,P} = 0 \qquad (\text{pH} = \text{p.z.s.e.}) \tag{1.48}$$

to show that the crossover point of two σ_H versus pH curves must also be that of two Δq versus pH curves,[51] a fact that can be used to check illusory p.z.s.e. values inferred from the crossover point of $\sigma_{H,titr}$ curves instead of that of σ_H curves.

To see this point in detail, consider the data in the table on p. 29, which lists components of surface charge for a California Alfisol measured at two ionic strengths (NaCl background electrolyte). The crossover point for Δq evidently occurs at pH $\lesssim$ 4.2, whereas that for $\delta\sigma_{H,titr}$ is around pH 7.5. This large discrepancy results from the dependence of the initial value of σ_H (σ_H^{init} in column 5) on ionic strength, making each curve of $\delta\sigma_{H,titr}$ offset significantly from σ_H by a *different* datum value, thus leading to an illusory crossover point. The value of σ_H^{init} can be calculated with eq. 1.44 and the definition, $\delta\sigma_{H,titr} \equiv \sigma_H - \sigma_H^{init}$, given $\sigma_o = -64.5$ mmol$_c$ kg^{-1} for this vermiculite-containing soil. (The 10% variability in σ_H^{init} values calculated at each titration point in the table below is typical of the data as a whole.) Adding the average value of σ_H^{init} to $\delta\sigma_{H,titr}$ we obtain σ_H (column 6), which goes to zero around pH 4.3. Thus, p.z.n.p.c. $\approx$ 4.3 $\approx$ p.z.s.e. in this example. Note that $\delta\sigma_{H,titr} = 0$ near pH 7.3, which is nowhere near the true p.z.n.p.c.

The p.z.n.c. is the pH value at which the intrinsic surface charge is equal to zero. Comparison of eqs. 1.39 and 1.44 shows that surface charge balance requires the condition:

$$\sigma_{in} + \Delta q = 0 \tag{1.49}$$

that is, *the net adsorbed ion charge vanishes at the p.z.n.c.* (fig. 1.12). It is common practice to utilize index ions, such as K^+ and Cl^-, in the determination of p.z.n.c. from a Δq versus pH curve.[42] Evidently, the value of p.z.n.c. will depend on the choice of "index" ions, although this dependence is small if the ions are chosen from the following group: Li^+, Na^+, K^+, ClO_4^-, and NO_3^-. Standard methods of measuring p.z.n.c. for soils and sediments have been reviewed by Zelazny et al.,[42] who also provide tabulations of representative p.z.n.c. values for specimen minerals. As a broad rule, p.z.n.c. values for silica, humus, clay minerals, and most manganese oxides or oxyhydroxides are below pH 4, whereas those for aluminum

[NaCl] ($mol\ dm^{-3}$)	pH	Δq	$\delta\sigma_{H,titr}$ ($mmol_c\ kg^{-1}$)	σ_H^{init} ($mmol_c\ kg^{-1}$)	σ_H ($mmol_c\ kg^{-1}$)
0.02[a]	4.3	81	56	−72[b]	−7[b]
	4.8	(84)[b]	37	(−63)[c]	−25
	5.4	103	25	−63	−38
	5.8	108	17	−61	−46
	6.5	124	7.5	−67	−55
	7.1	128	0.24	−54	−63
	7.6	122	−0.17	−58	−63
	7.9	128	−0.19	−63	−63
				−63 ± 6[d]	
0.05[a]	4.2	63	48	−46[b]	4[b]
	4.6	72	35	−44	−9
	5.5	94	16	−46	−28
	5.8	91	12	−39	−32
	6.5	102	2.5	−41	−41
	7.0	104	0.03	−39	−44
	7.3	117	−0.08	−53	−44
	7.8	115	−1.50	−48	−46
				−44 ± 5[d]	

[a] Data from S. J. Anderson and G. Sposito.[48]
[b] $-\sigma_H^{init} \equiv \sigma_o + \delta\sigma_{H,titr} + \Delta q$ for each titration point, then $\sigma_H \equiv \delta\sigma_{H,titr} + \sigma_{Have}^{init}$.
[c] q_{Cl} not determined, so $\sigma_H^{init} \equiv \sigma_{Have}^{init}$
[d] $\sigma_{Have}^{init} \pm$ std. dev.

and iron oxides or oxyhydroxides and for calcite are above pH 7. Thus, *the p.z.n.c. tends to increase as chemical weathering of a soil proceeds, with attendant loss of humus and silica.*

The p.z.c. is the pH value at which the net total particle charge is equal to zero. Thus, by eq. 1.43, at the p.z.c., there is no surface charge to be neutralized by ions in the diffuse swarm and any adsorbed ions that exist must be bound in surface complexes. Therefore, the p.z.c. could be measured by ascertaining the pH value at which a perfect charge balance exists among the ions in an aqueous solution with which particles have been equilibrated. More commonly, p.z.c. is inferred from the pH value at which a suspension flocculates rapidly,[49] or from that at which the particle electrophoretic mobility vanishes, the i.e.p.[1] Equality between i.e.p. and p.z.c., however, is problematic, because it requires demonstrating that no part of the diffuse ion swarm will be advected with a particle as it moves steadily in response to a uniform, constant electric field. Otherwise, the i.e.p. corresponds to the vanishing of σ_p plus some poorly defined fraction of the diffuse swarm charge. Fair and Anderson[52] have performed a very illuminating model calculation of the electrophoretic mobility for ellipsoidal particles that carry zero net total charge, that is, $\sigma_p = 0$, but have a heterogeneous surface charge distribution (zero net particle charge because of mutually canceling patches of positive and negative surface charge), much like natural particles should. These model particles, whose shape can range broadly from prolate to oblate, also were assumed to have a "no-slip" boundary condition at the solid–liquid interface, thus obviating the problem of diffuse swarm advection. With any heterogeneous distribution of surface charge on the particles that yields a nonzero electric quadruple moment, there also will be a nonzero electrophoretic mobility, despite the existence of zero net surface charge overall. Fair and Anderson[52] illustrate this conclusion for an ellipsoid of revolution that has a band of positive charge sandwiched between two canceling bands of negative charge distributed along the ellipsoid symmetry axis. If its shape is either prolate or oblate, this electrically neutral particle will have a nonzero electrophoretic mobility. Thus, $\sigma_p = 0$ would not be sufficient to imply that the i.e.p. has been achieved for nonspherical, charge-heterogeneous particles.

The charge balance conditions in eqs. 1.43 and 1.44, together with the constraints imposed in eqs. 1.46 and 1.50 (see the *Thermodynamic Stability* inset), lead to a set of broad statements about the points of zero charge known as *PZC Theorems.*[53] The first of these theorems concerns the relationship between p.z.n.p.c. and p.z.n.c. At the latter point of zero charge, eq. 1.44 reduces to the condition:

$$\sigma_o + \sigma_H = 0 \qquad (\text{pH} = \text{p.z.n.c.}) \tag{1.51}$$

If p.z.n.c. > p.z.n.p.c., σ_H must have a negative sign in eq. 1.51 because of the stability condition on σ_H in eq. 1.50 and the fact that $\sigma_H \equiv 0$ at p.z.n.p.c. (table 1). It follows that $\sigma_o > 0$ if p.z.n.c. > p.z.n.p.c. Similarly, if p.z.n.c. < p.z.n.p.c., $\sigma_o < 0$. Therefore,

1. *The sign of the difference (p.z.n.c. − p.z.n.p.c.) is the sign of the structural charge,* σ_o.

QED

For example, p.z.n.c. < p.z.n.p.c. typically for kaolinitic Oxisols (fig. 1.12) and specimen kaolinite samples,[41,47] indicating at once that structural charge exists in these materials, likely from the presence of 2:1 layer type clay minerals, given the lack of isomorphic substitutions in kaolinite. More generally, natural particles whose surface chemistry is dominated by 2:1 clay minerals ($\sigma_o < 0$) must always have p.z.n.c. values below their p.z.n.p.c. values.

A corollary of PZC Theorem 1 is that, *for natural particles without 2:1 clay minerals* (and, strictly speaking, without oxide or oxyhydroxide minerals having structural charge[39]) *p.z.n.c. = p.z.n.p.c.* Equality of the two points of zero charge means that the pH value at which σ_H is equal to zero can be determined through ion adsorption measurements alone.[51] On the other hand, natural particles for which $|\sigma_H/\sigma_o| \ll 1$ at any pH value permit the determination of σ_o through conventional ion adsorption measurements as a maximal CEC value (eq. 1.33).[45]

The difference between p.z.n.c. and p.z.c. is that a diffuse swarm can exist at the former pH value, whereas it cannot at the latter pH value. The use of suspension flocculation to infer p.z.c.[49,50] is compromised by the fact that flocculation usually occurs in the presence of a small—but nonzero—electrostatic repulsive force which is not strong enough to preclude van der Waals attraction from inducing flocculation. The use of electrokinetic methods to infer p.z.c. is problematic, as described above, because there may be no unique relationship between electrokinetic motion and surface charge for heterogeneous nonspherical particles. Surface charge balance, as expressed by combining eqs. 1.42 and 1.43,

Thermodynamic Stability

A necessary and sufficient condition for an aqueous system comprising natural particles to be in a state of equilibrium is that the Gibbs energy of the system be a minimum under constraints of controlled T, P, and mole numbers of all components. This condition is expressed mathematically by a zero value of the first-order and a positive value of the second-order infinitesimal variation of the Gibbs energy, under the three constraints mentioned, for all changes between equilibrium states. The vanishing of the first-order variation leads to the uniformity of the chemical potential of each component among all phases in the system [e.g., equality of the chemical potential of a component in the aqueous solution phase (adsorptive) with that in the solid phase (adsorbate)].

A positive second-order variation of the Gibbs energy in turn requires positive values of certain partial derivatives of the chemical potentials of the components, one of which is (at fixed T, P):

$$\left(\frac{\partial \mu_i}{\partial n_i}\right)_{\mathrm{T,P}} > 0$$

where μ_i is the chemical potential of a component i whose mole number is n_i. For the case of an equilibrated adsorbate, n can be replaced with the relative surface excess, and μ_i can be equated to the chemical potential of the corresponding adsorptive in aqueous solution,

$$\mu_i = \mu_i^0 + \mathrm{RT}\ln(i)$$

where μ_i^0 is the standard-state chemical potential of the adsorptive with which the adsorbate has come into equilibrium, R is the molar gas constant, and (i) is the activity of the adsorptive in the aqueous solution contacting the adsorbent. It follows that thermodynamic stability imposes the following constraint:

$$\left(\frac{\partial \ln(i)}{\partial n_i^{(w)}}\right)_{\mathrm{T,P}} > 0$$

Therefore, increases in the amount of an adsorbate must follow increases in the aqueous solution activity of the adsorptive with which it is equilibrated, if the system is thermodynamically stable.

An important application is to the response of net adsorbed proton surface charge

$$\sigma_{\mathrm{H}} \equiv n_{\mathrm{H}}^{(w)} - n_{\mathrm{OH}}^{(w)}$$

to changes in pH [$\equiv -\log(\mathrm{H}^+)$]:

$$\left(\frac{\partial \sigma_{\mathrm{H}}}{\partial \mathrm{pH}}\right)_{\mathrm{T,P}} < 0 \qquad (1.50)$$

showing that *decreases in σ_{H} must follow increases in pH* in a stable adsorbent-aqueous solution system at fixed T and P. Equation 1.50 prescribes an essential stability condition to which each curve of σ_{H} versus pH must adhere.

A useful discussion of the thermodynamic stability concepts used here can be found in Chapter 6 of A. Münster, *Classical Thermodynamics*, John Wiley, New York, 1970.

$$\sigma_{\text{in}} + \sigma_{\text{S}} + \sigma_{\text{d}} = 0 \tag{1.52}$$

however, yields a relationship between p.z.n.c. and p.z.c. Suppose that $\sigma_S = 0$ at the p.z.n.c. Then σ_d must also vanish because of eq. 1.52 and the fact that $\sigma_{\text{in}} = 0$ at the p.z.n.c. But $\sigma_d = 0$ means pH = p.z.c. Therefore, p.z.c. = p.z.n.c. if $\sigma_S = 0$ at the p.z.n.c. Conversely, if p.z.c. = p.z.n.c., then $\sigma_{\text{in}} = 0 = \sigma_d$ and, again by eq. 1.52, $\sigma_S = 0$ of necessity. The general conclusion to be drawn is in the theorem:

2. *The p.z.c. is equal to the p.z.n.c. if and only if the Stern layer charge, σ_S, is zero at the p.z.n.c.*

QED

Note that PZC Theorem 2 is trivially true if the only adsorbed species are those in the diffuse ion swarm. If surface complexes exist, Theorem 2 will not hold unless the ions adsorbed in them (other than H^+ or OH^-) meet a condition of zero net charge at the p.z.n.c. This might occur for monovalent ions adsorbed "indifferently" in outer-sphere surface complexes by largely electrostatic interactions (e.g., Li^+ and Cl^-). Electrolytes for which $\sigma_S = 0$ at the p.z.n.c. are indeed termed *indifferent electrolytes*, in the sense that relatively weak electrostatic interactions dominate their adsorption. The p.z.c. of natural particles contacting solutions of indifferent electrolytes thus can be determined by ion adsorption measurements.

As originally conceived,[54] the Stern layer comprises both inner-sphere and outer-sphere surface complexes. If these species do not combine to yield zero net charge at the p.z.n.c., p.z.c. $\neq$ p.z.n.c., according to Theorem 2. The close relationship between p.z.c. and σ_S can be exposed further by consideration of the charge balance constraint in eq. 1.46 at the p.z.c.:

$$\delta\sigma_{\text{H}} + \delta\sigma_{\text{S}} = 0 \quad (\text{pH} = \text{p.z.c.}) \tag{1.53}$$

which refers to changes under which σ_p remains equal to zero. If the Stern layer charge is made to increase, say, by increasing the amount of surface-complexed cation charge, then the net proton charge must compensate this change by decreasing, which, according to the stability condition in eq. 1.50, requires the p.z.c. also to increase: the pH value at which $\sigma_H + \sigma_S$ balances σ_o must be higher, as σ_S becomes higher, in order that σ_H will be negative enough to meet the condition of charge balance. In the same way, the pH value at which $\sigma_d = 0$ must be lower, as σ_S becomes lower, in order that σ_H will become positive enough to compensate exactly the decrease in σ_S. This line of reasoning is epitomized in the theorem:

3. *If the Stern layer charge increases, the p.z.c. also increases, and vice versa.*

QED

Theorem 3 indicates the role of cation surface complexation in increasing p.z.c., and that of anion surface complexation in decreasing p.z.c. It does *not* imply, however, that shifts in p.z.c. necessarily signal the effect of *strong* ion adsorption, since changes in the number of outer-sphere surface complexes are sufficient to change p.z.c.

These properties of the points of zero charge, although useful in the characterization of particle charge, also illustrate the essential ambiguity of macroscopic chemical measurements when faced with adducing molecular mechanisms. The lack of uniqueness among points of zero charge as signatures of underlying molecular behavior is poignantly expressed in eq. 1.47, which provides a rigorous macroscopic connection between p.z.s.e. and a true point of zero charge. If Δq and its first derivative with respect to ionic strength both vanish at the same pH value, then p.z.s.e. = p.z.n.c. If $\sigma_o = 0$ also, then p.z.s.e. = p.z.n.p.c., as desired. Little more can be said without some detailed model of the electrical double layer, which, of course, may turn out to be incomplete as a molecular description.

1.5 Ion Adsorption Trends

The three categories of adsorbed ion species identified in sections 1.2 and 1.3, and further epitomized as the contributors to adsorbed ion surface charge on the right side of eq. 1.40, can be related as well to conventional parlance in the study of natural particle reactions with aqueous electrolyte solutions.[39,50] The diffuse ion swarm and outer-sphere surface complex mechanisms of adsorption are likely to involve electrostatic bonding, whereas inner-sphere surface complex mechanisms are likely to involve bonding in which electron configurations become shared. Since this latter type of bonding depends significantly on the electron distribution of both the surface functional group and the complexed ion, it is appropriate to consider inner-sphere surface complexation as a molecular basis for the term *specific adsorption*. Correspondingly, diffuse ion adsorption and outer-sphere surface complexation are a molecular basis for the term, *nonspecific adsorption*, the "nonspecificity" referring to a very weak dependence on the electron configuration of the surface functional group and adsorbed ion, as would be expected for the reactions of solvated species.

Readily exchangeable ions are adsorbed ions that can be displaced easily from natural particles by leaching with an indifferent electrolyte solution of prescribed composition, concentration, and pH value. Despite the empirical nature of this concept, there is consensus[55] that ions adsorbed specifically are not to be considered readily exchangeable. From this point of view, *only fully solvated adsorbed ions* are readily exchangeable ions. Operationally, these ions will therefore exhibit adsorption edges or envelopes that are shifted significantly by changes in the concentration of the background electrolyte solution.[56]

Additional insight into the differences in behavior between readily exchangeable and specifically adsorbed ions can be obtained through the use of *Schindler diagrams*.[57] A Schindler diagram is a banded rectangle in which the charge properties of an adsorbent and an adsorptive are compared as a function of pH in the range normally observed for natural particles, say, pH 3–9.5. The top band contains a vertical line denoting the p.z.n.c. of the adsorbent (fig. 1.13). The central band contains vertical lines denoting either the p*K value for hydrolysis (that based on water as a reactant, for metal cations) or the log K value for protonation (that based on the proton as a reactant, for ligands) of the adsorptive. The bottom

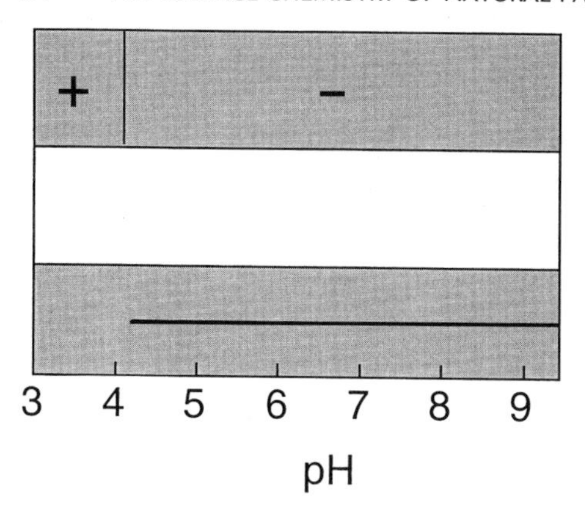

Fig. 1.13 Prototypical Schindler diagram for an adsorbent with low p.z.n.c. (clay minerals, humus, silica) reacting with monovalent cations, showing its broad pH range of functioning as a cation exchanger.

band depicts with a horizontal line the range of pH over which adsorption is to be expected when based solely on unlike-charge attraction for the adsorbent and the adsorptive. *This pH range, therefore, indicates a condition under which the adsorbent can surely function as a cation or anion exchanger. If adsorption is observed experimentally to occur at pH values outside this range, specific adsorption mechanisms are implied.*

As a first very simple example of a Schindler diagram, consider an adsorbent composed of clay minerals and humus primarily, with the adsorptive being an ion of an indifferent electrolyte (e.g., Li^+, Na^+, Cl^-, or NO_3^-). The p.z.n.c. of the adsorbent will not likely exceed 4.0,[39,41] and the p*K value for metal hydrolysis as well as log K for anion protonation of the indifferent adsorptives always will lie outside the range between 3 and 9. Therefore, the Schindler diagram will feature a top band with a vertical line at pH 4 (or to its left), a central band that has no vertical lines, and a bottom band with either a horizontal line extending to the right of pH 4 (cations) or one extending to the left of pH 4 (anions). It follows that adsorbents comprising principally humus and clay minerals (e.g., soils from temperate grassland regions) will function effectively as cation exchangers under most conditions. Conversely, adsorbents comprising principally iron and aluminum oxides (e.g., uncultivated soils from the humid tropics), for which p.z.n.c. > 7 typically,[39,42] will function effectively as anion exchangers. These trends are illustrated for adsorptive cations in fig. 1.13.

A second example can be developed for the particle-aqueous solution system described in fig. 1.1, in which the adsorbent is amorphous $Fe(OH)_3$ and the adsorptives are the trace metal cations Pb^{2+}, Cu^{2+}, and Cd^{2+}. The relevant p.z.n.c. value is 7.9,[58] and the respective p*K values are 7.7, 8.1, and 10.1. Therefore, the Schindler diagram for this system features a top band with a vertical line at pH 7.9, a central band with vertical lines at pH 7.7 and 8.1, and a bottom band with a horizontal line from pH 7.9 to 9.5 (fig. 1.14). The rather narrow range of pH over which the adsorbent can function as a cation exchanger is apparent, as is the conclusion that specific adsorption mechanisms must be operating in the reactions of Pb^{2+} and Cu^{2+} with $Fe(OH)_3$ (*am*), since their adsorption edges (fig. 1.1) plateau at pH $\lesssim$ p.z.n.c. while the adsorbent surface

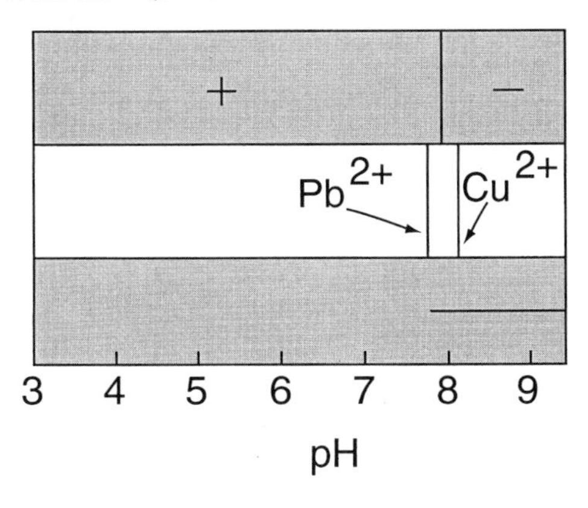

Fig. 1.14 A Schindler diagram for the system described in fig. 1.1, showing the typical functioning of iron oxides as anion exchangers over a broad range of pH. The pH values for the onset of metal cation hydrolysis (reduction of cation charge) are indicated for Pb^{2+} and Cu^{2+}. The narrow range of pH for which purely electrostatic cation adsorption mechanisms are possible ($pH > 8$) is noteworthy.

is still positively charged. The adsorption edge for Cd^{2+}, on the other hand, only commences for pH > p.z.n.c. and, therefore, is consistent with a cation exchange mechanism.

The same approach can be used to analyze the adsorption envelope in fig. 1.2, which refers to a calcareous soil reacting with borate in a NaCl background electrolyte solution. The relevant p.z.n.c. value is 9.5,[58] and log K for $B(OH)_4^-$ is 9.23. Therefore, the corresponding Schindler diagram has a top band with uniformly positive adsorbent surface charge indicated, a central band with a vertical line at pH 9.2, and a bottom band with a horizontal line extending over the very narrow range of pH between 9.2 and 9.5 (fig. 1.15). Quite clearly, then, specific adsorption mechanisms are involved in the reaction of borate with this soil. The relatively sharp peak in the adsorption envelope at a pH value approximately equal to log K for borate protonation, however, bears scrutiny (fig. 1.2). At pH values below 9.2, borate anions exist to some degree and are attracted to the positively charged adsorbent in increasing numbers as pH increases from 7 to 9. At pH values above 9.2, the adsorptive is predominantly anionic, but now the adsorbent is also becoming increasingly negatively charged, leading to a sharp fall-off in the adsorption envelope at pH > p.z.n.c. Thus, the *resonance feature* in fig. 1.2 can be interpreted as the net effect of interplay

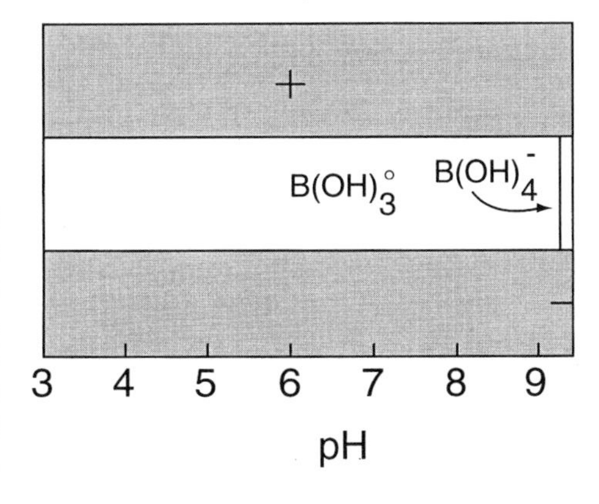

Fig. 1.15 A Schindler diagram for the system described in fig. 1.2, showing the broad range of pH over which particles in a calcareous soil can attract anions, and the extremely narrow range of pH over which borate adsorption can be governed by purely electrostatic mechanisms.

between ligand charge and adsorbent charge. Because specific adsorption mechanisms play a strong role in the reaction between ligand and absorbent, however, the resonance feature will be broadened relative to that resulting from charge relationships alone.

The Schindler diagram is not restricted to inorganic adsorptives, of course. Adsorption envelopes for aromatic carboxylic acids reacting with γ-Al_2O_3, for example, exhibit shapes that are similar to that for borate in fig. 1.2.[59] An interesting example is phthalic acid, whose two carboxyl groups have log K values of 2.95 and 5.41 for protonation. Given the p.z.n.c. value of 8.7 for the aluminum oxide adsorbent,[58] a Schindler diagram for this system would indicate that anion exchange is possible up to pH 8.7. The observed "ligand-like" adsorption envelope (fig. 1.16) is consistent with this prediction and can be deconvoluted conceptually into a pair of broad resonances that represent the adsorption of the two possible anionic species.[59] If the values of log K for a diprotic ligand adsorptive are even more widely separated (e.g., glycine, with log K values of 2.35 and 9.78), then the pair of corresponding resonances will appear as two distinct peaks in the adsorption envelope.[57] The overall conclusion to be drawn is that *the interplay of ligand and adsorbent charge can produce resonances in an adsorption envelope.*

A comparison of the sequence of pH_{50} values for the adsorption edges in fig. 1.1 with the sequence of p*K values for the three adsorptive metal cations shows quickly that the two parameters are correlated positively, that is, low selectivity implies a high pH value for hydrolysis. This kind of positive correlation, noted long ago,[60] has been apparent in many studies of metal cation adsorption by oxide or oxyhydroxide minerals.[61] It is not surprising. In conceptual terms, it amounts to the general rule, that *metal cations which hydrolyze readily will adsorb strongly* (i.e., adsorb at pH values well below their p*K value). From a local coordination perspective, the complexation of a metal cation with OH^- in aqueous solution is

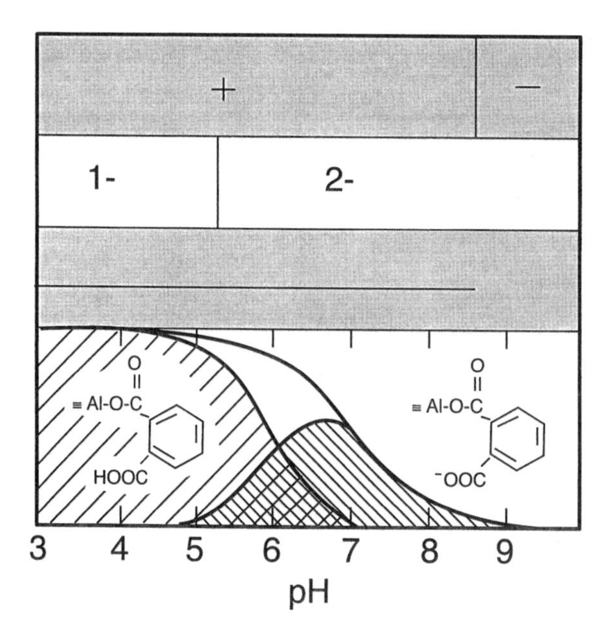

Fig. 1.16 A Schindler diagram for the reactions of γ-Al_2O_3 with phthalic acid anions, showing a close correspondence with the observed[59] adsorption edge. A chemical model of the reactions based on the two postulated surface species indicated was used to deconvolute the adsorption edge into a broad resonance, with a plateau between pH 3 and 5, and a narrower resonance with a maximum near pH 6.7.

analogous to the inner-sphere complexation of the metal cation to an ionized surface hydroxyl group, with the role of the proton in solution now being played by the metal in the adsorbent structure to which the surface hydroxyl is bonded. Similarly, the strength of adsorption of an anion to an oxide mineral has been observed to be correlated positively with its log K for protonation.[62] From a local coordination perspective, *anions that protonate strongly will adsorb strongly*, the complexation of a proton in aqueous solution being analogous to that with a proton on a positively charged adsorbent.

Notes

1. For a discussion of IUPAC terminology in colloid and surface chemistry, see D.H. Everett, *Pure App. Chem.* **31**:577 (1972).
2. Experimental methods for investigating adsorption reactions on natural particles are discussed by M.C. Amacher, pp. 19–59 in *Rates of Soil Chemical Processes*, D.L. Sparks and D.L. Suarez (Eds.), Soil Science Society of America, Madison, WI, 1991. See also N.J. Barrow and T.C. Shaw, *J. Soil Sci.* **30**:67 (1979), for a useful discussion of the effects on adsorption reactions from varying either θ_w (in eq. 1.4) or the vigor of physical mixing.
3. Quantitation of chemical species in aqueous media is described critically in D.L. Sparks (Ed.), *Methods of Soil Analysis, Part 3, Chemical Methods*, Soil Science Society of America, Madison, WI, 1996.
4. See pp. 219–237 in *The Scientific Papers of J. Willard Gibbs*, Vol. I, Longmans, Green, and Co., New York, 1906. As pointed out by Gibbs, the concept of relative surface excess obviates any need to specify the location of an adsorbent surface.
5. For this and other basic chemical concepts, see I. Mills, T. Cvitaš, K. Homann, N. Kallay, and K. Kuchitsu, *Quantities, Units and Symbols in Physical Chemistry*, Blackwell Scientific Publications, Oxford, 1993.
6. See, e.g., G. Sposito and R. Prost, *Chem. Rev.* **82**:553 (1982).
7. C. Amrhein and D.L. Suarez, *Soil Sci. Soc. Am. J.* **54**:999 (1990). The data for this example appear in table 2 of this article, but the present calculation does not consider the negative adsorption of anions other than chloride.
8. See, e.g., chapter 8 in G. Sposito, *The Chemistry of Soils*, Oxford University Press, New York, 1989 for discussions of these basic concepts.
9. The empirical classification of adsorption isotherms is developed fully by C.H. Giles, T.H. MacEwan, S.N. Nakhwa, and D. Smith, *J. Chem Soc., London,* 3973 (1960); C.H. Giles, D. Smith, and A. Huitson, *J. Colloid Interface Sci.* **47**:755 (1974); C.H. Giles, A.P. D'Silva, and I.A. Easton, *J. Colloid Interface Sci.* **47**:766 (1974).
10. Useful working discussions are given by D.G. Kinniburgh, *Environ. Sci. Technol.* **20**:895 (1986) and D.A. Ratowsky, *J. Soil Sci.* **37**:183 (1986).
11. M.H. Kurbatov, G.B. Wood, and J.D. Kurbatov, *J. Phys. Chem.* **55**:1170 (1951). The term "Kurbatov plot" was coined by D.G. Kinniburgh and M.L. Jackson, pp. 91–160 in *Adsorption of Inorganics at Solid-Liquid Interfaces*, ed. by M.A. Anderson and A.J. Rubin, Ann Arbor Science, Ann Arbor, MI, 1981.
12. See pp. 32f in W. Stumm, *Chemistry of the Solid-Water Interface,* Wiley-Interscience, New York, 1992.
13. M.M. Benjamin and J.O. Leckie, *Environ. Sci. Technol.* **15**:1050 (1981), **16**:162 (1982).
14. G. Sposito, *Soil Sci. Soc. Am. J.* **46**:1147 (1982). See also I.M. Klotz, *Science* **217**:1247 (1982). In the literature of biophysical chemistry, figure 1.4 is known as a *Scatchard plot*.

15. For a comprehensive review of the many different adsorption isotherm equations used in chemical transport studies, see C.C. Travis and E.L. Etnier, *J. Environ. Qual.* **10**:8 (1981). None of the isotherm models surveyed by these authors has any unique chemical significance, of course.
16. J.A. Veith and G. Sposito, *Soil Sci. Soc. Am. J.* **41**:697 (1977).
17. D.G. Kinniburgh, J.A. Barker, and M. Whitfield, *J. Colloid Interface Sci.* **95**:370 (1983).
18. G. Sposito, pp. 217–228 in *Geochemical Processes at Mineral Surfaces*, J.A. Davis and K.F. Hayes (Eds.), American Chemical Society, Washington, DC, 1986.
19. A brief quantitative discussion of homogeneous coprecipitation reactions is given in chapter 3 of G. Sposito, *Chemical Equilibria and Kinetics in Soils*, Oxford University Press, New York, 1994.
20. Useful reviews of approaches to discerning adsorption from precipitation reactions on natural particles are given by L. Charlet, pp. 273–305 in *Chemistry of Aquatic Systems*, G. Bidoglio and W. Stumm (Eds.), Kluwer Academic, Boston, 1994; and by K.F. Hayes and L.E. Katz, pp. 147–223 in *Physics and Chemistry of Mineral Surfaces*, P.V. Brady (Ed.), CRC Press, Boca Raton, FL, 1996. Working accounts of both experiment and modeling are given by D.A. Dzombak and F.M.M. Morel, *J. Colloid Interface Sci.* **112**:588 (1986); J.M. Zachara, J.A. Kittrick, L.S. Dake, and J.B. Harsh, *Geochim. Cosmochim. Acta* **53**:9 (1989); P. Wersin, L. Charlet, R. Karthein, and W. Stumm, *Geochim. Cosmochim. Acta* **53**:2787 (1989); S.E. Fendorf, D.L. Sparks, M. Fendorf, and R. Gronsky, *J. Colloid Interface Sci.* **148**:295 (1992); L. Charlet and A. Manceau, *J. Colloid Interface Sci.* **148**:443 (1992); S.K. Lower, P.A. Maurice, S.J. Traina, and E.H. Carlson, *Am. Mineral.* **83**:147 (1998).
21. *High adsorptive concentration*: M.B. McBride, *Soil Sci. Soc. Am. J.* **43**:693 (1979); M.M. Benjamin, *Environ. Sci. Technol.* **17**:686 (1983). *Long reaction time*: J. Torrent, U. Schwertmann, and V. Barron, *Clays Clay Min.* **40**:14 (1992); S.L. Stipp, M.F. Hochella, G.A. Parks, and J.O. Leckie, *Geochim. Cosmochim. Acta* **56**:1941 (1992); C.C. Ainsworth, J.L. Pilon, P.L. Gassman, and W.G. van der Sluys, *Soil Sci. Soc. Am. J.* **58**:1615 (1994).
22. P.W. Schindler and W. Stumm, pp. 83–110 in *Aquatic Surface Chemistry*, W. Stumm (Ed.), John Wiley, New York, 1987.
23. P.A. O'Day, G.A. Parks, and G.E. Brown, *Clays Clay Miner.* **42**:337 (1994).
24. L. Spadini, A. Manceau, P.W. Schindler, and L. Charlet, *J. Colloid Interface Sci.* **168**:73 (1994).
25. Adsorbate structures are reviewed carefully by P.A. O'Day, *Rev. Geophys.* **37**:249 (1999).
26. A. Manceau, L. Charlet, M.C. Boisset, B. Didier, and L. Spadini, *Appl. Clay Sci.* **7**:201 (1992).
27. These concepts were introduced in G. Sposito, *Soil Sci. Soc. Am. J.* **45**:292 (1981).
28. G. Sposito, N.T. Skipper, R. Sutton, S.-H. Park, A.K. Soper, and J.A. Greathouse, *Proc. Natl. Acad. Sci. USA* **96**:3358 (1999).
29. F.-R.C. Chang, N.T. Skipper, K. Refson, J. A. Greathouse, and G. Sposito, pp. 88–106 in *Mineral-Water Interfacial Reactions*, D.L. Sparks and T.J. Grundl (Ed.), American Chemical Society, Washington, DC, 1998.
30. See, e.g., chapter 2 in G. Sposito, *The Chemistry of Soils*, Oxford University Press, New York, 1989.
31. J.R. Bargar, S.N. Towle, G.E. Brown, and G.A. Parks, *Geochim. Cosmochim. Acta* **60**:3541 (1996).
32. J.R. Bargar, S.N. Towle, G.E. Brown, and G.A. Parks, *J. Colloid Interface Sci.* **185**:473 (1997).
33. L. Spadini, A. Manceau, P.W. Schindler, and L. Charlet, *J. Colloid Interface Sci.* **168**:73 (1994).
34. J.R. Rustad, A.R. Felmy, and B.P. Hay, *Geochim. Cosmochim. Acta* **60**:1563 (1996).

35. B.A. Manning, S.E. Fendorf, and S. Goldberg, *Environ. Sci. Technol.* **32**:2383 (1998).
36. A. Manceau and W.P. Gates, *Clays Clay Miner.* **45**:448 (1997).
37. I.D. Brown, *Acta Cryst.* B**48**:553 (1992). For a comprehensive introduction to the bond valence concept, see I.D. Brown, pp. 1–30 in *Structure and Bonding in Crystals*, Vol. II, M. O'Keeffe and A. Navrotsky (Eds.), Academic Press, New York, 1981.
38. G. Sposito, pp. 291–314 in *Environmental Particles*, vol. 1, J. Buffle and H.P. van Leeuwen (Eds.), CRC Press, Boca Raton, FL, 1992.
39. See, e.g., chapter 7 in G. Sposito, *The Chemistry of Soils*, Oxford University Press, 1989. Layer charge also exists in some manganese oxides from isomorphic substitution of Mn^{3+} for Mn^{4+} and from vacancies, yielding σ_o values as low as -2 mol$_c$ kg^{-1}. [See B. Lanson, V.A. Drits, E. Silvester, and A. Manceau, *Am. Mineral.* **85**:826 (2000).]
40. S.J. Anderson and G. Sposito, *Soil Sci. Soc. Am. J.* **55**:1569 (1991); J. Chorover, M.J. DiChiaro, and O.A. Chadwick, *Soil Sci. Soc. Am. J.* **63**:169 (1999). See also L.W. Zelazny, L. He, and A. Vanwormhoudt, pp. 1231–12553 in D.L. Sparks (Ed.), *Methods of Soil Analysis*, Soil Science Society of America, Madison, WI, 1996.
41. See the discussion in B.K. Schroth and G. Sposito, *Clays Clay Miner.* **45**:85 (1997).
42. See, e.g., L.W. Zelazny et al., in D.L. Sparks (Ed.), *Methods of Soil Analysis*, Soil Science Society of America, Madison, WI, 1996.
43. A titration curve measures the *response* of an adsorbent to an input of acid or base and, therefore, cannot give information about the initial state of an adsorbent *before* the input. For a discussion of this issue, see G. Sposito, *Environ. Sci. Technol.* **32**:2815 (1998); **33**:208 (1999).
44. L. Charlet, P. Wersin, and W. Stumm, *Geochim. Cosmochim. Acta* **54**:2329 (1990).
45. See, e.g., M.E. Sumner and W.P. Miller, pp. 1201–1229 in D.L. Sparks, (Ed.), *Methods of Soil Analysis*, Soil Science Society of America, Madison, WI, 1996. Bivalent cations are selected so as to achieve maximal values of q_+ in eq. 1.27.
46. See, e.g., G. Sposito, pp. B-241 – B-263, in *Handbook of Soil Science*, M. Sumner (Ed.), CRC Press, Boca Raton, FL, 2000.
47. J. Chorover and G. Sposito, *Geochim. Cosmochim. Acta* **59**:875 (1995).
48. This approach was introduced by S.J. Anderson and G. Sposito, *Soil Sci. Soc. Am. J.* **56**:1437 (1992). Applications of the approach to Spodosols, including "Chorover plots," are presented in T.A. Polubesova, J. Chorover, and G. Sposito, *Soil Sci. Soc. Am. J.* **59**:772 (1995) and in U. Skyllberg and O.K. Borggaard, *Geochim. Cosmochim. Acta* **62**:1677 (1998).
49. See, e.g., L. Liang and J.J. Morgan, *Aquat. Sci.* **52**:32 (1990) for a useful discussion of this more general concept of a point of zero charge.
50. A useful discussion of all aspects of fundamental colloidal behavior can be found in R.J. Hunter, *Introduction of Modern Colloid Science,* Oxford University Press, New York, 1993.
51. L. Charlet and G. Sposito, *Soil Sci. Soc. Am. J.* **51**:1155 (1987).
52. M.C. Fair and J.L. Anderson, *J. Colloid Interface Sci.* **127**:388 (1989).
53. G. Sposito, *Chimia* **43**:169 (1989).
54. O. Stern, *Z. Elektrochem.* **30**:508 (1924).
55. See, e.g., chapter 4 in W. Stumm, *Chemistry of the Solid-Water Interface*, Wiley-Interscience, New York, 1992.
56. J.A. Davis and D.B. Kent, *Rev. Mineral.* **23**:177 (1990).
57. P.W. Schindler, *Rev. Mineral.* **23**:281 (1990).
58. S. Goldberg, S.M. Lesch, and D.L. Suarez, *Soil Sci. Soc. Am. J.* **64**:1356 (2000).
59. R. Kummert and W. Stumm, *J. Colloid Interface Sci.* **75**:373 (1980).

60. P.W. Schindler, B. Fürst, R. Dick, and P.U. Wolf, *J. Colloid Interface Sci.* **55**:469 (1975).
61. P.W. Schindler and G. Sposito, pp. 115–145 in *Interactions at the Soil Colloid-Soil Solution Interface*, G.H. Bolt et al. (Eds.), Kluwer Academic, Boston, 1991. F. Wang, J. Chen, and W. Forsling, *Environ. Sci. Technol.* **31**:448 (1997).
62. L. Balistrieri and T.T. Chao, *Soil Sci. Soc. Am. J.* **51**:1145 (1987).

For Further Reading

W. Stumm, *Geoderma* **38**:19 (1986) and P. W. Schindler, *Pure Appl. Chem.* **63**:1697 (1991). These position papers, by two of the founders of natural particle surface chemistry, provide brief overviews of the role of coordination phenomena in ion adsorption reactions. An equally brief (and compelling) update on the seminal ideas of Stumm and Schindler is provided by L. Charlet, (see note 20).

P. A. O'Day, *Rev. Geophys.* **37**:249 (1999). This remarkable review presents a comprehensive guide to the concepts and methodology of molecular-scale mineral surface geochemistry, a most useful companion to the present chapter.

G. E. Brown et al., *Chem. Rev.* **99**:77 (1999). This encylopedic review of metal oxide structure and surface reactivity represents the consensus of fourteen leading scientists on the major achievements and trends in an important area of natural particle geochemistry, including microbial interactions with mineral surfaces.

P. M. Huang, N. Senesi, and J. Buffle, Eds., *Structure and Surface Reactions of Soil Particles*, John Wiley: New York, 1998. An IUPAC-sponsored volume that offers detailed insights into soil particle surface chemistry in chapters 1, 8, 10, 11, and 12.

D. Langmuir, *Aqueous Environmental Geochemistry*, Prentice Hall: Upper Saddle River, NJ, 1997. Chapters 9 and 10 in this advanced textbook provide brief, applications-oriented surveys of the concepts discussed in the present chapter.

Research Matters

1. E. M. Murphy and J. M. Zachara [*Geoderma* 67:103 (1995)] have compared typical "metal-like" adsorption edges for trace metals (e.g., Cu^{2+}) with typical "ligand-like" adsorption envelopes for humic substances (e.g., humic acid) on metal oxide surfaces. They predicted that the metal adsorption edge would be shifted upward at low pH but downward at high pH, thereby exhibiting a maximum at some intermediate pH, if the metal binds strongly to the humic substance and the latter is present along with the metal oxide absorbent (see fig. 8 in their article). Given that eq. 1.12 applies simultaneously to (1) metal adsorption by the oxide, (2) metal binding by the humic substance, and (3) humic substance adsorption by the oxide, develop an explanation for this conclusion that is related to the parameters, b and pH_{50}, based on three separate applications of eq. 1.12.
2. R. Kummert and W. Stumm [*J. Colloid Interface Sci.* 75:373 (1980)] investigated the adsorption of four aromatic organic acids by an aluminum oxide (γ-Al_2O_3) suspended in 100 mol m^{-3} $NaClO_4$ solution.
   (a) Evaluate their method by which a σ_H versus pH curve (their fig. 3a) for the adsorbent was determined (their eq. [9] is the same as eq. 1.32, with their a the same as c_s in the present chapter). Estimate the p.z.n.c. using the data presented in their fig. 3.
   (b) Evaluate the shift of the titration curve as a result of phthalic acid addition (0.5 mol m^{-3}), that is, the graph of pH versus $|c_A - c_B|$ in their fig. 4a, in terms of the stability condition in eq. 1.46. N.B. This evaluation requires careful scrutiny of their eq. [10].

(c) Prepare Schindler diagrams for the four organic acids investigated using the graphs in their fig. 7. Compare the predictions of these diagrams with the adsorption envelopes shown in the figure.

(d) Estimate pH_{50} for each adsorption envelope in their fig. 7 using the "ligand-like" portion (ignore that for benzoic acid, since $D_{ben} < 1$ for all pH values). Examine your results for positive correlation with the sum of log K values for protonation of the organic acids:

Organic Acid	log K_1	log K_2
Catechol	9.4	12.80
Phthalic	2.95	5.41
Salicylic	2.97	13.74

N.B. Correlation with the *sum* of log K values is indicated by the participation of both aqueous solution species in the proposed surface speciation of the adsorbates.

3. J. R. Bargar, S. N. Towle, G. E. Brown, and G. A. Parks [*J. Colloid Interface Sci.* 185:473 (1997)] have applied the bond valence concept and Pauling Rule 2 (eq. 1.23) to evaluate the Brønsted acidity of surface hydroxyl groups on aluminum oxides. Using eqs. 1.24 and 1.25, prepare a table of surface species like those in tables 2 and 3 in their article for surface oxygen ions coordinated to 1, 2, or 3 Al^{3+} and, for each type of coordination to Al^{3+}, coordinated as well to 0, 1, or 2 H^+. Take $|Z_O| = 2.0$ in making your evaluation of how closely eq. 1.23 is satisfied. Ignore hydrogen bonding and, for the species with protonated surface oxygen ions, replace the terms "most stable," "plausible," and "doesn't occur" with your own terms indicating their relative Brønsted acidity.
4. J. Chorover and G. Sposito [*Geochim. Cosmochim. Acta* 59:875 (1995)] have measured the three components of surface charge in eq. 1.44 for several kaolinitic soils from the humid tropics. The principal data are presented in table 1 and in figs. 1 and 2 of their article.
   (a) Estimate the p.z.n.p.c. for each soil at each background electrolyte (LiCl) concentration. What correlation exists between p.z.n.p.c. and LiCl concentration? Explain this correlation in terms of the stability condition in eq. 1.47 by considering both positive and negative shifts in the LiCl concentration, at pH = p.z.n.p.c., from 0.005 mol kg^{-1} taken as the initial value.
   (b) Why is it not generally appropriate to use p.z.n.p.c. to determine the range of pH over which soils can function as cation exchangers?
5. D. G. Kinniburgh, J. A. Barker, and M. Whitfield [*J. Colloid Interface Sci.* 95:370 (1983)] compare the "two-site Langmuir" isotherm (eq. 1.13) with the Tóth isotherm, $n_i^{(w)} = bKc_1/[1 + (Kc_i)^\beta]^{1/\beta} (0 < \beta < 1)$ where b, K, and β are empirical parameters, in respect to modeling the sorption of Ca^{2+} and Zn^{2+} by ferrihydrite.
   (a) Derive an equation for K_{di} as a function of $n_i^{(w)}$ analogous to eq. 1.14 by making use of eq. 1.8. Then plot K_d versus n for $0 \le n \le b$ and compare your result to the trend in the data plotted in fig. 1.4. Are the two conditions met in your plot for fitting Tóth adsorption isotherm data to eq. 1.13?
   (b) Show that the Tóth isotherm parameters b and K are related to the "two-site Langmuir" isotherm parameters according to the equations $b = b_1 + b_2$, $K = (b_1K_1 + b_2K_2)/b$.
   (c) Show that the Tóth parameter β is the solution to the optimization problem of finding the value of the exponent that makes the equation $y^\beta + x^\beta = 1$ true for all $0 \le y \le 1$, $0 \le x \le 1$, where $y \equiv K_d/bK$, $x \equiv n/b$ and y is the function of x that follows from the relation between K_d and n derived in (a).
   (d) Kinniburgh et al. show that the "two-site Langmuir" and Tóth isotherms correspond to dramatically different "site affinity distribution functions" (SADFs; see their fig. 6), which describe the distribution of Langmuir affinity parameters in a heterogeneous adsorbent whose sites have been grouped into classes, each of which is assumed to adsorb according to the classic Langmuir

isotherm, $n_i^{(w)} = bKc_i/(1 + Kc_i)$ where b is the maximum value of the surface excess and K is the Langmuir affinity parameter. Evidently, eq. 1.13 corresponds to a SADF with two sharp spikes, at $K = K_1$ and K_2. The Tóth SADF is, by contrast, a continuous distribution with strong "negative skew" toward very small values of K. Review the commentary on the interpretation of SADFs given by Kinniburgh et al., then develop your own position on the issue of the chemical interpretation of a SADF.

2

The Spectroscopic Detection of Surface Species

2.1 Strategy

Surface complexation models have been applied successfully to soil colloids to bring matters full circle; but, like their predecessors, they rely solely on prior molecular concepts and are tested only by goodness-of-fit to adsorption data. Since the model assumptions are so different and the models so plausible, one is left to wonder what physical truth they bear. One fears that the fog will lift only to reveal a Tower of Babel.

It is inevitable that methodologies not equipped to explore molecular structure will produce ambiguous results in the study of surface speciation. The method of choice for investigating molecular structures is spectroscopy. Surface spectroscopy, both optical and magnetic, is the way to investigate surface species, and thus to verify directly the molecular assumptions in surface speciation models. When the surface species are detected they need not be divined from adsorption data, and the choice of a surface speciation model from the buffet of available software becomes a matter unrelated to goodness-of-fit.[1]

As in all subdisciplines of chemical science, the surface chemistry of natural particles has as its ultimate goal a *mechanistic* understanding of the surficial reactions these particles undergo in aqueous media.[2] Taking for example the adsorption of Cd^{2+} by an hydroxylated mineral surface, one can envision at least two different chemical equations on which to base the mathematical modeling of the adsorption edge (illustrated in fig. 1.1):

$$\equiv \mathrm{FeOH} + \mathrm{Cd}^{2} \leftrightarrows \equiv \mathrm{FeO}^{-} \cdots \mathrm{Cd}^{2+} + \mathrm{H}^{+} \qquad (2.1a)$$

$$\equiv \mathrm{FeOH} + \mathrm{Cd}^{2+} \leftrightarrows \equiv \mathrm{FeOCd}^{+} + \mathrm{H}^{+} \qquad (2.1b)$$

where $\equiv$ FeOH represents a reactive hydroxyl group bound to Fe(III) on the surface of $Fe(OH)_3$ (*am*) and the adsorbed Cd^{2+} products on the right side are either outer-sphere (eq. 2.1a) or inner-sphere (eq. 2.1b) surface complexes, respectively. (Left out of this simple example are the possible formation of diffuse layer species and hydrolyzed Cd^{2+} species bound in surface complexes.) The *thermodynamic* equilibrium constants[2] for the two alternative adsorption reactions in eq. 2.1 have the same mathematical form in terms of activities because both reactions exhibit the same stoichiometry:

$$K_{ads}^{OS} = \frac{(\equiv FeO^{-} \cdots Cd^{2+})\,(H^{+})}{(\equiv FeOH)\,(Cd^{2+})} \tag{2.2a}$$

$$K_{ads}^{IS} = \frac{(\equiv FeOCd^{+})\,(H^{+})}{(\equiv FeOH)\,(Cd^{2+})} \tag{2.2b}$$

where the superscript on the left side of each equation identifies the type of surface complex formed and ( ) represents thermodynamic activity.[2] But any theoretical model of the corresponding *conditional* equilibrium constant,[2] which has to provide a calculation of the activity coefficient of adsorbed Cd^{2+}, must then differentiate significantly between the mathematical forms of K_{adsC}^{OS} and K_{adsC}^{IS}, because the molecular structures of outer-sphere and inner-sphere surface complexes differ considerably. (See figs. 1.7–1.10, in which the dipolar nature of outer-sphere complexes, indicated on the right side of eq. 2.1a, is clearly apparent.) Accordingly, the pH dependence of Cd^{2+} adsorption—manifest in fig. 1.1 and rooted theoretically in the pH-dependent behavior of the conditional equilibrium constant for the adsorption reaction—would reflect the details of the model used to calculate an adsorbate activity coefficient.[3]

At least this would be true if the model featured no adjustable parameters, requiring as input only molecular properties determined independently of adsorption data (e.g., ionic radii). Unfortunately, extant models of adsorbate activity coefficients suitable for natural adsorbents *do* contain parameters that are difficult to measure independently (e.g., surface densities of reactive functional groups or capacitance densities determined by the degree of local charge separation in surface complexes) and, therefore, are treated as adjustable.[4] Current mathematical models thus have sufficient flexibility through parameter ambiguity to provide an accurate representation of adsorption edges, envelopes, or isotherms.[5] For example, the adsorption edge for Pb^{2+} on the hydroxylated adsorbent, γ-Al_2O_3, has been modeled equally well with either an outer-sphere or an inner-sphere surface complexation mechanism assumed,[6] and the net adsorbed proton charge on this adsorbent can be described accurately as a function of pH with a variety of models that assume *any* mechanism of background electrolyte adsorption, from diffuse layer to inner-sphere surface complex.[7] This kind of mechanistic ambiguity calls directly for molecular-level experimentation designed to detect the surface species that actually have formed.[8]

Spectroscopic techniques suitable for the detection of surface species on natural particles have undergone much development in instrumentation and accessibility over the past decade.[9] Out of this development has come a sharper understanding of the four key physical criteria necessary to an accurate spectroscopic identification of adsorption reaction products. A spectroscopic method must be: *selective*, able to distinguish adsorbate from both adsorbent and adsorptive when all three contain the same ion or molecule; *sensitive*, able to detect surface species at environmentally relevant concentrations; *noninvasive*, able to characterize adsorbate structure in the presence of an aqueous solution, and with minimal sample preparation; *interpretable*, amenable to accurate deconvolution when complex spectra result, and to realistic modeling in terms of a few molecular parameters. The spectroscopic approaches to be surveyed in the present chapter have, in the consensus of experts,[8] proven to meet these four criteria reasonably well.

Spectroscopic methods are designed to yield data on the transitions among the energy levels accessible to a chemical species, in the present case, an adsorbed species. These data (the spectral response of an adsorbate over a range of input photon energy) are then interpreted to elucidate molecular structure, often by comparison with related spectra of suitable model compounds or, more quantitatively, by a calculation of molecular properties (coordination number, ligation, bond length and orientation, oxidation state, etc.) based on a theoretical model of adsorbate molecular structure. This kind of interpretation is possible because spectral data are assumed to provide information about *local* bonding environments in an adsorbate through their effects on the transitions among adsorbate energy levels. Thus, the ion, atom, or molecule of an adsorbate is expected to show a different spectral response from what it does when it is an adsorptive not interacting with an adsorbent surface.

Two important, long-standing variations on this strategy to deduce molecular structures of adsorbates should be mentioned at this juncture.[10] One is the use of *adsorbate surrogates* as in situ probes to explore adsorption mechanisms. The concept here is to select a chemical species whose spectral response in a variety of molecular environments is well characterized, then utilize this species as an adsorbate to determine the bonding environment for homologous adsorptives. For example, *o*-phosphate ions could be reacted with an hydroxylated mineral surface and their spectral response as an adsorbate (e.g., vibrational or nuclear magnetic resonance spectra) could be used to determine the generic characteristics of inner-sphere surface complexation of oxyanions by the mineral.[11] Similarly, Cu^{2+}, which has well-characterized electron-spin-resonance and X-ray absorption spectra, or $^{113}Cd^{2+}$, which exhibits nuclear magnetic resonance, could be doped into a 2:1 clay mineral-water system primarily adsorbing Ca^{2+} to explore adsorption mechanisms for bivalent cations.[12] A second useful variation on the direct application of spectroscopy to an adsorbate is the use of *reporter units* in an adsorbent to signal the presence and bonding characteristics of an adsorbate. Examples of this approach include monitoring of: the vibrational spectra or ^{13}C, ^{1}H nuclear magnetic resonance spectra of carboxylate groups exposed on the surface of humus; the vibrational spectra of reactive structural hydroxyls in mineral adsorbents; and the nuclear magnetic resonance spectra of structural $^{29}Si^{4+}$ or $^{27}Al^{3+}$ in mineral adsorbents.[13] The information provided by these two common alternatives to direct probing of an adsorbate can add much to the credibility of a proposed adsorption mechanism.

Irrespective of the transition processes investigated by a spectroscopic method, the spectral data produced always represent the *spatially integrated* response of an adsorbate, adsorbate surrogate, or reporter unit, with the averaging lengthscales ranging from micrometers to centimeters. This inherent feature of spectroscopic methods, a facet of instrumental design and sensitivity, has significant implications for the interpretation of spectral data. The first is that these data will not distinguish any effects of *molecular-scale* heterogeneity in the structure of an adsorbent surface, unless these effects result in a major shift of adsorption mechanism. For example, variability in layer charge among single-layer particles of 2:1 clay minerals will be averaged in the spectral response of adsorbed metal cations, since their mode of adsorption does not change dramatically as layer charge increases.[14] On the other hand, variability in the distribution and reactivity of surface hydroxyl groups on different crystallographic planes of mineral adsorbents often is dramatic and can be detected in the spectral response of adsorbed metal cations.[15]

A second important implication of spatial averaging in spectral data is their resultant *convolution* of spectra from several different kinds of adsorbed chemical species associated with different adsorbent phases in the admixtures characteristic of natural particles. Deconvolution of the spectrum of an adsorbate to infer speciation in this case depends sensitively on whether the number of different adsorbed species is sufficiently small to ensure a unique resolution into component spectra; whether the spectroscopic method is sufficiently sensitive to detect all adsorbed species present; whether the spectral response of each species is truly a measure of its population size; and whether the database of model compound spectra available is sufficiently large to permit accurate comparisons with in situ sample spectra.[16]

Spectroscopic techniques are legitimate probes of adsorbate molecular structure, not because of their spatial resolution of molecular behavior, but because they explore adsorbate states over *molecular time scales* (fig. 2.1). Transitions among energy levels corresponding to different accessible states are provoked by the photons input to an adsorbate from a spectral source, the type of transition (or elementary excitation) being dependent on the energy borne by the photon and, therefore, on the frequency of the electromagnetic radiation it constitutes.

Figure 2.1 illustrates a broad range of electromagnetic radiation categories (gray boxes in the second column) and the components of atomic or molecular structure that adsorb them to undergo transitions among energy levels. These transitions are known to be affected by the *local coordination environment* of the atom or molecule in which they occur, a point that will emerge clearly in subsequent sections of the present chapter. Thus, the spectral response associated with each type of excitation in fig. 2.1 provides useful information about the local molecular structure surrounding the nucleon, electron, or chemical bond that is perturbed by an input of photons to engage in transitions among accessible energy levels. With each category of input photon there is an associated frequency of electromagnetic radiation, and with each frequency there is an associated time scale (inverse of frequency), shown in the first column of fig. 2.1. This time scale, during which excitation occurs and molecular behavior is probed by a spectroscopic method, ranges over 18 orders of magnitude.

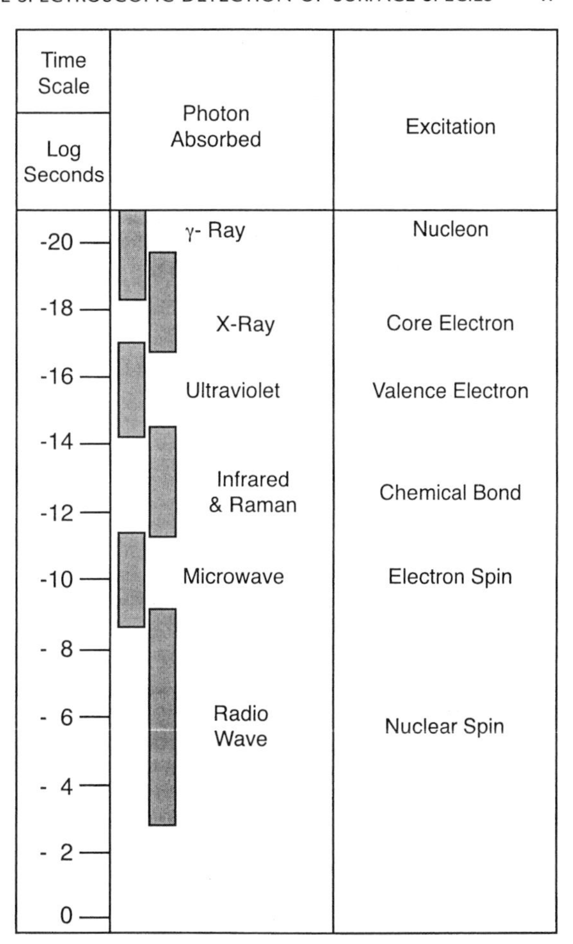

Fig. 2.1 Time scales associated with spectroscopic probes of adsorbate molecular structure. The second column indicates the spectral regions that are used to excite resonance conditions in the atomic structure components listed in the third column. After O'Day.[2]

The physical significance of the timescales in fig. 2.1 can be appreciated more fully once a set of convenient "time markers" has been introduced. This is usually done in terms of the molecular behavior of liquid water.[17] The time scale for the vibration of a hydrogen bond in liquid water under ambient conditions is about 10^{-13} s (or 0.1 ps), approximately in the center of the "Infrared & Raman" box in the second column of fig. 2.1. Therefore, the transitions provoked in an adsorbate by γ-ray, X-ray, UV-visible, and some infrared photons will occur on time scales smaller than that over which the bond between two water molecules in the bulk liquid can vibrate once. In particular, X-ray and UV-visible photons cause transitions that are effectively *instantaneous* on the vibratory time scale. This disparity implies that chemical bonds in the vicinity of an adsorbate molecule undergoing transitions from X-ray or UV-visible photons will be in a variety of instantaneous states corresponding to the panoply of configurations that can occur for chemical bonds (e.g., many different stretched or bent configurations of the bonds). By contrast, microwave and radio wave photons cause transitions that occur over rather long timescales compared to the vibratory one and, therefore, they will be affected only by *vibrationally averaged states* of an adsorbate that correspond to the average vibratory structure of its chemical bonds.

Another useful time marker is the time scale for the diffusion of a "tagged" water molecule in the bulk liquid. This time scale is about 5 ps—say 10^{-11} s for simplicity. It follows that, for example, X-ray, UV-visible, and infrared photons induce transitions that are largely unaffected by translational motions in the vicinity of an adsorbate, whereas microwave and especially radio wave photons induce transitions that will be affected by translational motions of neighboring molecules. Thus, these latter transitions reflect *diffusionally averaged* molecular structure to some degree; that is, the coordination environment probed is more nearly like the "equilibrium structure" deduced with techniques, such as X-ray or neutron diffraction, which explore the ordered arrangements of atoms on long time scales.[18] What should be apparent from this brief summary is that *different spectroscopic methods operate over different intrinsic time scales and, therefore, give complementary information about adsorbate molecular structure.*

These perspectives can be illustrated by a brief consideration of two examples drawn from opposite extremes of the range of time scale in fig. 2.1. On the shortest timescales indicated in the figure, surface Mössbauer spectroscopy can provide information about the instantaneous molecular structure of an adsorbate. This technique involves the absorption of γ-ray photons by the nuclei of certain isotopes of a small group of chemical elements, most importantly ^{57}Fe, with interpretation of the resulting spectral data as to the oxidation state [Fe(III) versus Fe(II)] and the local coordination environment (octahedral versus tetrahedral, degree of structural disorder, etc.).[19] Surface Mössbauer spectroscopy extends the approach taken to characterize bulk mineral structures by examining the interfacial region of an Fe-bearing adsorbent which has been ^{57}Fe enriched.[20] Enrichment of an adsorbate provides direct insight as to the configurations of adsorbed metal cations, whereas surface enrichment of the adsorbent permits its structural ^{57}Fe to serve as *reporter units*.

This latter methodology[21] is optimal for use with specimen Fe-oxide minerals that have been synthesized with ^{56}Fe only—thus making them transparent to Mössbauer spectroscopy—which then are reacted with an $FeCl_3$ solution containing only ^{57}Fe ions. For reaction times of minutes to hours, enrichment of the structural Fe at or near the adsorbent surface occurs by isotopic exchange. The ^{57}Fe-enriched adsorbent then can be reacted with an adsorptive of choice and the Mössbauer spectra of the ^{57}Fe reporter units can be used to characterize adsorbate structure. Application of this protocol to strong arsenate adsorption by ferrihydrite[21] gave the interesting result that arsenate coordination to a surface site decreased the degree of disorder in the vicinal Fe(III) bonding environment; that is, arsenate adsorption (by inner-sphere surface complexation[22]) stabilized nearby Fe(III) octahedra at the adsorbent surface. Prior to reaction with arsenate ions, comparison of the ^{57}Fe Mössbauer spectra of a surface-enriched and a natural-abundance sample of ferrihydrite revealed significant disorder in the surface Fe(III) octahedra (elongated and weakened Fe-O bonds), as would be expected for polyhedra not ensconced in a bulk crystalline structure. Because of the very short time scale over which the nuclear transitions producing Mössbauer spectra take place, these differences in disorder must reflect all possible vibrational configurations of the Fe-O bonds under conditions in which the participating ions are effectively stationary. The reduction in distortion of the surficial Fe(III) octahedra induced by arsenate adsorption then can be interpreted broadly as additional evidence for the concept of inner-sphere surface complexation being an organizing process akin to crystal growth (section 1.2).

At the other extreme of time scale relative to Mössbauer spectroscopy are the techniques of X-ray and neutron diffraction. These methodologies determine the angular distribution of photons or neutrons which have been scattered elastically and coherently by either electrons or nuclei, respectively, in an adsorbate.[17] Since coherent scattering phenomena refer to macroscopic time intervals, diffraction methods provide information about the diffusionally averaged molecular structure of an adsorbate. Recent studies of concentrated aqueous solutions[23] have made use of isotopic-difference neutron diffraction to examine molecular structure in significant detail. This technique involves the subtraction of diffraction patterns of samples whose only difference in composition and state is in enrichment with an isotope of a metal cation that has a significantly different coherent scattering probability from the most abundant isotope (e.g., replacement of $^{7}Li^{+}$, which constitutes 92.5% of natural Li^{+} and is effectively incapable of coherent neutron scattering, by $^{6}Li^{+}$, which scatters neutrons coherently quite well). This approach has been applied with great success to a variety of aqueous electrolyte solutions[23] and currently is proving useful in elucidating adsorbate molecular structure on the basal planes of hydrated 2:1 clay minerals.[24] For example, $^{6}Li/^{7}Li$ neutron diffraction experiments have revealed outer-sphere surface complexes between Li^{+} and sites on the siloxane surface of vermiculite hydrated by two layers of water molecules.[25] These surface complexes, whose distorted octahedral character was revealed by $^{2}H/^{1}H$ neutron diffraction experiments on the interlayer water,[25] represent the equilibrium or static structure of the adsorbate which survives in spite of exchange among both adsorbed Li^{+} and adsorbed water molecules over time scales on the order of hundreds of picoseconds.[17]

2.2 X-ray Absorption Spectroscopy

X-ray absorption spectroscopy (XAS) involves the detection and quantification of X-ray attenuation by a sample containing chemical elements of interest whose atoms have electrons in energy levels that are sufficiently low-lying ("inner-core") to respond to input X-ray photons. In practice, given the four criteria for optimal spectroscopic studies of natural particles, as stated in section 2.1, chemical elements with atomic numbers above 20 are the most accessible to XAS, although the lower limit can be placed even at 5 if the four criteria are relaxed somewhat.[26] As indicated in fig. 2.1, XAS probes molecular structure on a time scale that is very short when compared to that for the vibration of a hydrogen bond in liquid water. Therefore, XAS gives information about the "instantaneous" configurations of adsorbed species and, of course, it averages their spatial distribution in a sample over length scales of micrometers and above.

The principal applications of XAS in surface geochemistry characterize surface complexes,[26] although coherent X-ray scattering techniques can be applied to detect both diffuse layer species and surface complexes.[26, 27] The two most widely used XAS methods investigate the near-edge structure (XANES) and the extended fine structure (EXAFS) of an X-ray absorption spectrum for a selected chemical element, as illustrated in fig. 2.2.[26] The XANES portion of the spectrum arises from the absorption of incident photons whose energies form a narrow band around that at which the chemical element begins to attenuate X-rays (termed the *absorption edge*, E_0, whose value increases with atomic number). These photons excite bound electrons into unoccupied energy levels that still represent bound electronic states in the absorber atom (fig. 2.2). Eventually the electrons return to their initial states through the emission of X-ray photons or through the transfer of energy—without radiation—to another electron, which then is ejected from the atom.[28] The spectral features observed in XANES are particularly indicative of the oxidation state of an absorber atom because of the strong effect of valence and of electron delocalization on the "inner-core" electrons that can absorb radiation at X-ray frequencies. The symmetry properties of the local coordination environment of the absorber atom also influence XANES, with high coordination symmetry restricting the classes of participating electronic orbitals and, therefore, decreasing the likelihood of the necessary compatibility between initial and final states of an electronic transition. For example, both the oxidation state and coordination symmetry for Mn (II versus III or IV; tetrahedral versus octahedral) can be adduced from XANES analysis.[26]

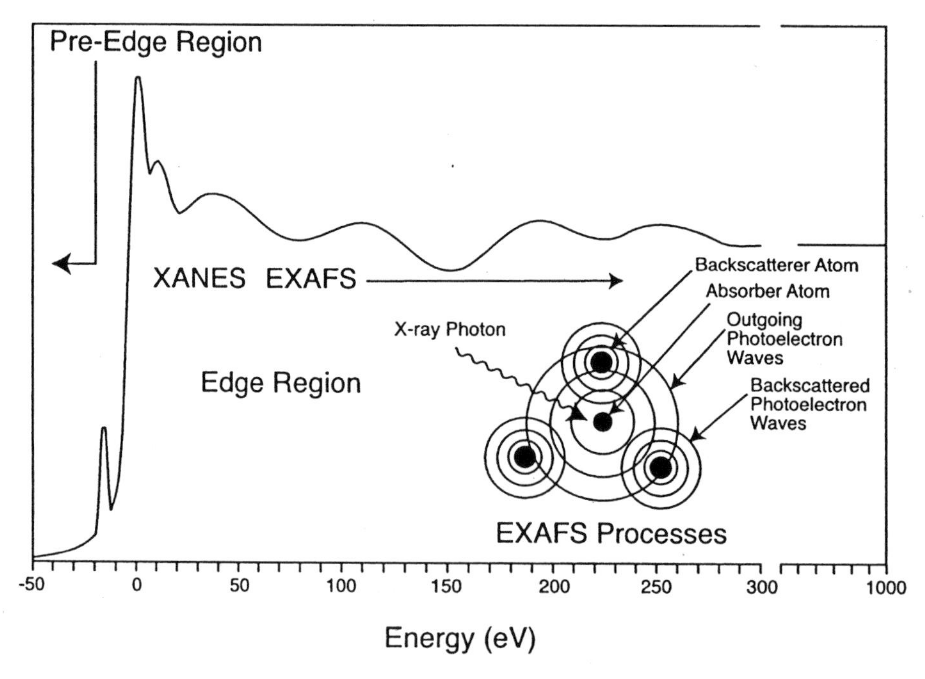

Fig. 2.2 Schematic drawing of an X-ray absorption spectrum, showing the XANES and EXAFS portions with energy measured relative to E_0 (the absorption edge). XANES spectra result from excitation into bound states, whereas EXAFS spectra result from the ejection of photoelectrons. After fig. 8 in Schulze and Bertsch.[26]

The EXAFS portion of the spectrum, by contrast, arises from the absorption of photons energetic enough to eject electrons completely from the absorber atom (fig. 2.2). These photoelectrons can be scattered coherently by neighboring atoms positioned within a few tenths of a nanometer from the absorber atom. This molecular-scale distance is comparable in magnitude to the de Broglie wavelength of the photoelectron (λ_B),[28] which means that interference is possible between the matter waves associated with the ejected and scattered electrons. The interference phenomenon, in turn, modulates the X-ray absorption spectrum in predictable ways, with peak amplitude affected by the type and population of the scattering atoms, and peak separation affected by their distance from the absorber atom.[26] Therefore, EXAFS analysis can give information about the ligation, coordination number, and bond distances in the local molecular environment of an absorber atom. Because XAS is a multielement technique, this information can be obtained about more than one atom in an adsorbate (and about atoms in the adsorbent selected to serve as reporter units) by tuning the spectrometer to different incident photon energies. The broad applicability and the inherent sensitivity of XAS have thus made it a spectroscopic method of choice in many studies of the surface chemistry of natural particles.[26, 29]

An EXAFS portion of an X-ray absorption spectrum [$\chi(k)$] is obtained by subtracting from an experimental spectrum that for an isolated absorber atom (as approximated by suitable fitting polynomials) and plotting the result against the de Broglie wavenumber of the photoelectron k [$\equiv 2\pi/\lambda_B$; $(\hbar k)^2 \equiv 2m_e (E - E_0)$].[28] Often the product $k^3\chi(k)$ or $k^2\chi(k)$ is plotted instead of the unweighted EXAFS spectrum to emphasize high-wavenumber oscillations.[30] If the EXAFS spectrum is a superposition of spectra associated with each different type of scattering atom near an absorber atom, then it is reasonable to deconvolute it using Fourier transformation to yield, by definition,[26] a one-dimensional representation of the relative probability of encountering a scattering atom at a molecular distance R (the Fourier transform variable conjugate to k) from the absorber atom. This mathematical representation of local structure around the absorber is a *radial distribution function* whose peaks indicate the significant presence of nearby atomic neighbors. Each peak is then separately back-transformed to yield a partial EXAFS spectrum. This piece of the original $\chi(k)$ is examined carefully for its compatibility with a suitable mathematical model, based on a selected ligation and coordination geometry, in order to identify the type, number, and distance of separation of the scattering atoms. The entire effort is facilitated by having at hand a catalog of experimental EXAFS spectra for the absorber atom studied in a variety of coordination environments, including those of differing crystalline structure but with the same pairing of absorber and scattering atoms as the one under investigation. As a rule, EXAFS analysis along these lines can provide accurate pictures of ligation; coordination number estimates good within a few percent; and bond distance estimates good within a few picometers for adsorbates containing a wide variety of chemical elements.[26]

These concepts of data analysis can be illustrated by a consideration of the experimental underpinnings[31] for the outer-sphere surface complex of Pb(II) that is depicted in fig. 1.8. The adsorbent is corundum (α-Al_2O_3), obtained in single-crystal form with its (0001) plane exposed preferentially (fig. 2.3). Examination of

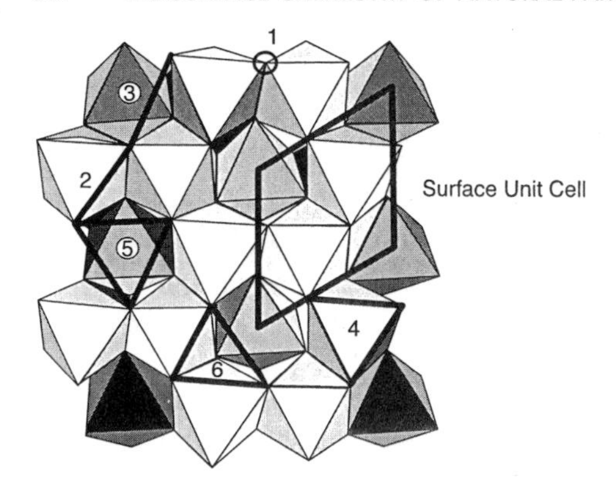

Fig. 2.3 The (0001) plane in α-Al_2O_3, showing six possible sites for Pb^{2+} adsorption: (1) bridging surface OH; (2), (3) octahedral edges; (4) octahedral faces; (5), (6) octahedral interstices. After fig. 1 in Bargar et al.[33]

the adsorbent surface using conventional methods showed it to be molecularly flat and clean, and free of dramatic effects of structural relaxation after exposure. This surface was reacted under CO_2-free N_2 gas with dilute $Pb(NO_3)_2$ in a 0.1 m $NaNO_3$ background electrolyte solution at pH 7 for 1.5 hours, then checked for adsorbate surface excess ($\Gamma_{Pb} = 0.10 \pm 0.05$ μmol m^{-2}) and composition (no nitrate present, 2 to 8 water monolayers estimated) using X-ray photoelectron spectroscopy before its investigation by XAS while under the N_2 atmosphere.

The spectrometer was tuned to the Pb(II) L_{III}-adsorption edge ($E_0 \approx 13.055$ keV) to produce fluorescence-yield XANES spectra with energy resolution between 6 and 16 eV.[32] The XANES portion of the spectrum occupied a band of energies running to about 100 eV above E_0, whereas the EXAFS portion of the spectrum ranged to about 1 keV above E_0. The latter spectral component was extracted after the background contribution (extrapolation of the pre-E_0 part of the spectrum to energies in the EXAFS region) and the single-absorber component [a quartic polynomial function of λ_B optimized for Pb(II) L_{III}-edge XAS] were removed numerically from the experimental spectrum.[30] The resulting EXAFS oscillations were then plotted as the $k^3\chi(k)$ form of weighted spectrum.

A small catalog of Pb(II) XANES spectra is provided in fig. 2.4.[31] The spectrum of Pb(II) adsorbed on the (0001) plane appears to be similar to that for the aquo-ion of Pb^{2+}, whereas it differs from the XANES spectra for Pb^{2+} adsorbed in known inner-sphere surface coordination; hydrolyzed in aqueous solution; bonded to oxygen in an oxide mineral; or bonded to oxygen in a lead nitrate precipitate. The feature in the adsorbed Pb(II) spectrum around 13.09 keV does differ from the broad second peak in the Pb^{2+} spectrum, suggesting that spectral effects of the adsorbent are present. It may be noted in passing that Pb^{2+} aquo-ions in aqueous solution or in crystalline hydrates are known to be coordinated to five or more oxygen atoms. The principal peak near 13.06 keV in their XANES spectra corresponds to a 2p $\rightarrow$ 6d electronic transition in the metal cation,[32] which has the orbital configuration $(5d)^{10}(6s)^2$.

The EXAFS portion of the spectrum was assumed to be a superposition of terms having the sinusoidal form[26]

$$\chi_j(k) = A_j(k)\sin[2kR_j + \phi_j(k)] \tag{2.3}$$

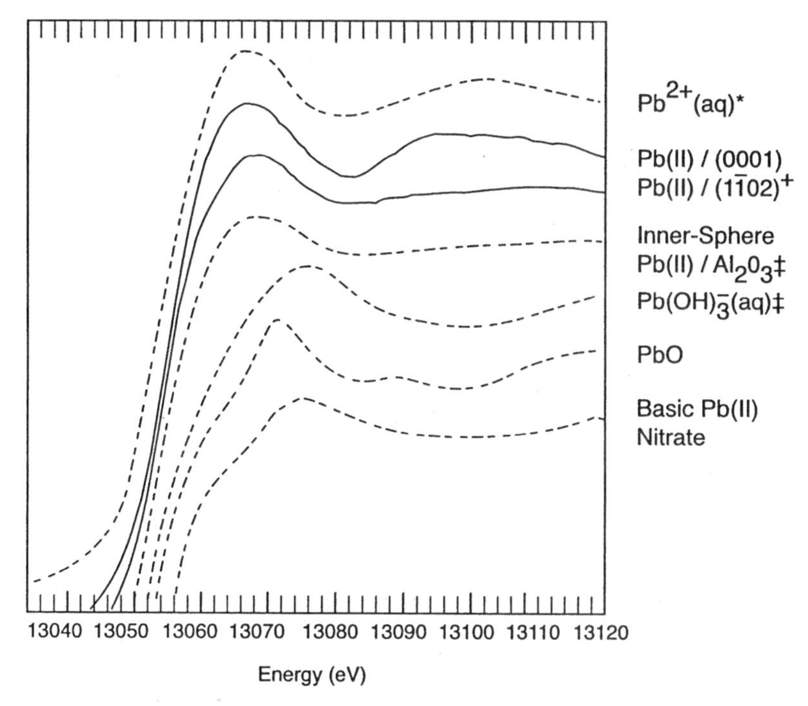

Fig. 2.4 A catalog of Pb(II) XANES spectra showing, from the top down, $Pb(ClO_4)_2$ in aqueous solution (*); Pb(II) adsorbed on the (0001) plane of α-Al_2O_3 bearing two to eight monolayers of water ("ex situ"); Pb(II) adsorbed on the $(1\bar{1}02)$ plane of α-Al_2O_3 in aqueous suspension ("in situ,"); Pb(II) in an inner-sphere surface complex on α-Al_2O_3; aqueous $Pb(OH)_3^-$; and two Pb(II) solids. After fig. 2 in Bargar et al.[33] with additional data.[32]

where

$$A_j(k) = F_j(k)\exp\{-2[\sigma_j^2 k^2 + (R_j/\lambda_j)]\}N_j/kR_j^2 \tag{2.4}$$

is the amplitude of a partial EXAFS spectrum contributed by a scattering atom j located at a distance R_j from the absorber atom and $\phi_j(k)$ is the phaseshift of the oscillation in $\chi_j(k)$. The first two factors on the right side of eq. 2.4 are the scattering amplitude (F_j) and its damping, which is caused by thermal motions and structural disorder among the scattering atoms (σ_j), as well as comparability between R_j and the mean free path (λ_j) of the photoelectron. These factors are either calculated theoretically or deduced from EXAFS spectra of known structural provenance which describe the same absorber-scattering atom pair, leaving only the coordination number (N_j) and R_j as adjustable parameters.[26,32] The phaseshift ϕ_j comprises terms for both the absorber and the scattering atom that are usually computed with theoretical models.[26,30,32] Experience indicates that the estimates of N_j and R_j based on this approach are accurate to within 20% and 0.005 nm, respectively.

The weighted EXAFS spectrum and corresponding Fourier transform to yield a radial distribution function are shown in fig. 2.5 for Pb(II) adsorbed on the

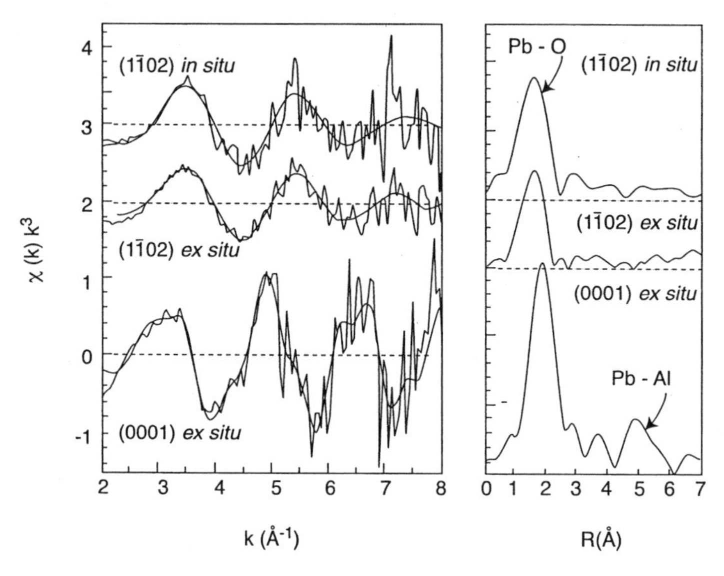

Fig. 2.5 Weighted EXAFS spectra (left) and their Fourier transforms (right, uncorrected for phase shift) for Pb(II) adsorbed on α-Al_2O_3. The peak at 0.49 nm (4.9Å) in the (0001) ex situ spectrum corresponds to 0.578 nm after phase-shift correction. The smooth curves through the EXAFS spectra correspond to model fits based on eq. 2.3. After fig. 2 in Bargar et al.[31]

(0001) and $(1\bar{1}02)$ planes in α-Al_2O_3.[33] The spectra for the inner-sphere surface complex on the latter plane differs markedly from that for the (0001) plane, which features a major peak at about 0.2 nm (2 Å) and a minor peak at about 0.5 nm (5 Å). (The ripple at about 0.38 nm is an artifact of the Fourier transform.) Independent back-transformation of these two peaks followed by the least-squares fitting of the resultant partial spectra to eq. 2.3 indicated that the layer peak can be attributed to 2.7 ± 0.5 O at 0.251 ± 0.003 nm from Pb^{2+}, while the smaller peak represents 3.5 ± 0.7 Al at 0.578 ± 0.003 nm from Pb^{2+}. The Pb-O distance is too large for direct coordination to a surface O (0.22 – 0.23 nm), but is consistent with those determined for Pb^{2+} aquo-ions (0.247 – 0.253 nm). The Pb-Al distance and coordination number are consistent with a solvated adsorbate cation that is bound to "site 5" indicated in fig. 2.3. (Sites 1 to 4 and 6 can be eliminated because their Pb-Al coordination number must be < 2 and/or their Pb-Al distances must be > 0.6 nm.)

An outer-sphere surface complex on an octahedral interstice is suggested, therefore, by the EXAFS analysis. (Strictly speaking, the Pb-Al coordination number for this complex should be exactly 3.0. The somewhat larger value observed experimentally indicates that structural relaxation of the adsorbent surface may have taken place, allowing some next-nearest neighbor Al^{3+} also to approach the adsorbed Pb^{2+}.)[33] The visualization in fig. 1.8 is based on the formation of hydrogen bonds between the three solvating water molecules (the observed Pb-Al distance permits only one solvation shell) and OH on the exposed (0001) plane of corundum. In section 1.2, Pauling Rule 2 (eq. 1.23) was applied to show that the stable state of the doubly coordinated oxygen ions on this plane is single protonation to form an uncharged, hydroxylated surface. This surface can adsorb water through hydrogen bonding, and the configuration in fig. 1.8 can be viewed as an example that serves also to stabilize the outer-sphere complex. If the hydrogen bonds give typical O-H $\cdots$ O distances of 0.272 nm, the Pb^{2+} cation will be located about 0.42 nm above the centers of the surface OH, as shown in the figure. Two additional solvating water molecules depicted so as to conform to the usual coordination number ($\gtrsim 5$) observed for Pb^{2+} aquo-ions.[30–32]

2.3 Infrared Spectroscopy

Infrared absorption spectroscopy involves the detection and quantification of optical radiation with wavelengths in the range 1 to 100 μm, corresponding to a time scale in the range 10^{-15} to 10^{-12} s (fig. 2.1). This time scale includes that for vibrational motions of molecular structures and, therefore, infrared spectroscopy is applied quite broadly to investigate these motions both in adsorbates and for reporter units (e.g., surface OH groups) in adsorbents.[34] However, it must be added that this spectroscopic approach does not fare so well as does XAS in respect to the four criteria for optimal spectroscopic studies of natural particles, described in section 2.1. In particular, the presence of water, a very strong absorber of infrared radiation, is not accommodated as well.[35]

Infrared spectrometers operate as Michelson interferometers in order to enhance their sensitivity. The interference pattern created by an absorbing sample that has been placed in the path of one beam in the matched pair produced by a beamsplitting mirror is detected in the time domain, then Fourier-transformed into the frequency domain.[34] Fourier-transform infrared spectroscopy (FTIR) is applied to study adsorbates in two principal ways, as diffuse reflection (DRIFT) or as attenuated total reflection (ATR) spectroscopy. The former methodology examines infrared radiation that emerges from a sample, after absorption, refraction, and scattering by its constituent particles, as a diffusely reflected beam. This beam typically has explored a depth below the sample surface ranging from 1 to 10 μm, with preferential spectral weighting accorded to components that are weak absorbers of infrared radiation.[36] A DRIFT spectrum thus can yield information about adsorbate structure readily, but it can also be confounded by absorption from liquid water, if the sample is wet, and by specular reflection, if the sample is high in clay content. The ATR methodology, on the other hand, does not suffer from these problems because the multiply reflected beam explores a much smaller depth below the sample surface, thus making ATR the technique selected for most FTIR studies of adsorbates. Sample presentation is usually as a wet paste which is deposited on the crystal reflection element, then isolated to mitigate evaporation. Companion FTIR spectra of the supernatant solution and the adsorbent without adsorbate are also obtained and used to bring out features of the adsorbate spectrum by digital subtraction. Careful control of the spectrometer settings, supernatant solution composition, and sample temperature are necessary to ensure a reliable subtraction of spectra, particularly in reference to liquid water.

The ATR methodology has been applied extensively to examine the configurations of organic and inorganic anions adsorbed on hydroxylated mineral surfaces.[37] Figure 2.6 shows ATR-FTIR spectra of salicylate [$C_6H_4OHCOO^-$, pK_a = 2.97] adsorbed on the edge surface of the clay mineral, illite,[38] evidently in surface complexes with exposed Al^{3+} in Lewis acid sites, by analogy with its adsorption on alumina.[39] Comparison spectra of salicylic acid in solid form as a crystal and in the dried supernatant solution also are shown. The adsorbent itself has no significant absorption peaks in the wavenumber range of the spectra.

Conventional interpretation of FTIR spectra is based on the assumption that similar vibrational frequencies for the same bond between two atoms will be observed regardless of the molecule in which the bond occurs. Thus, for example, the peaks in fig. 2.6 near 1600, 1550–1570, and 1450–1470 cm^{-1} may be assigned to stretching vibrations of the carbon–carbon bonds in the benzene ring of salicylate, whereas those near 1510–1530 and 1320–1340 cm^{-1} may be assigned to stretching vibrations of the COO^- group.[40] Differences in peak shape, as well as

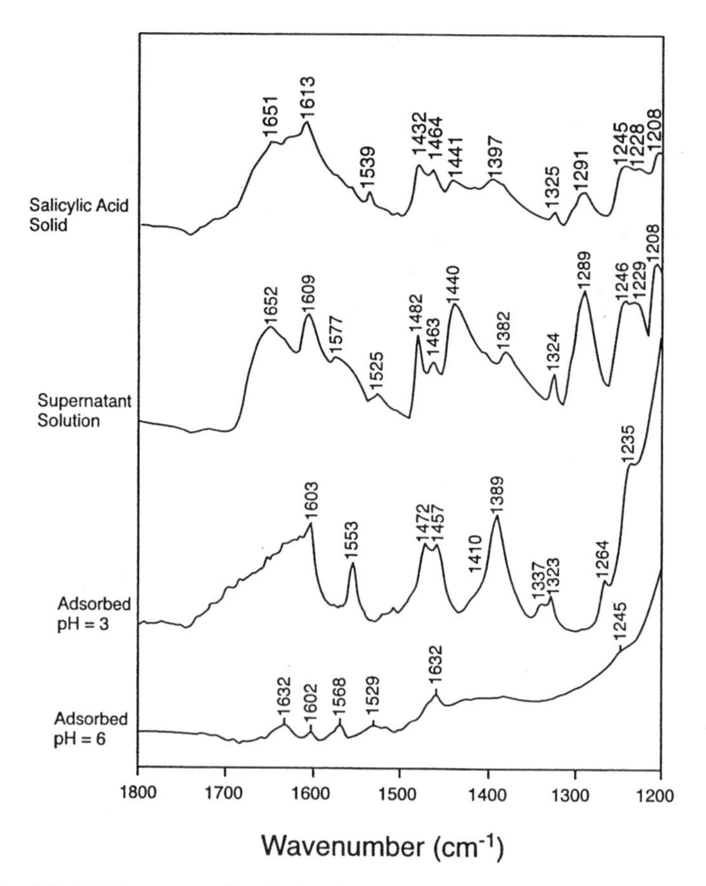

Fig. 2.6 ATR-FTIR spectra of salicylate in crystalline form; in a precipitate from a supernatant solution at pH 3; and adsorbed on the edge surface of illite at pH 3 or pH 6. After fig. 1 in Kubicki et al.[38]

wavenumber between spectra of salicylate in aqueous solution or in an adsorbate often have been useful guides to deducing the molecular configuration in the latter. Recently, this comparative approach has been replaced by molecular modeling of candidate structures for surface complexes with predictions of the corresponding vibrational frequencies. Some results of this approach are shown in fig. 2.7.[38] The x-axis of the graph at the bottom of the figure refers to the peak wavenumbers in the ATR-FTIR spectrum of salicylate adsorbed on illite at pH 3 (fig. 2.6), which matches the pK_a of the adsorptive and, therefore, should be near a resonance in its adsorption envelope (see section 1.5). The y-axis of the graph refers to the peak wavenumbers in the observed ATR-FTIR spectrum of salicylate adsorbed on Al_2O_3 at pH 5 (filled circles)[39] and to the vibrational frequencies calculated for the $Al(H_2O)_5$-salicylate complex illustrated above the graph (open squares). The slope of both plots in the graph is equal to 1 and the correlations between peak wavenumbers are very strong. The model complex features bonds between both O in the COO^- group and two Al^{3+}, together with a hydrogen bond between the phenolic OH group and one of the carboxyl O. This configuration results in a strong redshift of the carbonyl stretching vibration from near 1700 cm^{-1} in aqueous solution to near 1520 cm^{-1} in the model complex, thus providing a useful criterion for interpreting the ATR-FTIR spectrum. Moreover, the strong peak at 1389 cm^{-1} in the adsorbate spectrum (fig. 2.6) was found to correspond to bending vibrations of the CCH and COH benzene ring substituents, as would be supposed on the basis of aqueous solution spectra.[40]

Figure 2.8 shows ATR-FTIR spectra of phthalate [C_6H_4 $(COO^-)_2$, $pK_{a1} = 2.95$, $pK_{a2} = 5.41$] adsorbed on goethite (upper two spectra) along with the spectrum of the adsorptive in aqueous solution.[41] The two spectra of the adsorbate were obtained by deconvolution of ATR-FTIR spectra of adsorbed phthalate obtained over a broad range of pH and background electrolyte ionic strength.[41,42] The uppermost spectrum in fig. 2.8 is essentially that of the adsorbate at low pH and high ionic strength, whereas the middle spectrum is that of the adsorbate at high pH and low ionic strength. The upper spectrum has a prominent peak near 1420 cm^{-1} that has been assigned to a blueshifted C-O stretching vibration on the basis of a comparison to the FTIR spectrum of the inner-sphere $Al(H_2O)_4$-phthalate complex in aqueous solution.[43] The adsorbate spectrum is accordingly concluded to reflect inner-sphere surface complexation.[41–43] The lower spectrum, which is quite similar to that of phthalate in solution (bottom spectrum in fig. 2.8), shows a C-O stretching vibration near 1405 cm^{-1} and is attributed to an outer-sphere surface complex on the basis of a comparison to the spectrum of the outer-sphere Ca-phthalate complex in aqueous solution.[43] The observed ATR-FTIR spectrum of adsorbed phthalate over a broad range of pH and ionic strength can be fit well by a linear combination of these two "end-member" spectra, and the adsorption edge for phthalate can be reproduced by a linear combination of suitably normalized peak areas for the two C-O stretching vibrations.[41,42] These conclusions have been extended to other carboxylic acids and adsorbents.[41–43] Figure 2.9 illustrates the molecular configuration of the outer-sphere surface complex between phthalate and goethite.[41] Its similarity to the inner-sphere surface complex of Cd^{2+} with

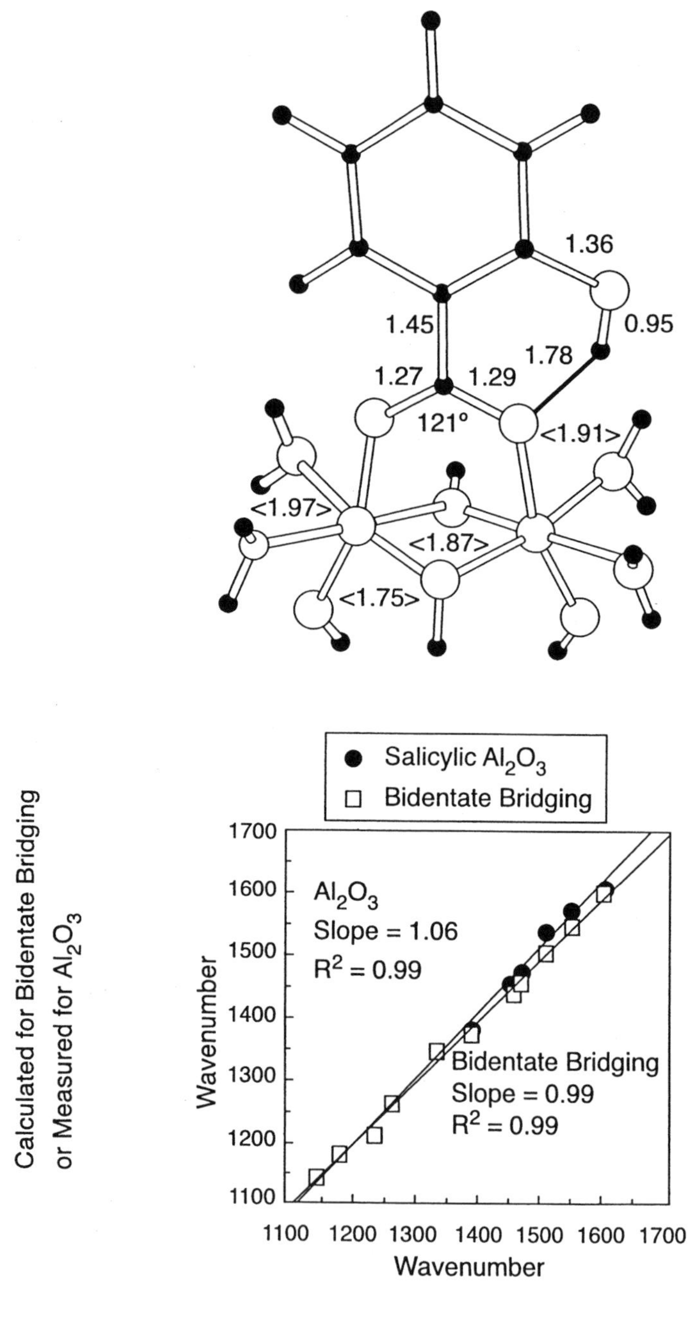

Fig. 2.7 Correlations between peak wavenumbers in the ATR-FTIR spectrum of salicylate adsorbed on Al_2O_3 at pH 5 (filled circles),[39] or of a model $Al(H_2O)_5$-salicylate complex (open squares, with the model complex itself shown above the graph), and peak wavenumbers in the ATR-FTIR spectrum of salicylate adsorbed on illite at pH 3. Numbers on the cartoon of the model complex are bond lengths (Å) or angles (°). After fig. 4 in Kubicki et al.[38]

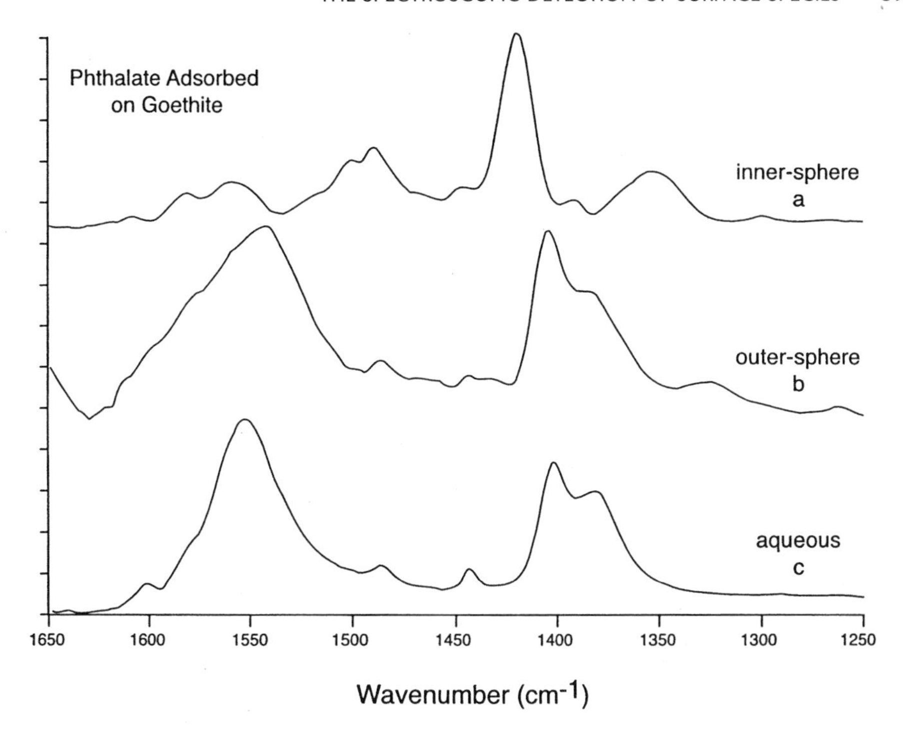

Fig. 2.8 ATR-FTIR spectra of phthalate (a) adsorbed on goethite at low pH and high ionic strength, (b) adsorbed on goethite at high pH and low ionic strength, and (c) in aqueous solution. After Boily et al.[41]

goethite (fig. 1.10) is notable. This surface complex is proposed to involve coordination to two adjacent $FeO_3(OH)_3$ octahedra through protonated surface OH groups, with stabilization by hydrogen bonding and a range of bond distances, consistent with the breadth of the C-O stretching peak for this surface species in fig. 2.8.[41]

Figure 2.10 shows ATR-FTIR spectra of borate [$B(OH)_4^-$, $pK_a = 9.23$] adsorbed on amorphous $Fe(OH)_3$ at two pH values and for several total boron concentrations.[44] On the basis of the ATR-FTIR spectrum of aqueous boric acid as a function of pH, the peaks in the wavenumber range 1200–1400 cm^{-1} were assigned to $B(OH)_3^0$, whereas that near 960–990 cm^{-1} was assigned to $B(OH)_4^-$. Their simultaneous presence in the spectrum indicates that surface complexes with both $-B(OH)_2$ and $-B(OH)_3$ units bonded through an O ion to Fe(III) in the adsorbent have formed on $Fe(OH)_3$ (*am*). The former surface complex would be favored for pH < 9.2, thus explaining the rising portion of the borate adsorption envelope in fig. 1.2. Indeed, inclusion of this neutral species in a chemical model of the adsorption envelope based on inner-sphere surface complexation leads to a better description than an approach involving only the charged species.[45] Similar interplay between ATR-FTIR spectroscopy and chemical modeling is useful for other oxyanion adsorptives (see section 4.2).[46]

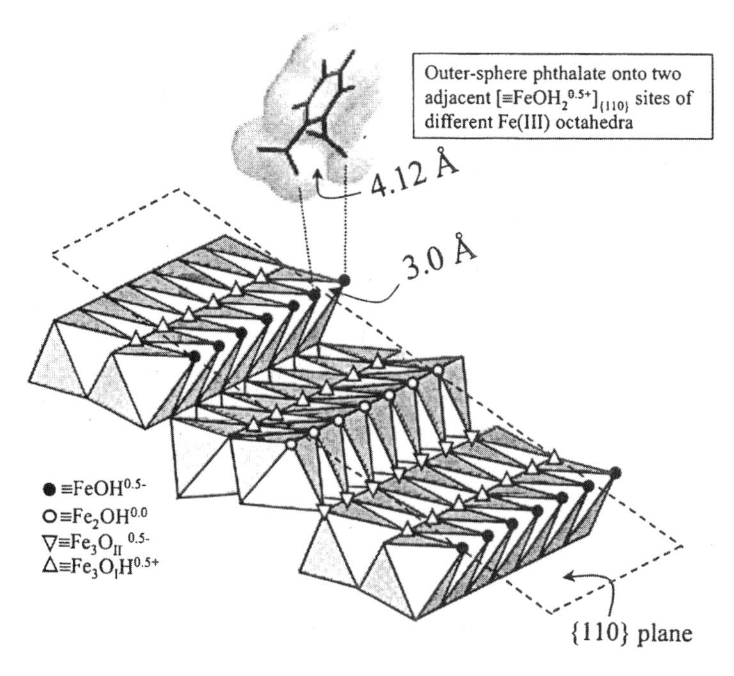

Fig. 2.9 Cartoon of an outer-sphere surface complex of phthalate on the (110) plane of goethite. Reprinted with permission from Boily et al.[41]

2.4 Electron Spin Resonance Spectroscopy

Electron spin resonance (ESR) spectroscopy involves the detection of optical radiation in the microwave region that is produced as a result of transitions between electron spin states in a paramagnetic ion.[47] These transitions have characteristic time scales between 10^{-11} and 10^{-9} s (fig. 2.1), long enough to be affected by translational and rotational molecular motions of an adsorbate and reflect only vibrationally averaged states of the molecular structure containing a paramagnetic species. In the study of adsorption by natural particles, ESR spectroscopy is applied mainly to in situ probes of the local coordination environment in an adsorbate, with the principal choices of probe being paramagnetic transition metals in certain oxidation states [V(IV), Cr(III), Mn(II), Fe(III), Mo(V), and, most importantly, Cu(II)] and stable organic radicals.[48] This approach fares reasonably well with respect to the four criteria of quality for spectroscopic studies of natural particles (section 2.1), its principal limitation being the relatively small set of paramagnetic species that is suitable for use. Especially successful are the many investigations of adsorbed water structure based on inferences from the ESR spectra of surrogate adsorbates on hydrated adsorbents.[49]

An ESR spectrum is generated by transitions between electron spin states when a sample, after placement in a static magnetic field, $\mathbf{B}_o$, is irradiated with microwave photons polarized at a 90° angle to the field direction.[47] The spectral lineshape depends on the local coordination environment (e.g., complexing ligands)

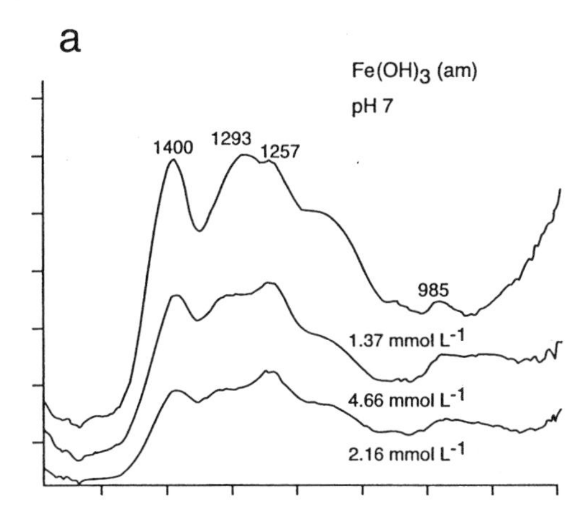

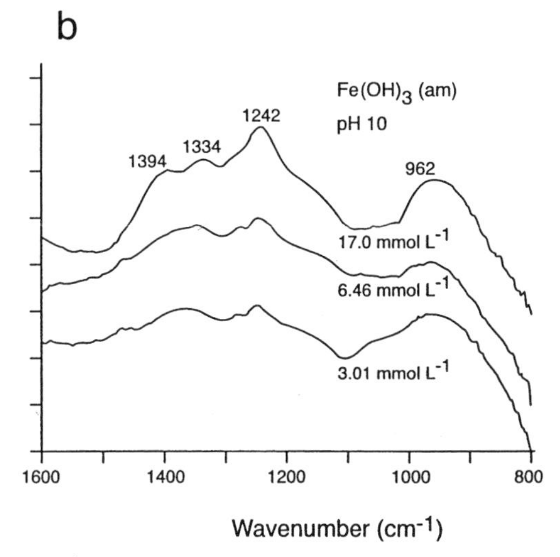

Fig. 2.10 ATR-FTIR spectra of borate adsorbed on $Fe(OH)_3$ (*am*) in the presence of NaCl solution at 25°C. (a) pH 7 and three borate supernatant solution concentrations, (b) pH 10 and three borate supernatant solution concentrations. After Su and Suarez.[44]

of the electrons undergoing the transitions, including the coupling of their spin magnetism with that of the nucleus of the atom in which they reside. These transitions are illustrated in fig. 2.11 for the single unpaired electron in Cu^{2+}, the exemplar probe species. An applied magnetic field in the kG range lifts the twofold degeneracy of the spin state $|1/2>$, so that resonance transitions can be induced by photons of frequency υ in the GHz range. Thus the ESR spectrum reports the energy required to reverse the spin direction of an electron ("spin flip") in an applied magnetic field. Fine structure (termed "hyperfine splitting") occurs in an ESR spectrum because of interactions between the electron spin and the nuclear spin (^{63}Cu or ^{65}Cu, $I = \frac{3}{2}$, fourfold degeneracy). The lineshape for Cu^{2+} in its typical tetragonal coordination (octahedral coordination, but with longer metal-ligand distances along the tetrad symmetry axis, around which, by definition, rotations through 90° leave the complex looking the same as originally) thus comprises a quadruplet of closely spaced peaks (fig. 2.11), if the tetrad symmetry

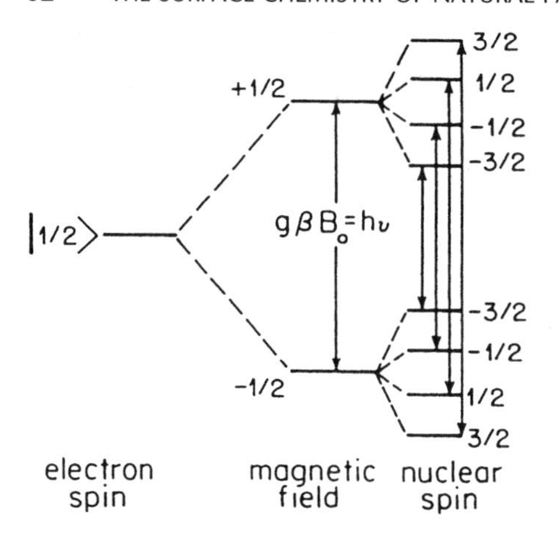

Fig. 2.11 Energy-level diagram for ESR "spin-flip" transitions (arrows) by the unpaired electron in Cu^{2+} (spin state $|1/2>$) in the presence of an applied magnetic field $\mathbf{B}_o$ and a magnetic nucleus (spin state $|3/2>$). The resonance equation involves the *g*-factor ($g \approx 2$) and the Bohr magneton ($\beta \equiv \mu_B = 9.274 \times 10^{-24}$ J T^{-1}) along with the Planck constant ($h = 6.626 \times 10^{-34}$ J s).

axis of the complex is either parallel or perpendicular to the direction of $\mathbf{B}_o$ over the *ca.* 100 ps time scale of the ESR transitions. This latter condition arises because the electron-nuclear interaction has different components along or at right angles to the tetrad symmetry axis.[47] Studies of adsorbates on natural particles usually involve powder samples, in which a broad range of orientations of the symmetry axis relative to $\mathbf{B}_o$ is likely, and, therefore, the resonance condition in fig. 2.11 will occur over a range of $\mathbf{B}_o$ values (for fixed υ, the usual convention in ESR spectroscopy). In this case, detection of the orientationally broadened peaks and measurement of the components of the interaction are facilitated by recording the first derivative of the lineshape with respect to B_o while the magnetic field is varied. The derivative-ESR spectrum exhibits "ripple" features at the positions of peaks in the absorption spectrum (see fig. 2.12, upper right side).[47,48]

An illustrative application of ESR spectroscopy is summarized in fig. 2.12. In the center of the figure is a visualization of an inner-sphere surface complex formed between Cu^{2+} and two oxygen sites on the surface of δ-Al_2O_3. The coordination of Cu^{2+} to oxygen ligands is tetragonal, with four of the ligands supplied by water molecules, two of which are positioned on the tetrad axis (*z*-axis) of the complex. The upper right corner of the figure shows a first-derivative ESR spectrum obtained at 25°C.[50] The left side of this spectrum is resolved into the four characteristic "ripples" that typify the electronic transitions in fig. 2.11. Their positions are marked by a line of tics, below which is given the value of the spectroscopic g-factor (fig. 2.11) associated with a parallel alignment between $\mathbf{B}_o$ and the tetrad symmetry axis.[47,50] The right side of the spectrum has only a single large "ripple" because the four characteristic transitions are not resolved. It corresponds to $\mathbf{B}_o$ directed perpendicularly to the tetrad symmetry axis. That $g_{||} > g_{\perp}$ for this ESR spectrum is a signal of a Cu^{2+} complex with tetragonal symmetry, as depicted in the visualization.[47,50]

The evident appearance of spectral signatures for both orientations of the surface complex demonstrates that *it is not rotating* at 25°C on the ESR time scale (*ca.* 100 ps). By contrast, in liquid water at 25°C, the $Cu(H_2O)_6^{2+}$ solvation complex produces a first-derivative ESR spectrum that features only a single large "ripple" at a g-value intermediate between $g_{||}$ and $g_{\perp}$.[50] This is because the solva-

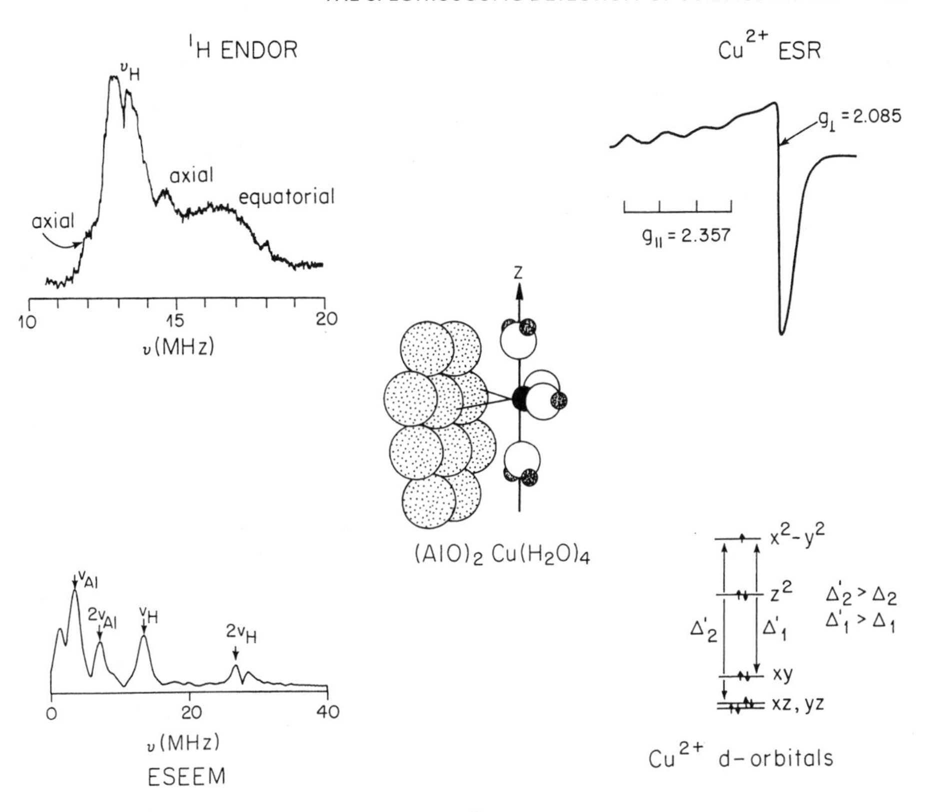

Fig. 2.12 Inner-sphere surface complex of Cu^{2+} on α-Al_2O_3 (center), with the corresponding *d*-orbital ESR transitions diagrammed in the lower right corner [Δ' is an orbital level separation in the surface complex, whereas Δ is that in the solvation complex, $Cu(H_2O)_6^{2+}$]. The first-derivative ESR spectrum of the surface complex is shown in the upper right corner, while its proton ENDOR and two-pulse ESEEM spectra are shown on the left side.

tion complex *is tumbling* on the ESR time scale. Reduction of the temperature to cryogenic levels is required in order to suppress the tumbling motion on the ESR time scale and resolve this spectrum into its two main components. This difference between the surface complex and the solvation complex of Cu^{2+} is reflected in the sketch of d-orbital energy levels that appears in the lower right corner of fig. 2.12. The measured separation of the levels between which the "spin-flip" transitions occur (Δ'_1, Δ'_2) is larger in the surface complex than in the solvation complex (Δ_1, Δ_2), indicating that some of the oxygen ligands in the former are bound more strongly to Cu^{2+} than in the latter.[47,50] Evidently these ligands are the adsorbent surface oxygens, which should constrain the surface complex against rotation.[51]

The ESR spectrum of Cu^{2+} adsorbed on δ-Al_2O_3 provides definitive evidence for surface complex formation, as opposed to adsorption in the diffuse swarm, which would yield a single-"ripple" lineshape like that for the solvation complex.[49] However, no quantitative information about the structure of the surface complex is provided, and the issue of whether it is truly inner-sphere is not settled unambiguously. One approach to resolving this conundrum is *electron-nuclear double resonance* (ENDOR) spectroscopy.[52] In this technique, both microwave (GHz) and radiowave (MHz) photons irradiate a sample simultaneously, the former being of high enough intensity to saturate an ESR transition (i.e., fully populate an upper energy level), while the latter induces transitions between nuclear spin states [*nuclear magnetic resonance* (NMR); see section 2.5] in a magnetic nucleus proximate to the paramagnetic ion undergoing ESR transitions. (As a general rule, this latter nucleus must be within 0.5 nm of the paramagnetic ion.) The result of this "double resonance" condition is a perturbation of the ESR lineshape that depends on frequencies of the NMR transitions and the strength of the electron spin-nuclear spin interactions induced.[52] The ESR spectral lineshape is thus altered in the radiofrequency range, such that pairs of NMR peaks appear with separations determined mainly by the distance between the interacting ion and nucleus.[53]

For the example of Cu^{2+} adsorbed on δ-Al_2O_3, the obvious choice of proximate magnetic nucleus is the proton, and a proton ENDOR spectrum obtained at cryogenic temperature is shown in the upper left corner of fig. 2.12.[52] The prominent doublet in the spectrum is split around the *Larmor frequency* for ^1H (13.3 MHz, denoted ν_H in the figure), which occurs because of the induced NMR transition. This frequency is defined analogously to the frequency ν in fig. 2.10, but for induced transitions between nuclear spin-$\frac{1}{2}$ energy levels (see section 2.5). The ENDOR pairs of NMR peaks are separated equidistantly from this frequency.[52] The very narrow doublet itself represents protons that are far away in the aqueous environment surrounding the surface complex ("matrix protons"). This conclusion follows on the basis of the inverse-cube dependence of an ENDOR peak separation on the distance between the paramagnetic ion and the stimulated magnetic nucleus (magnetic dipole-dipole interaction).[53] For the solvation complex, $Cu(H_2O)_6^{2+}$, the two peaks corresponding to the NMR spectrum of protons in the water molecules located on the tetrad axis have a separation of about 3.3–3.4 MHz.[53] These protons on the tetrad axis are the farthest removed from the Cu^{2+} ion in the solvation complex and, therefore, the tiny peak separation in the prominent doublet in fig. 2.12, 0.5 MHz, indicates that the "matrix protons" associated with it must be rather far from the adsorbed Cu^{2+} ion. Indeed, the two Cu-H distances should scale as $(3.35/0.5)^{\frac{1}{3}} = 1.9$, indicating that the narrow-doublet protons are about twice the distance from Cu^{2+} as are the water protons along the tetrad axis of the $Cu(H_2O)_6^{2+}$ complex. The two peaks to be associated with these latter protons in the surface complex are marked "axial" in fig. 2.12. Their separation is about 2.6 MHz, corresponding to a distance ratio of $(3.35/2.6)^{\frac{1}{3}} = 1.1$. Thus, the two water molecules on the tetrad axis of the surface complex are about 10% further away from adsorbed Cu^{2+} than they are in the solvation complex; that is, the surface complex is more elongated along its tetrad axis than is the solvation complex. Their distance from Cu^{2+} works out to be around 0.26 nm.[53] The protons in the two water molecules coordinated to Cu^{2+} in the same plane as the surface oxygen ions give rise to the broad feature in the ENDOR spectrum that extends from about 16 MHz to about 18.5 MHz (marked "equatorial"), which is also where it appears in the

ENDOR spectrum of the solvation complex.[53] Therefore, these water molecules must reside at a distance from Cu^{2+} that is similar to that for the four "in-plane" H_2O in the solvation complex, *ca.* 0.2 nm.[53] This latter distance is quite typical of Cu-O separations that are deduced from diffraction experiments on solvation complexes in aqueous Cu solvations.[17]

The proton ENDOR spectrum for the surface complex formed by Cu^{2+} on δ-Al_2O_3 has provided data consistent with the visualization of the complex in fig. 2.12, but no conclusive evidence for *inner-sphere* surface complexation was forthcoming, since nothing emerged that distinguished among the equatorial oxygen ligands binding to adsorbed Cu^{2+}. A technique that can add insight to this situation is *electron spin-echo envelope modulation* (ESEEM). In this technique, two or three microwave pulses are applied to a paramagnetic species to generate an "echo" (an induced magnetization that decays like the magnetization produced from a single pulse) whose amplitude, as a function of the time interval between pulses, is modulated by (dipolar) interactions between the paramagnetic electron and neighboring magnetic nuclei (e.g., protons) for which "spin flips" have been induced by the pulsed input.[52] The modulation pattern is sensitive to the type and number of nearest-neighbor nuclei and to their distance from the paramagnetic species. By computer simulation of the modulation pattern, one can estimate the coordination number of the species in terms of the nuclei.[54] Fourier transformation of the modulation pattern produces a spectrum whose principal peaks correspond to the Larmor frequencies of the vicinal nuclei and permit their identification.[52] In a logical sense, this technique is analogous to EXAFS spectroscopy (section 2.2), in that near-neighbor species modulate the amplitude of a signal induced by a perturbation of a target species in such a way as to reveal their number and identity.

The lower left corner of fig. 2.12 shows a two-pulse ESEEM spectrum of Cu^{2+} adsorbed on δ-Al_2O_3, obtained at cryogenic temperature.[55] Both the Larmor frequency and its first harmonic occur in the spectrum for ^1H ($\nu_H = 13.3$ MHz) and ^{27}Al ($\nu_{Al} = 3.4$ MHz) nuclei, showing that they are indeed near-neighbors of the paramagnetic Cu^{2+} species. Unfortunately, however, the difference in lineshape produced by inner-sphere versus outer-sphere complexation of Cu^{2+} to the surface oxygen ions appears only subtly, beyond the resolution of the spectrum.[55] Other ESEEM studies of Cu^{2+} adsorbed by clay minerals,[54,56] silica,[57] bacteria,[58] and calcium carbonate,[59] however, have been more successful in demonstrating the existence of inner-sphere surface complexation through direct simulation of the modulation pattern,[54–57] comparison of the Fourier-transform spectrum to that of suitable model complexes,[58,59] or resolution of unique features in the Fourier-transform spectrum that can be attributed to equatorial water molecules in the surface complex.[59]

2.5 Nuclear Magnetic Resonance Spectroscopy

Nuclear magnetic resonance (NMR) spectroscopy involves the detection of radiofrequency optical radiation produced by transitions between nuclear spin states in an atomic nucleus having either an odd number of nucleons or an odd number of protons.[60] These transitions occur on time scales between 10^{-9} s and 10^{-7} s (fig. 2.1), which means that they reflect not only vibrationally averaged states, but also rotationally and translationally averaged states of the molecular structure in which a magnetic nucleus is imbedded. The populations of these two latter states, however, are very sensitive to temperature, such that lowering the sample temperature to cryogenic levels (say, 173 K) can lengthen the time scales for rotations and translations sufficiently to free the NMR spectral lineshape from their effects. In this way, the details of atomic exchange between chemical species that differ in local molecular environment can be explored on time scales as long as milliseconds.[61] A variety of nuclei that figure prominently in the surface chemistry of natural particles is accessible to NMR spectroscopy,[60] among them ^{1}H, ^{2}H, ^{7}Li, ^{13}C, ^{15}N, ^{17}O, ^{19}F, ^{23}Na, ^{27}Al, ^{29}Si, ^{31}P, ^{35}Cl, ^{39}K, ^{113}Cd, ^{133}Cs. The proton and alkali metals among these elements are suitable in situ probes of adsorbates, as are the first-row nonmetals, P, and Cd; whereas Al and Si are useful as reporter units for oxide and silicate minerals. Thus NMR spectroscopy does well in meeting the four criteria of quality for spectroscopic studies of natural particles mentioned in section 2.1, the only exception being sensitivity. Adsorbate concentrations near millimolar (along with sample enrichment in the magnetic isotope of interest) usually are required in order to obtain spectra with good resolution.

A NMR spectrum provides information on the transitions between nuclear spin states that are promoted when a static magnetic field $\mathbf{B}_o$ is applied, with gradually increasing intensity, to a sample irradiated by radiofrequency photons polarized at a 90° angle to the field direction.[60] These nuclear spin transitions are analogous to that diagrammed in fig. 2.11 for two electron-spin energy levels if the nuclear spin quantum number also is $\frac{1}{2}$ (e.g., ^{1}H or ^{29}Si), with the difference for nuclei being that the factor β ($\equiv \mu_N = 5.051 \times 10^{-27}\ J\ T^{-1}$) in the resonance equation is about three orders of magnitude smaller than β for electron spin transitions ($\equiv \mu_B = 9.274 \times 10^{-24}\ J\ T^{-1}$), a reduction that is equal to the ratio of electron to proton mass.[60] It follows that the photon frequency leading to magnetic resonance (Larmor frequency) will similarly be reduced, and the molecular time scale probed will be increased inversely to this reduction (cf. fig. 2.1).

The NMR spectral lineshape is sensitive to the local molecular environment of a magnetic nucleus because neighboring electrons respond to $\mathbf{B}_o$ by reducing its magnitude, and, therefore, shielding the nucleus from the full effect of $\mathbf{B}_o$.[62] This diminishing of the field intensity acting on the magnetic nucleus is accounted for by introducing a factor $(1-\sigma)$ next to B_o in the resonance equation, where σ is the *mean principal chemical shielding*, a quantity of the order of 10^{-4} or less. It is equal to the average of the three elements along the principal diagonal of the *chemical shielding tensor*, $\bar{\bar{\sigma}}$, whose nine elements describe the dependence of chemical shielding on the relative orientation of the magnetic nucleus.[60] The magnitudes of these tensor elements depend on the detailed nature of the electron density vicinal to the magnetic nucleus and, therefore, yield information about its local molecular environment. This information is conventionally summarized in the *chemical shift* (in units of ppm),[60,61]

$$\delta \equiv \left[\frac{\nu_{\text{sample}} - \nu_{\text{ref}}}{\nu_{\text{ref}}}\right] \times 10^{6} \tag{2.5}$$

where ν is the resonance frequency (Larmor frequency) determined by adjusting the value of B_o and ν_{ref} is the resonance frequency for the magnetic nucleus in a sample selected as a convenient reference for interpreting NMR spectra. For most metal nuclei, the reference sample is an aqueous solution of known concentration, whereas for ^1H, ^{13}C, and ^{29}Si, tetramethylsilane [$Si(CH_3)_4$] is often chosen. The factor 10^6 appears in eq. 2.5 to anticipate the otherwise very small value of δ, which is simply proportional to the difference in σ values for the reference and sample nuclei, determined at a fixed B_o.

If the effects of the local molecular environment on a magnetic nucleus are isotropic, $\overline{\overline{\sigma}}$ can be replaced by the scalar quantity, σ. This situation is expected in aqueous solutions because of rapid translational motions. It is not expected, therefore, in adsorbates. For these compounds, a broadening of the NMR lineshape from the single sharp peak characteristic of isotropicity will occur in powder samples because each of the three principal elements of $\overline{\overline{\sigma}}$ yields a different resonance frequency and a nearly random set of orientations of the magnetic nucleus exists in these samples.[61] (There is an additional broadening, caused by interactions of the magnetic nucleus with other magnetic nuclei or with paramagnetic electrons and by structural disorder,[61] analogous to the effect of "hyperfine splitting" in fig. 2.11 in that a multiplicity of transitions is promoted.)

Broadening of the NMR lineshape (including the contributions from magnetic interactions) can be obviated to a large extent by an experimental technique, known as *Magic-Angle Spinning* (MAS), in which a sample is rotated about an axis making an angle of 54.7° with $\mathbf{B}_o$ at a rate larger than the spread of resonant frequencies across the broadened NMR peak.[60,61] (For example, if the resonant frequency is 80 MHz and the broadening, as measured by the range of δ-values, is 150 ppm, the rate of rotation must be larger than $1.5 \times 10^{-4} \times 80 \times 10^6$ Hz $= 12$ kHz.) This technique is effective because the orientation dependence of the chemical shielding (and that of most of the magnetic interactions, but not the structural disorder) goes as $1{-}3\cos^2\theta$, where θ is the angle between a principal axis characteristic of the anisotropy and $\mathbf{B}_o$, a quantity which equals zero when $\theta = 54.7°$. Rotation forces the magnetic nuclei in the sample to experience this "magic angle" preferentially, eliminating most of the causes of broadening and yielding a NMR peak corresponding to σ (as opposed to the elements of $\overline{\overline{\sigma}}$) alone.[63]

Static disorder in the molecular environment of the magnetic nucleus (or, equivalently, the existence of more than one distinct adsorbate species) will still contribute to broadening the MAS NMR peak, as will molecular motions whose time scales are comparable to the inverse of the resonance frequency. This latter "dynamic" disorder can be explored by lowering the sample temperature from a regime in which a narrow peak appears because molecular motions are rapid, to one in which broadening and eventually multiple peaks appear because molecular motions are slow on the NMR time scale.[61] Variable-temperature-MAS experiments can be especially illuminating when dynamic disorder is prevalent.[64]

Anisotropic chemical shielding is illustrated by the NMR spectrum in fig. 2.13,[65] which shows the chemical shift for $^{113}Cd^{2+}$ (resonance frequency of 79.88 MHz for $B_o = 8.4T$) adsorbed on the two-layer hydrate of montmorillonite (1.5 nm layer spacing). The dashed curve through the spectrum is calculated by convoluting a gaussian broadening function (conventionally used to simulate

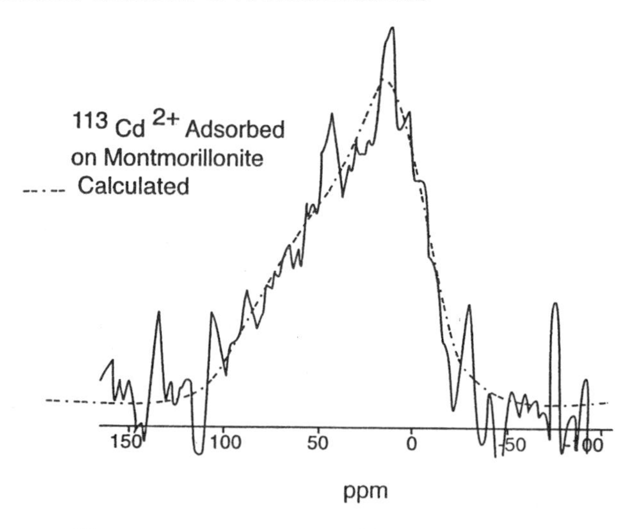

Fig. 2.13 A static ^{113}Cd NMR spectrum for Cd^{2+} adsorbed on montmorillonite, with the chemical shift referenced to a 100 mol m^{-3} solution of $CdSO_4$. The dashed curve through the spectrum is a model fit based on a chemical shielding tensor with three nonzero elements, $\sigma_{||} = -92$ ppm and two $\sigma_{\perp} = 8$ ppm. (Note that the chemical shift and the chemical shielding have opposite sign.) After Tinet et al.[65]

spectral broadening caused by static disorder and magnetic interactions[61]) with a theoretical NMR spectrum corresponding to $\overline{\overline{\sigma}}$ with three principal diagonal elements, $\sigma_{||} = -92 \pm 5$ ppm and two $\sigma_{\perp} = 8 \pm 5$ ppm, where $||$ refers to parallel alignment of the crystallographic *c*-axis of the clay mineral (the direction perpendicular to its siloxane surfaces) and $\mathbf{B}_o$. These elements of $\overline{\overline{\sigma}}$ are referenced to the spectrum of $^{113}Cd^{2+}$ in a 100 mol m^{-3} solution of $CdSO_4$[65] and, therefore, $\sigma_{||}$ is smaller, while $\sigma_{\perp}$ is larger than σ for ^{113}Cd in the reference aqueous solution. The mean principal value of the relative chemical shielding is $\sigma = \frac{1}{3}(-92 + 8 + 8) = -25 \pm 9$ ppm for ^{113}Cd adsorbed on the clay mineral. According to eqs. 2.4 and 2.5, the NMR spectrum in fig. 2.13 reflects a resonance near $\delta = 92$ ppm for chemical shielding contributed by $\sigma_{||}$, whereas a resonance near –8 ppm, contributed by the two $\sigma_{\perp}$ elements, gives rise to the pronounced absorption near 0 ppm. This interpretation is substantiated by ^{113}Cd NMR spectra obtained for a clay–film sample oriented at several angles relative to the direction of $\mathbf{B}_o$.[64] A modest peak corresponding to $\sigma_{||}$ appears when the film (with the clay particles made to lie in the plane of the film) is perpendicular to $\mathbf{B}_o$, whereas a gradual shift in position toward 0 ppm, and a large increase in peak intensity, occurs as the film is rotated to lie in a plane parallel to $\mathbf{B}_o$ and $\sigma_{\perp}$ begins to dominate the lineshape.

The negative value of $\sigma_{||}$ indicates decreased shielding while the slightly positive value of $\sigma_{\perp}$ indicates slightly increased shielding of adsorbed $^{113}Cd^{2+}$ relative to $^{113}Cd^{2+}$ in 100 mol m^{-3} $CdSO_4$ solution. Moreover, the existence of just two nonzero elements of $\overline{\overline{\sigma}}$ for adsorbed $^{113}Cd^{2+}$ implies a local molecular environment that is symmetric about an axis perpendicular to the siloxane surface of the clay mineral. In the two-layer hydrate of montmorillonite, this environment comprises a solvation shell with six or more water molecules and two opposing layers of the clay mineral (cf. fig. 1.9). This kind of local configuration is expected generally for bivalent metal cations, such as Ca^{2+}, adsorbed in the interlayer region of montmorillonite, for which Cd^{2+} can serve as a NMR-active in situ probe.[66] The chemical shielding data for $^{113}Cd^{2+}$ suggest that the bonding of a bivalent cation to the two opposing clay layers through only two solvating water molecules reduces the shielding electron density in its environment relative to the solvation complex formed in aqueous solution.[67] Consistently with this interpretation, the chemical shift associated with $\sigma_{||}$ was observed to decrease from 92 ppm to near 0 as the water content was increased, while that associated with $\sigma_{\perp}$ did not change. This trend suggests that the influence of proximate clay layers is the major cause of the chemical shift.

The effect of dynamic disorder on NMR spectra is illustrated in fig. 2.14,[68] which portrays the chemical shift (relative to a 100 mol m^{-3} solution of CsCl at 25°C) for ^{133}Cs (resonance frequency at 65.5 MHz for $B_o = 11.7$ T) adsorbed on illite exposed to water vapor at 100% relative humidity. Magic-angle spinning at 4–5 kHz was used to suppress anisotropic chemical shielding and magnetic inter-

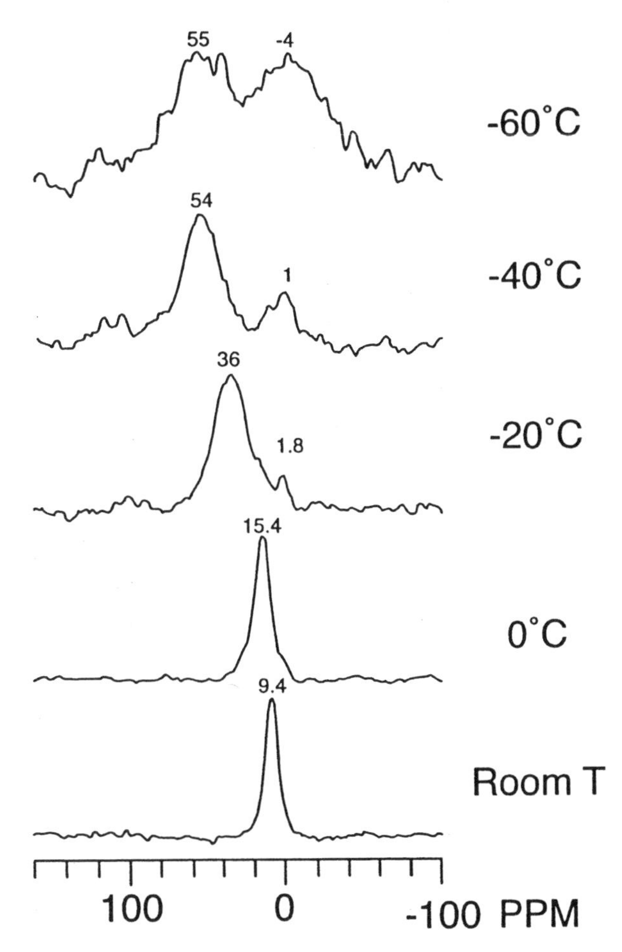

Fig. 2.14 A MAS NMR spectrum for $^{133}Cs^{+}$ adsorbed on illite exposed to 100% relative humidity and variable temperature. After Kim et al.[68]

actions, such that the NMR peak corresponds to the mean principal value of $\overline{\overline{\sigma}}$ for the adsorbate. At 25°C ("Room T") a single peak appears at $\delta = 9.4$ ppm, indicating a molecular environment for $^{133}Cs^+$ that is slightly less shielded than in a 100 mol m^{-3} solution of CsCl at the same temperature. Below 0°C, however, two peaks are observed, one corresponding to $^{133}Cs^+$ in a deshielded molecular environment ($\delta \approx 36$–55 ppm) and the other corresponding to a molecular environment similar to that in aqueous solution ($\delta \approx 0$). The impact of decreasing temperature is much greater on the former NMR peak. These peaks can be assigned to Cs^+ in inner-sphere and outer-sphere surface complexes, respectively.[69] Like K^+ (cf. fig. 1.9), Cs^+ does not solvate strongly and, therefore, is likely to coordinate directly with negatively charged sites on the adsorbent. This species can exchange with solvated Cs^+ adsorbed on the clay mineral and, at 25°C, this exchange is sufficiently rapid to produce only a single, motionally

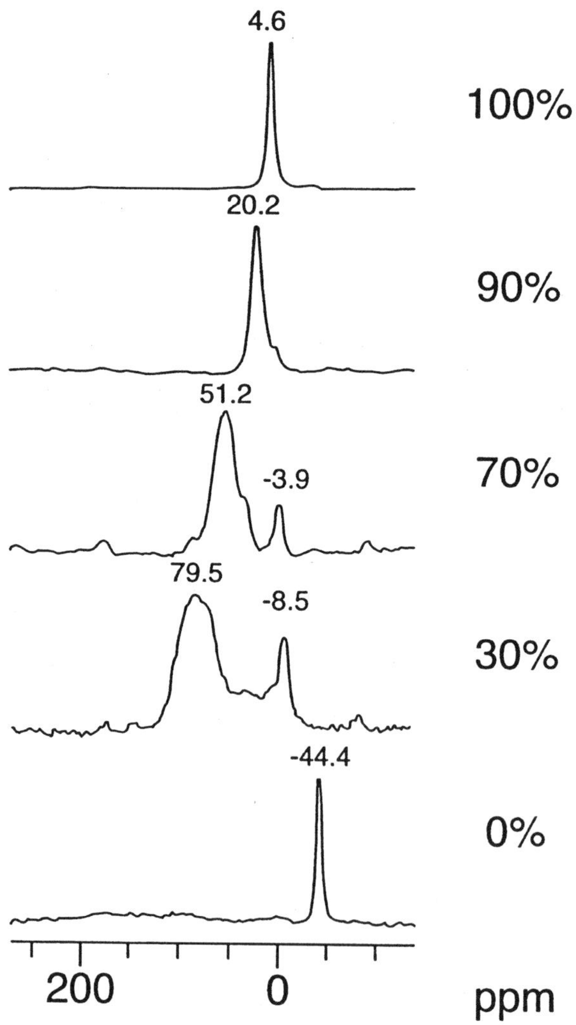

Fig. 2.15 A MAS NMR spectrum for $^{133}Cs^+$ adsorbed on illite at 25°C after exposure to variable relative humidity. After Kim et al.[68]

averaged NMR peak. The rate of this exchange can be estimated approximately by measuring the width of the narrower of the two peaks after conversion to frequency units.[70] Taking this width to be about 20 ppm, one estimates a rate of exchange in excess of several kHz, in agreement with the rate of MAS required for the NMR experiment. A simple model calculation[70] suggests that this rate may be as high as 100 kHz.

Figure 2.15 shows ^{133}Cs NMR MAS spectra for the same system as in fig. 2.14, but this time the spectra were collected at 25°C for varying relative humidity (RH).[69] (The asterisks denote experimental artifacts created in the spectra from modulation of the applied radiofrequency signal by the spinning of the sample.[60]) At 100% RH, a motionally averaged peak near $\delta = 0$ ppm is observed, but this peak separates into two as the RH decreases (i.e., at lower water content of the illite sample) toward ambient RH ($\approx$ 35%). As in the case of the MAS spectra in fig. 2.14, the peak at $\delta > 0$ is assigned to strongly bound Cs^+, whereas that at $\delta \lesssim 0$ is assigned to solvated adsorbed Cs^+. (The spectrum at 0% RH comprises two peaks as well, but the low-intensity, broad peak ranging over $0 < \delta < 200$ ppm is not readily observable in the figure.[68])

Notes

1. G. Sposito, pp. 33–57 in *Aquatic Chemistry*, C.P. Huang, C.R. O'Melia, and J.J. Morgan (Eds.), American Chemical Society, Washington, DC, 1995.
2. The emphasis here is on *elementary* chemical reactions, which take place on a molecular level exactly as they are written, as opposed to *overall* chemical reactions, in which stoichiometry and molecularity are not necessarily related. This difference is discussed in more detail in section 3.1. Equations 2.1a and 2.1b are intended to be elementary reactions describing the adsorption of Cd^{2+} by $Fe(OH)_3$. See chapter 1 in G. Sposito, *Chemical Equilibria and Kinetics in Soils*, Oxford University Press, New York, 1994, for a discussion of these basic concepts. The rationale for considering elementary reactions in preference to overall reactions in chemical modeling is expressed cogently by P.A. O'Day, *Rev. Geophys.* **37**:249 (1999).
3. This conundrum of which surface complexation model to select often is set aside, for practical purposes, by a description of adsorption with an overall reaction whose conditional equilibrium constant is calculated empirically and then correlated with chemical properties, such as pH, type of adsorbent, adsorptive hydrolysis or protonation constant, etc. Good examples of this approach, which bridges between the strictly empirical Kurbatov plot (eq. 1.7) and fundamental models based on molecular concepts, are in S.-Z. Lee, H.E. Allen, C.P. Huang, D.L. Sparks, P.F. Sanders, and W.J.G.M. Peijnenburg, *Environ. Sci. Technol.* **30**:3418 (1996); F. Wang, J. Chen, and W. Forsling, *Environ. Sci. Technol.* **31**:448 (1997); H. Radovanovic and A.A. Koelmans, *Environ. Sci. Technol.* **32**:753 (1998); and J.A. Davis, J.A. Coston, D.B. Kent, and C.C. Fuller, *Environ. Sci. Technol.* **32**:2820 (1998).
4. For good working discussions of the continuing saga of parameter ambiguity, see K.F. Hayes, G. Redden, W. Ela, and J.O. Leckie, *J. Colloid Interface Sci.* **142**:448 (1991); S. Goldberg, *J. Colloid Interface Sci.* **145**:1 (1991); L. Katz and K.F. Hayes, *J. Colloid Interface Sci.* **170**:447 (1995); S. Pivovarov, *J. Colloid Interface Sci.* **196**:321 (1997); and J.A. Davis et al. *Environ. Sci. Technol.* **32**:2820 (1998).
5. S. Goldberg, *Advan. Agron.* **47**:233 (1992).

6. H. Hohl and W. Stumm, *J. Colloid Interface Sci.* **55**:281 (1976); J.A. Davis and J.O. Leckie, *J. Colloid Interface Sci.* **67**:90 (1978).
7. J. Westall and H. Hohl, *Advan. Colloid Interface Sci.* **12**:265 (1980).
8. Brief apologies for the spectroscopic approach to adsorption reactions are given by G. Sposito, *Aquatic Chemistry*, C.P. Huang, C.R. O'Melia, and J.J. Morgan (Eds.), American Chemical Society, Washington, DC, 1995; P.A. O'Day, *Rev. Geophys.* **37**:249 (1999); and S. Goldberg, pp. 337–412 in *Structure and Surface Reactions of Soil Particles*, P.M. Huang, N. Senesi, and J. Buffle (Eds.), John Wiley, New York, 1998.
9. Introductory reviews of surface spectroscopic techniques are in chapters 1 and 2 of *Environmental Particles*, Vol. 2, ed. by J. Buffle and H.P. van Leeuwen (Eds.), Lewis Publishers, Boca Raton, FL, 1993; chapters 2–5 of *Clay Mineralogy*, M.J. Wilson (Ed.), Chapman and Hall, London, 1994; chapters 6, 7, and 9 in *Structure and Surface Reactions of Soil Particles*, P.M. Huang, N. Senesi, and J. Buffle (Eds.), John Wiley, New York; and chapters 2 and 8 in *Mineral-Water Interfacial Reactions*, D.L. Sparks and T. Grundl (Eds.), American Chemical Society, Washington, DC, 1999. For a thorough survey of spectroscopic approaches, see *Spectroscopic Methods in Mineralogy and Geology*, F.C. Hawthorne, (Ed.), Mineralogical Society of America, Washington, DC, 1988.
10. C.T. Johnston, G. Sposito, and W.L. Earl, pp. 1–36 in J. Buffle and H.P. van Leeuwen (Eds.), *Environmental Particles*, vol. 2, Lewis Publishers, Boca Raton, FL, 1993.
11. See, e.g., M.I. Tejedor-Tejedor and M.A. Anderson, *Langmuir* **6**:602 (1990) and W.F. Bleam, P.E. Pfefter, S. Goldberg, R.W. Taylor, and R. Dudley, *Langmuir* **7**:1702 (1991).
12. See, e.g., D.R. Brown and L. Kevan, *J. Am. Chem. Soc.* **110**:2743 (1988) and J.D. Morton, J.D. Semrau, and K.F. Hayes, *Geochim. Cosmochim. Acta* **65**:2709 (2001) for Cu^{2+}, D. Tinet, A.M. Fangere, and R. Prost, *J. Phys. Chem.* **95**:8804 (1991) for Cd^{2+}.
13. See, e.g., J.A. Leenheer, G.K. Brown, P. MacCarthy, and S.E. Cabaniss, *Environ. Sci. Technol.* **32**:2410 (1998); G. Sposito, R. Prost, and J.-P. Gaultier, *Clays Clay Miner.* **31**:9 (1983); and W. Möhl, A. Schweiger, and H. Motschi, *Inorg. Chem.* **29**:1536 (1990).
14. See, e.g., S. Xu and J.B. Harsh, *Clays Clay Miner.* **40**:567 (1992), for a review of evidence supporting an effect of layer charge on the mode of surface complexation by 2:1 clay minerals.
15. See, e.g., C. Papelis and K.F. Hayes, *Colloids Surf.* **107A**:89 (1996); X.-G. Wang, W. Weiss, Sh. K. Shaikhutdinov, M. Ritter, M. Petersen, F. Wagner, R. Schlöge, and M. Scheffler, *Phys. Rev. Lett.* **81**:1038 (1998); M.L. Schlegel, A. Manceau, D. Chateigner, and L. Charlet, *J. Colloid Interface Sci.* **215**:140 (1999); D.G. Strawn and D.L. Sparks, *J. Colloid Interface Sci.* **216**:257 (1999); and S.R. Randall, D.M. Sherman, K.V. Ragnarsdottir, and C.R. Collins, *Geochim. Cosmochim. Acta* **63**:2971 (1999) for good working discussions of the effects of molecular-scale heterogeneity on adsorbate spectra.
16. A careful discussion of spectral deconvolution is presented in A. Manceau, B. Lanson, M.L. Schlegel, J.-C. Hargé, M. Musso, L. Eybart-Bérard, J.-L. Hazemann, D. Chateigner, and G.M. Lamble, *Am. J. Sci.* **300**:289 (2000).
17. G. Sposito and R. Prost, *Chem. Rev.* **82**:553 (1982). See also H. Ohtaki and T. Radnai, *Chem. Rev.* **93**:1157 (1993), for a compilation of molecular time scales related to ion solvation in aqueous solutions.
18. X-ray and neutron diffraction methods are discussed briefly by H. Ohtaki and T. Radnai, *Chem. Rev.* **93**:1157 (1993); and by A.K. Soper, *J. Phys. Condens. Matter* **9**:2717 (1997) in the context of water and aqueous solutions.
19. Useful introductions to Mössbauer spectroscopy are given by F.C. Hawthorne, *Rev. Mineral.* **18**:255 (1988) and by B.A. Goodman, pp. 68–119 in M.J. Wilson (Ed.), Clay Mineralogy, Chapman and Hall, London, 1994.

20. A.M. van der Kraan, *Phys. Stat. Sol. A* **18**:21 (1973); S. Ambe and F. Ambe, *Langmuir* **6**:644 (1990).
21. B.A. Rea, J.A. Davis, and G.A. Waychunas, *Clays Clay Miner.* **42**:23 (1994).
22. S. Fendorf, M.J. Eick, P. Grossl, and D.L. Sparks, *Environ. Sci. Technol.* **31**:315 (1997) and references cited therein.
23. See, e.g., G.W. Neilson and J.E. Enderby, *Advan. Inorg. Chem.* **34**:195 (1989) for a brief review of isotopic-difference neutron diffraction by aqueous electrolyte solutions. Detailed experimental results are tabulated by H. Ohtaki and T. Radnai, *Chem. Rev.*. **93**:1157 (1993).
24. The application to 2:1 clay minerals is described by N.T. Skipper, A.K. Soper, and J.D. C. McConnell, *J. Chem. Phys.* **94**:5751 (1991).
25. N.T. Skipper, M.V. Smalley, G.D. Williams, A.K. Soper, and C.H. Thompson, *J. Phys. Chem.* **99**:14201 (1995). See also D.H. Powell, H.E. Fischer, and N.T. Skipper, *J. Phys. Chem. B.* **102**:10899 (1998).
26. D.G. Schulze and P.M. Bertsch, *Advan. Agron.* **55**:1 (1995); L. Charlet and A. Manceau, pp. 117–164 in *Environmental Particles*, Vol. 2, ed. by J. Buffle and H.P. van Leeuwen (Eds.), Lewis Publishers, Boca Raton, FL, 1993; D.G. Schulze, J.W. Stucki, and P.M. Bertsch, eds., *Synchrotron X-ray Methods in Clay Science*, The Clay Minerals Society, Aurora, CO, 1999. See also chapters 4 and 5 in *Mineral Surfaces*, D.J. Vaughan and R.A.D. Pattrick (Eds.), Chapman and Hall, London, 1995, for brief surveys of XAS and X-ray scattering methods as applied to adsorbates.
27. Good working discussions of coherent X-ray scattering techniques are in Y. Qian, N.C. Sturchio, R.P. Chiarello, P.F. Lyman, T.-L. Lee, and M.J. Bedzyk, *Science* **265**:1555 (1994) and N.C. Sturchio, R.P. Chiarello, L. Chang, P.F. Lyman, M.J. Bedzyk, Y. Qian, H. You, D. Yee, P. Geissbuhler, L.B. Sorensen, Y. Liang, and D.R. Baer, *Geochim. Cosmochim. Acta* **61**:251 (1997). Because it probes a time scale much shorter than that for diffusion, XAS cannot readily distinguish a diffuse layer species from an outer-sphere surface complex, whereas X-ray scattering methods can do so because of their very small scales of spatial resolution in the electrical double layer on mineral surfaces. For recent applications, see, e.g., P. Fenter, P. Geissbühler, E. Demasi, G. Srajer, L.B. Sorensen, and N.C. Sturchio, *Geochim. Cosmochim. Acta* **64**:1221 (2000); P. Fenter, L. Cheng, S. Rihs, M. Machesky, M.J. Bedzyk, and N.C. Sturchio, *J. Colloid Interface Sci.* **225**:154 (2000); and L. Cheng, P. Fenter, K.L. Nagy, M.L. Schlegl, and N.C. Sturchio, *Phys. Rev. Lett.* **87**:156103 (2001).
28. These fundamental processes underlying the spectroscopy of atoms and molecules are discussed accessibly in a classic little book, by G. Herzberg, *Atomic Spectra and Atomic Structure*, Dover, New York, 1944.
29. Applications of XAS to determine the chemical forms of environmentally important elements in situ are discussed in detail by A. Manceau et al., *Am. J. Sci* **300** (Zn); P.M. Bertsch, D.B. Hunter, S.R. Sutton, S. Bajt, and M.L. Rivers, *Environ. Sci. Technol.* **28**:980 (1994) (U); I.J. Pickering, G.E. Brown, and T.K. Tokunaga, *Environ. Sci. Technol.* **29**:2456 (1995) (Se); D.E. Morris, P.G. Allen, J.M. Berg, C.J. Chisolm-Branse, S.D. Conradson, R.J. Donohoe, N.J. Hess, J.A. Musgrave, and C.D. Tait, *Environ. Sci. Technol.* **30**:2322 (1996) (U); A Manceau, M.-C. Boisset, G. Sarret, J.-L. Hazemann, M. Mench, P. Cambier, and R. Prost, *Environ. Sci. Technol.* **30**:1540 (1996) (Pb); M.D. Szulczewski, P.A. Helmke, and W.F. Bleam, *Environ. Sci. Technol.* **31**:2954 (1997) (Cr); A.L. Foster, G.E. Brown, T.N. Tingle, and G.A. Parks, *Am. Mineral.* **83**:553 (1998) (As); G.E. Brown, A.L. Foster, and J.D. Ostergren, *Proc. Natl. Acad. Sci. USA* **96**:3388 (1999) (Pb, As, Se); M.L. Schlegel, A. Manceau, D. Chateigner, and L. Charlet, *J. Colloid Interface Sci.* **215**:140 (1999), M.L. Schlegel, L. Charlet, and A. Manceau, *J. Colloid Interface Sci.* **220**:392 (1999) (Co), J.P. Fitts, G.E. Brown, and G.A. Parks, *Environ. Sci. Technol.* **34**:5122 (2000) (Cr); J.D. Morton et al., *Geochim. Cosmochim. Acta* **65**:2709 (2001) (Cu); B.C. Bostick, M.A. Vairavamurthy, K.G.

Karthikeyan, and J.D. Chorover, *Environ. Sci. Technol.* **36**:2670 (2002) (Cs). The penultimate article by M.L. Schlegel et al. describes the application of polarized EXAFS to detect surface species.

30. See, e.g., G.N. George and I.J. Pickering, *EXAFSPAK*, Stanford Synchrotron Radiation Laboratory, Stanford, CA, 1995, and S. Webb, *SixPACK,* www-ssrl.slac.standford.edu/~swebb/sixpack.htm, 2002, for useful discussions of XAS data analysis and software for data reduction.
31. J.R. Bargar, S.N. Towle, G.E. Brown, and G.A. Parks, *Geochim. Cosmochim. Acta* **60**:3541 (1996). Planes of atoms in α-Al_2O_3 are designated by four *Miller-Bravais indices* (hkil), where, by convention, h+k= $-i$ and an overbar refers to a plane cutting the negative portion of the coordinate axis perpendicular to the plane of the axes to which h and k refer. See, e.g., chapter 2 in L.V. Azároff, *Introduction to Solids*, McGraw-Hill, New York, 1960.
32. J.R. Bargar, G.E. Brown, and G.A. Parks, *Geochim. Cosmochim. Acta* **61**:2617 (1997).
33. J.R. Bargar, S.N. Towle, G.E. Brown, G.A. Parks, *J.Colloid Interface Sci.* **185**:473 (1997).
34. P.F. McMillan and A.M. Hofmeister, pp. 99–159 in F.C. Hawthorne (Ed.), *Spectroscopic Methods in Mineralogy*, Mineralogical Society of America, Washington, DC, 1988, A. Piccolo and P. Conte, pp. 183–250, in P.M. Huang et al. (Eds.), *Structure and Surface Reactions of Soil Particles*, John Wiley, New York, 1998; D.L. Suarez, S. Goldberg, and C. Su (Eds.), *Mineral-Water Interfacial Reactions*, pp. 136–178 in D.L. Sparks and T. Grundl, American Chemical Society, Washington, DC, 1999 offer useful technical introductions to FTIR, DRIFT, and ATR spectroscopy. Combined FTIR-visible sum-frequency generation spectroscopy as applied to mineral-water interfaces is described by M.S. Yeganeh, S.M. Dougal, and H.S. Pink, *Phys. Rev. Lett.* **83**:1179 (1999).
35. Raman spectroscopy has the advantage of less sensitivity to liquid water and to the presence of a silicate adsorbent, making it an alternative to FTIR spectroscopy if an adsorbate is Raman-active. See, e.g., P.F. McMillan and A.M. Hofmeister, F.C. Hawthorne (Ed.), *Spectroscopic Methods in Mineralogy.*, Mineralogical Society of America, Washington, DC, 1988.
36. See, e.g., A. Piccolo and P. Conte, in P.M. Huang et al. (Eds.), *Structure and Surface Reactions of Soil Particles*, John Wiley, New York, 1998; R.W. Parker and R.L. Frost, *Clays Clay Miner.* **44**:32 (1996); C.M. Koretsky, D.A. Sverjensky, J.W. Salisbury, and D.M. D'Aria, *Geochim. Cosmochim. Acta* **61**:2193 (1997).
37. For reviews, see J.D. Kubicki, L.M. Schroeter, M.J. Itoh, B.N. Nguyen, and S.E. Apitz, *Geochim. Cosmochim. Acta* **63**:2709 (1999) and D.L. Suarez et al. (Eds.), *Mineral-Water Interfacial Reactions*, American Chemical Society, Washington, DC, 1999.
38. J.D. Kubicki, M.J. Itoh, L.M. Schroeter, and S.E. Apitz, *Environ. Sci. Technol.* **31**:1151 (1997).
39. M.V. Biber and W. Stumm, *Environ. Sci. Technol.* **28**:763 (1994).
40. E.C. Yost, M.I. Tejedor-Tejedor, and M.A. Anderson, *Environ. Sci. Technol.* **24**:822 (1990).
41. J.-F. Boily, P. Persson, and S. Sjöberg, *Geochim. Cosmochim. Acta* **64**:3453 (2000).
42. P. Persson, J. Nordin, J. Rosenquist, L.-O. Öhman, and S. Sjöberg, *J. Colloid Interface Sci.* **206**:252 (1998).
43. J. Nordin, P. Persson, E. Laiti, and S. Sjöberg, *Langmuir* **13**:4085 (1997); J. Nordin, P. Persson, A. Nordin, and S. Sjöberg, *Langmuir* **14**:3655 (1998).
44. C.Su and D.L. Suarez, *Environ. Sci. Technol.* **29**:302 (1995).
45. S. Goldberg, *Soil Sci. Soc. Am. J.* **63**:823 (1999).

46. See the review by D.L. Suarez et al. (Eds.), *Mineral-Water Interfacial Reactions*, American Chemical Society, Washington, DC, 1999, and the discussion in Section 4.2.
47. G. Calas, *Rev. Mineral.* **18**:513 (1988); K. Dyrek and M. Che, *Chem. Rev.* **97**:305 (1997).
48. Lucid reviews of this approach are given by M.B. McBride, pp. 362–388 in *Geochemical Processes at Mineral Surfaces*, J.A. Davis and K.F. Hayes (Eds.), American Chemical Society, Washington, DC, 1986, and by M.V. Cheshire and N. Senesi, pp. 323–373 in P.M. Huang et al. (Eds.), *Surface Reactions of Soil Particles*, John Wiley, New York, 1999. Adsorbates on humus adsorbents can be investigated by ESR spectroscopy using stable free radicals in the adsorbent as reporter groups, if charge-transfer mechanisms are operative in the adsorption reaction. For a useful review of this technique, see N. Senesi, *Advan. Soil Sci.* **14**:77 (1990).
49. See M.B. McBride, in J.A. Davis and K.F. Hayes (Eds.), *Geochemical Processes at Mineral Surfaces*, American Chemical Society, Washington, DC, 1986, and G. Sposito, *Ion Exchange Solvent Extraction* **11**:211 (1993) for reviews. An innovative use of stable organic radical probes for characterizing adsorbed water on 2:1 clay minerals is described by M.B. McBride and P. Baveye, *Soil Sci. Soc. Am. J.* **59**:388 (1995).
50. H. Motschi, *Colloids Surf.* **9**:333 (1984), and pp. 111–125 in *Aquatic Surface Chemistry*, W. Stumm (Ed.), John Wiley, New York, 1987. In general, $g_{ave} = \frac{1}{3}g_{\|} + \frac{2}{3}g_{\perp}$.
51. Similar ESR spectra for Cu^{2+} adsorbed on oxygen surface sites are reported by M.B. McBride, *Clays Clay Miner.* **30**:21 (1982) (alumina); P. van Cutsem, M.M. Mestdagh, P.G. Rouxhet, and C. Gillet, *Reactive Polymers* **2**:31 (1984) (algae); J.B. Harsh, H.E. Doner, and M.B. McBride, *Clays Clay Miner.* **32**:407 (1984) (hydroxy-hectorite); W. Möhl, H. Motschi, and A. Schweiger, *Langmuir* **4**:580 (1988) (bacteria); F.J. Weesner and W.F. Bleam, *J. Colloid Interface Sci.* **196**:79 (1997) (boehmite, γ-AlOOH); E. Kiefer, L. Sigg, and P. Schosseler, *Environ. Sci. Technol.* **31**:759 (1997) (algae); and P.M. Schosseler, B. Wehrli, and A. Schweiger, *Geochim. Cosmochim. Acta* **63**:1955 (1999) (calcite, $CaCO_3$). The formation of Cu^{2+} surface clusters or nuclei (fig. 1.5) can be detected by a marked broadening of an ESR spectrum as a result of interactions between electrons in neighboring Cu^{2+} ions [see discussions by McBride (1982) and by Weesner and Bleam (1997) in the articles cited above]. Similar insight is possible through analysis of EXAFS spectra of adsorbed Cu^{2+}, as shown by J.D. Morton et al., *Geochim. Cosmochim. Acta* **65**:2709 (2001).
52. A comprehensive review of ENDOR and ESEEM spectroscopies is given by A. Schweiger, *Angew. Chem. Int. Ed. Engl.* **30**:265 (1991).
53. M. Rudin and H. Motschi, *J. Colloid Interface Sci.* **98**:385 (1984).
54. D.R. Brown and L. Kevan, *J. Am. Chem. Soc.* **110**:2743 (1988). See also K. Dyrek and M.Che, *Chem. Rev.* **97**:305 (1997).
55. W. Möhl, A. Schweiger, and H. Motschi, *Inorg. Chem.* **29**:1536 (1990).
56. D.R. Brown and L. Kevan, *J. Phys. Chem.* **92**:1971 (1988).
57. T. Ichiwaka, H. Yoshida, and L. Kevan, *J. Chem. Phys.* **75**:2485 (1981).
58. W. Möhl, H. Motschi, and A. Schweiger, *Langmuir* **4**:580 (1988).
59. P.M. Schosseler, B. Wehrli, and A. Schweiger, *Geochim. Cosmochim. Acta* **63**:1955 (1999).
60. Perhaps the most useful introduction to the applications of NMR spectroscopy to deduce adsorbate structure is given by W.L. Earl and C.T. Johnston, pp. 251–280, in P.M. Huang et al. (Eds.), *Structure and Surface Reactions of Soil Particles*, John Wiley, New York, 1998. Applications to natural organic matter are described by A. Piccolo and P. Conte, pp. 183–250 in P.M. Huang et al. (Eds.), Structure and Surface Reactions of Soil Particles, John Wiley, New York, 1998. More technical introductions are given by R.J. Kirkpatrick, *Rev.*

Mineral. **18**:341 (1988) and W.F. Bleam, *Advan. Agron.* **46**:91 (1991). Comprehensive monographs on NMR applications in environmental geochemistry include M.A. Wilson, *NMR Techniques and Applications in Geochemistry and Soil Chemistry*, Pergamon, New York, 1987, and M.A. Nanny, R.A. Minear, and J.A. Leenheer, *Nuclear Magnetic Resonance Spectroscopy in Environmental Geochemistry*, Oxford University Press, New York, 1997.

61. J.F. Stebbins, *Rev. Mineral.* **18**:405 (1988). A good working discussion of the use of temperature variation to reveal interspecies exchange rates is given by C.A. Weiss, R.J. Kirkpatrick, and S.P. Altaner, *Geochim. Cosmochim. Acta* **54**:1655 (1990).
62. This opposing response of electrons to an applied magnetic field, termed diamagnetism, is a universal phenomenon for any assembly of moving electric charges. See, e.g., chapter 2 in F.E. Low, *Classical Field Theory*, John Wiley, New York, 1997.
63. Good working discussions of MAS techniques are given by J.F. Stebbins, *Rev. Mineral.* **18**:405 (1988) and by W.L. Earl and C.T. Johnston, pp. 251–280 in P.M. Huang et al. (Eds.), *Structure and Surface Reactions of Soil Particles*, John Wiley, New York, 1998.
64. Besides temperature, a useful variable with which to explore dynamic disorder is adsorbent water content, since the presence of a fluid phase should both alter the molecular environment of an adsorbate and facilitate its diffusional motions. See, e.g., Y. Kim and R.J. Kirkpatrick, *Geochim. Cosmochim. Acta* **61**:5199 (1997).
65. D. Tinet et al., *J. Phys. Chem.* **95**:8804 (1991).
66. See, e.g., G. Sposito, *Ion Exchange Solvent Extraction* **11**:211 (1993), and V. Laperche, J.F. Lambert, R. Prost, and J.J. Fripiat, *J. Phys. Chem.* **94**:8821 (1990) for discussions of the utility of magnetic bivalent cations as surrogates for alkaline earth metal cations. This approach is exploited, with varying degrees of success, in S. Bank, J.F. Bank, and P.D. Ellis, *J. Phys. Chem.* **93**:4847 (1989) and P. Di Leo and P. O'Brien, *Clays Clay Miner.* **47**:761 (1999). Problems can arise if the surrogate metal cation forms complexes that are unlikely for Ca^{2+} or Mg^{2+} under the same conditions.
67. Studies of the NMR spectrum of Cd^{2+} referenced to aqueous solutions of $Cd(ClO_4)_2$, wherein no complexes are formed, indicate that coordination of Cd^{2+} with O-ligands leads to chemical shifts clustered around 0 ppm, whereas coordination with N-ligands leads to $40 < \delta < 300$ ppm and coordination with S-ligands leads to $400 < \delta < 800$ ppm, illustrating the deshielding effect of increasingly stronger chemical bonds. These trends are discussed for Cd complexes with humic substances by J. Li, E.M. Perdue, and L.T. Gelbaum, *Environ. Sci. Technol.* **32**:483 (1998), who infer from them predominant Cd^{2+} coordination with O-ligands. See also K.H. Chung, S.W. Rhee, H.S. Shin, and C.H. Moon, *Can. J. Chem.* **74**:1360 (1996).
68. Y. Kim, R.J. Kirkpatrick, and R.T. Cygan, *Geochim. Cosmochim. Acta* **60**:4059 (1996).
69. Y. Kim and R.J. Kirkpatrick, *Geochim. Cosmochim. Acta* **61**:5199 (1997). The spectral interpretation in this paper is that adopted in the present chapter. See also A. Labouriau, C.T. Johnston, and W.L. Earl, pp. 181–197 in M.A. Nanny et al., *Nuclear Magnetic Resonance Spectroscopy in Environmental Geochemistry*, Oxford University Press, New York, 1997; and B.C. Bostick et al., *Environ. Sci. Technol.* **36**:2670 (2002) for complementary insights.
70. C.A. Weiss et al., *Geochim. Cosmochim. Acta* **54**:1655 (1990).

For Further Reading

G. E. Brown and G. A. Parks, *Internat. Geol. Rev.* 43:963 (2001) and A. Manceau, B. Lanson, M. L. Schlegel, J. C. Hargé, M. Musso, L. Eybert-Bérard, J.-L Hazemann, D.

Chateigner, and G. M. Lamble, *Am. J. Sci.* 300:289 (2000). These recent position papers, authored by two leading research groups applying X-ray absorption spectroscopy to natural particles, provide excellent insights into the strategy behind spectroscopic approaches to adsorbate speciation.

F. C. Hawthorne, Ed., *Spectroscopic Methods in Mineralogy and Geology*, Mineralogical Society of America: Washington, DC, 1988. This edited volume contains in-depth, solid introductions to the practice and theory of the spectroscopic methods described in chapter 2.

P. A. Fenter, M. L. Rivers, N. C. Sturchio, and S. R. Sutton (eds.), *Applications of Synchrotron Radiation in Low-Temperature Geochemistry and Environmental Science*, Mineralogical Society of America: Washington, D.C., 2002. This edited volume of workshop lectures is an excellent application-oriented introduction to the use of X-ray absorption spectroscopy, with additional chapters on X-ray reflectivity and microscopy, as well as synchrotron-based IR spectroscopy.

M. J. Wilson, Ed., *Clay Mineralogy: Spectroscopic and Determinative Chemical Methods*, Chapman and Hall: London, 1994. This edited volume has useful applications chapters on IR, ESR, and NMR spectroscopy emphasizing clay minerals.

P. M. Huang, N. Senesi, and J. Buffle, Eds., *Structure and Surface Reactions of Soil Particles*, John Wiley: NY, 1998. This IUPAC-sponsored volume contains useful chapters on IR, ESR, and NMR spectroscopy as applied to soil materials.

Research Matters

1. Manceau et al. [*Environ. Sci. Technol.* 30:1540 (1996)] have obtained a Pb(II) EXAFS spectrum for an agricultural soil contaminated because of its proximity to an alkyllead production facility. This spectrum was analyzed as described in section 2.2, with the weighted EXAFS spectrum and corresponding radial distribution function graphed similarly to the example in fig. 2.5 (sample "P" in figs. 6 and 7 in the article by Manceau et al). A catalog of reference Pb(II) EXAFS spectra appears in fig. 8 of this article. Evaluate the method used by Manceau et al. to conclude that alkyllead contaminants in the soil have become transformed to Pb-humic substance complexes. As background for your evaluation, review the comparative analysis of fig. 2.5 given in section 2.2. and the "Background" section in the article of Manceau et al.
2. Xia et al. [*Geochim. Cosmochim. Acta* 61:2211 (1997)] present Pb(II) EXAFS spectra for Pb-organic matter complexes, the organic matter having been extracted by reaction of a chelating resin with a Mollisol soil in aqueous suspension. The EXAFS spectra of Pb-organic matter complexes prepared at pH 4, 5, and 6 were analyzed as described in section 2.2 (eq. 2.3), but with σ_j as well as R_j and N_j in eq. 2.4 as adjustable parameters. Figures 7 and 8 in the article by Xia et al. (1997) are exactly analogous to fig. 2.5, with the major peak near 0.2 nm (2 Å) in fig. 8 of Xia et al. (1997) assigned to 4 O at 0.230 to 0.244 nm from Pb^{2+}, while the minor peak near 0.27 nm (2.7 Å) is attributed to 2 C at about 0.33 nm from Pb^{2+}. These results suggest a complex in which Pb^{2+} is coordinated to two carboxyl groups and a pair of water molecules as nearest neighbors. In support of this concept, Xia et al. (1997) perform a bond valence calculation (Pauling Rule 2, discussed in section 1.2), assigning s_{PbO} values in the range 0.4 to 0.6 for any O in the complex. Evaluate the method used by Xia et al. (1997) to deduce the molecular structure of the Pb-organic matter complex, applying in your evaluation the discussion in sections 1.2 and 2.2, and that in the "Background" section of the article by Manceau et al. (1996), cited in problem 1.
3. Kubicki et al. [*Geochim. Cosmochim. Acta* 63:2709 (1999)] have obtained ATR-FTIR spectra of salicylate adsorbed on the edge surfaces of illite and kaolinite at pH 3. The positions of the absorption peaks observed in the spectra are listed below in cm^{-1}. Make assignments of these peaks to vibrational motions of the

salicylate molecule, then perform a linear regression analysis of the observed peak wavenumbers on the calculated peak wavenumbers for the 10 model salicylate surface complexes listed in table 2 in the article by Kubicki et al. (1999), following the example in fig. 2.7 while consulting Kubicki et al., *Environ. Sci. Technol.* 28:763 (1994). It will be necessary to make a table of corresponding wavenumbers between experimental and theoretical spectra in order to do the regression calculation. Be sure to report confidence intervals for the slope and intercept of the regression line. On the basis of your regression calculations, select the most probable surface complex model for each spectrum.

Illite	Kaolinite
1247	1249
1265	1390
1326	1464
1337	1485
1392	1529
1460	1554
1471	1575
1511	1610
1555	1745
1570	
1604	

4. Schlosseler et al. [*Geochim. Cosmochim. Acta* 63:1955 (1999)] have employed Cu^{2+} as an in situ probe of the local coordination environment on the surface of calcite ($CaCO_3$). Their ESR spectrum of Cu^{2+} adsorbed on the calcite surface [10 min reaction with micromolar $Cu(NO_3)_2$ at pH 8.5] appears very similar to that in the upper right corner of fig. 2.12, with $g_{\parallel} = 2.37$ and $g_{\perp} = 2.07$. Their two-pulse ESEEM spectrum of adsorbed Cu^{2+} resembles that in the lower left corner of fig. 2.12, except that peaks at the Larmor frequency and its first harmonic occur for ^{13}C ($\nu_C = 3.38$ MHz) instead of ^{27}Al. Their four-pulse ESEEM spectrum can be compared to the ^1H ENDOR spectrum in the upper left corner of fig. 2.12, in that there are shoulder features attributable to axial and equatorial protons in a tetragonal Cu^{2+} complex. These features are diminished sharply in adsorbed Cu^{2+} when compared to the solvation complex, $Cu(H_2O)_6^{2+}$. Basing your discussion on the analysis of fig. 2.12 given in section 2.4, evaluate the conclusion by Schlosseler et al. (1999), that Cu^{2+} experiences "strong adsorption at the mineral–water interface with rapid dehydration and formation of highly coordinated monodentate complexes in a thin, structured calcium carbonate surface layer."
5. Nordin et al. [*Geochim. Cosmochim. Acta* **63**:3513 (1999)] report MAS ^{19}F NMR spectra (resonance frequency at 376.45 MHz for $B_o = 9.6$ T; MAS at 16.5 kHz) for F^- adsorbed on alumina at pH 5 ("ligand-like" adsorption). Their spectra can be deconvoluted into two peaks whose relative intensities shift with increasing F^- surface excess [fig. 4 in the article by Nordin et al. (1999)]. One peak is centered at –131 ppm while the other is at –142 ppm, both chemical shifts being measured relative to $CFCl_3$. (A third peak, at $\delta = -151$ ppm, attributed to soluble Al-fluoride complexes in micropores, appeared in some of their spectra as well.) Given $\delta = -224$ ppm for solid-phase NaF; $\delta = -132$ ppm for F replacing OH bound to a pair of structural Al^{3+} in a dioctahedral sheet; and $\delta = -174$ ppm for solid AlF_3, explicate the two chemical shift values observed for F^- adsorbed on alumina in terms of chemical shielding. The chemical shift for F at an apex of two edge-sharing Al octahedra was measured by Huve et al. [*Clays Clay Miner.* **40**:186 (1992)].

3

Surface Chemical Kinetics

3.1 Phenomenology

Surface chemical kinetics encompass the time-dependent processes observed during equilibration between an adsorbate and the two phases it contacts. These processes are conditioned on pressure, temperature, composition, and whether the system observed is open (flowthrough reactor) or closed (batch reactor) with respect to the adsorptive. The time scales of observation vary widely, from microseconds to millennia, because the processes themselves equilibrate over a broad range of time scale, with ion exchange reactions generally being fastest and mineral dissolution reactions generally being slowest. A variety of experimental techniques and apparatus has been developed to address this variability and to provide quantitation of the kinetic species involved under controlled conditions.[1]

The principal laboratory measurements in surface chemical kinetics involve the time dependence of composition variables at fixed total volume, applied pressure, and temperature. For example, the decrease in adsorptive concentration with time during the equilibration of a surface complexation process can be monitored, or the time evolution of transformation products from a surface redox reaction can be followed, while pH or pE are controlled at constant temperature and pressure.[2] These time-dependent data are usually converted to reaction rates for purposes of conceptual interpretation. The *reaction rate* is defined as the time derivative of the *extent of reaction* ξ, which, in turn, is equal to the change in mole number of a substance as a chemical reaction proceeds, divided by the stoichiometric coefficient of the substance in the reaction. This coefficient is conventionally taken to be negative for reactants, positive for products.[3] Regardless of the sign of ξ, $d\xi/dt$, the reaction rate, is always non-negative. Since most surface reactions are investigated at fixed total volume (V), it proves to be convenient in practice to define the *rate of concentration change*, $d(\xi/V)/dt$, with ξ/V being the change in concentration (mol m^{-3}) of a substance as a chemical reaction proceeds.[4] The time derivative $d(\xi/V)/dt$ is then conventionally termed the *rate of reaction* for the sake of simplicity.[3] Note again that the rate of reaction is a positive-definite quantity, and that it depends on the stoichiometry of the reaction it characterizes.

Measured values of reaction rates are interpreted conceptually by *rate laws* in which the rate of reaction is expressed mathematically as a function of temperature, pressure, composition, and any other variables deemed essential to model rate behavior. There are no general constraints on the functional form of a rate law, except that it must yield a positive-definite reaction rate whenever the Gibbs energy of the reactants is larger than that of the products in the reaction whose rate is being modeled.[5] Thus, for example, a rate law need not be written as the difference between two terms ("forward rate" minus "backward rate"), nor does it have to be expressed in terms of powers of concentration variables. Even if a rate law is written as the difference between forward and backward rates, with each rate proportional to powers of the concentrations of reactants or products, respectively, there is no general requirement that, at equilibrium, the ratio of forward to backward rates can be factored to yield an equilibrium constant, but only that the ratio can be factored to yield some positive-valued function (otherwise left unspecified) of the equilibrium constant.[5] In short, *the formulation of rate laws is strictly an empirical exercise, subject only to very broad constraints.*

These concepts can be illustrated with the example of a redox reaction between ferrous iron and chromate [Cr(VI)] in aqueous solution,[6] a process whose rate law also will prove useful in sections 3.3 and 3.5. The redox reaction may be written in a schematic form:

$$\mathrm{Fe(II)} + \mathrm{Cr(VI)} \rightarrow \mathrm{Fe(III)} + \mathrm{Cr(V)} \tag{3.1}$$

in which Fe is oxidized and Cr is reduced. This overall redox reaction represents a one-electron transfer process that is believed to be the rate-limiting step in the three-electron transfer that ultimately produces Cr(III) from Fe(II) and perforce requires a 3:1 stoichiometric ratio overall between Fe and Cr.[7,8] The thermodynamic constraint on the reaction in eq. 3.1, with the direction of reaction as indicated, is that the pE value for the redox couple Cr(VI)/Cr(III) must be larger than that for the redox couple Fe(III)/Fe(II) under the conditions of the reaction.[9] The rate of the reaction is then expressed:

$$-\frac{d}{dt}[\mathrm{Cr(VI)}] = -\frac{d}{dt}[\mathrm{Fe(II)}] \tag{3.2}$$

where [ ] is a molar concentration and the minus sign indicates division of $d[\ ]/dt$ by the stoichiometric coefficient –1 according to convention.[4] This rate is positive-definite so long as pE{Cr(VI)/Cr(V)} $\geq$ pE{Fe(III)/Fe(II)} during the reaction in eq. 3.1.

A rate law for the redox reaction in eq. 3.1 can be formulated quite generally as:

$$-\frac{d}{dt}[\mathrm{Cr(VI)}] = \phi([\mathrm{Cr(VI)}], [\mathrm{Fe(II)}], \mathrm{env}) = -\frac{d}{dt}[\mathrm{Fe(II)}] \tag{3.3}$$

where $\phi(\ )$ is a positive-valued function of its arguments and "env" refers to all variables on which the rate may depend, other than the concentrations of the two reactants (e.g., temperature, ionic strength, pH, pE, etc.). The mathematical form of $\phi(\ )$ can be determined by fixing one of the reactant concentrations (say, [Cr(VI)]) and monitoring the decline of the other with time as the redox reaction proceeds from an initial state far from equilibration (fig. 3.1).[8] For example, at 25°C and pH 5.9,[8] the observed time dependence of either reactant concentration is exponential, i.e.,

$$-\frac{d}{dt}[\mathrm{Cr(VI)}] = \mathrm{K}([\mathrm{Fe(II)}], \mathrm{env})\ [\mathrm{Cr(VI)}] \tag{3.4a}$$

$$-\frac{d}{dt}[\mathrm{Fe(II)}] = \mathrm{K}'([\mathrm{Cr(VI)}], \mathrm{env})\ [\mathrm{Fe(II)}] \tag{3.4b}$$

where K and K′ are functions of all variables held fixed. These functions can be determined by applying eqs. 3.4 to experiments conducted at several values of the (excess) fixed reactant concentration, with the result that:[8]

$$-\frac{d}{dt}[\mathrm{Cr(VI)}] = k\ [\mathrm{Fe(II)}]^{\alpha}\ [\mathrm{Cr(VI)}]^{\beta} \tag{3.5}$$

where $\alpha = 0.9 \pm 0.3$ and $\beta = 1.2 \pm 0.3$.[7,8] The exponents α and β in eq. 3.5 are called *partial reaction orders* with respect to each reactant and their sum is termed the *overall reaction order* ($n \equiv \alpha + \beta$).[4] In general, these kinetics parameters will not necessarily have integer values, nor will they necessarily be equal to the stoichiometric coefficients in the reaction whose rate they describe. Moreover, since a rate law need not be expressed in terms of powers of concentrations, a reaction order need not even exist. (No reaction order exists for the Michaelis-

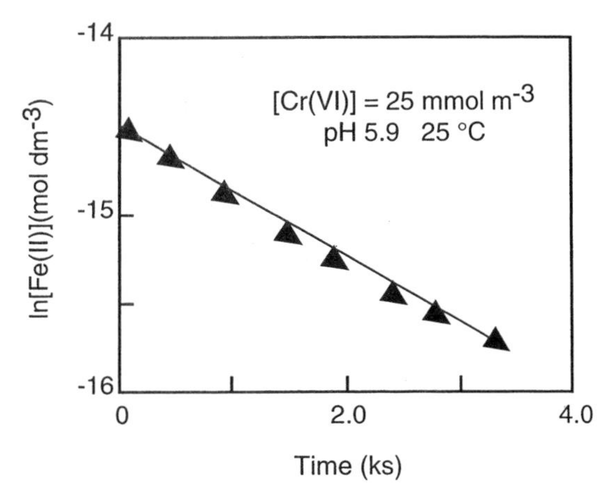

Fig. 3.1 Pseudo first-order time dependence of [Fe(II)] observed during the reduction of chromate by ferrous iron.[8] The initial concentration of Fe(II) (0.5 mmol m^{-3}) is well below that of Cr(VI), so that eq. 3.4b applies.

Menton rate law,[3] whereas the stoichiometry for the two-step reaction that the rate law describes is 1:1.)

Within experimental precision, the rate law in eq. 3.5 is *first order* in each of the reactant concentrations and is *second order* overall. The parameter k in eq. 3.5 is accordingly termed a *second order rate coefficient* ($dm^3\ mol^{-1}\ s^{-1}$) and the parameters K and K′ in eqs. 3.4 are termed *pseudo first-order rate coefficients* (s^{-1}), the adjective "pseudo" being added because they do not depend solely on "env" variables, as does k in eq. 3.5.[3] That the overall order of the reaction in this example is the same as the sum of stoichiometric coefficients of the reactants in eq. 3.1, and that there is equality between α or β and the stoichiometric coefficient of Fe(II) or Cr(VI), do *not* imply that the reaction mechanism involves the combination of one Fe(II) species with one Cr(VI) species. This latter property of the reaction refers to its *molecularity* (number of each type of reactant species that combine), whereas the *stoichiometry* refers to the moles of each reactant that combine, and, of course, *order* is a strictly empirical concept arising from kinetics.[3] When all three concepts do coincide, a reaction is said to be *elementary*; otherwise, it is *overall* or *composite*.[3] Most reactions investigated in the surface chemistry of natural particles are composite.

The pseudo first-order rate coefficient in eq. 3.4a shows a pronounced dependence on pH (fig. 3.2),[6] indicating that pH is an important "env" variable for the redox reaction in eq. 3.1. A pH dependence might be expected on the basis of the chemical speciation of Fe(II) and Fe(III), which is strongly affected by hydrolysis reactions. Three reduction half-reactions involving these species are known:

$$Fe^{3+} + e^- = Fe^{2+} \qquad \log K_R = 13.0 \tag{3.6a}$$

$$FeOH^{2+} + e^- = FeOH^+ \qquad \log K_R = 5.79 \tag{3.6b}$$

$$Fe(OH)_2^+ + e^- = Fe(OH)_2^0 \qquad \log K_R = -2.8 \tag{3.6c}$$

These half-reactions suggest that a reasonable model of the pH dependence of K([Fe(II)], pH) can be expressed by the equation:[7,8]

$$K([Fe(II)], pH) = k_0[Fe^{2+}] + k_1[FeOH^+] + k_2[Fe(OH)_2^0] \tag{3.7}$$

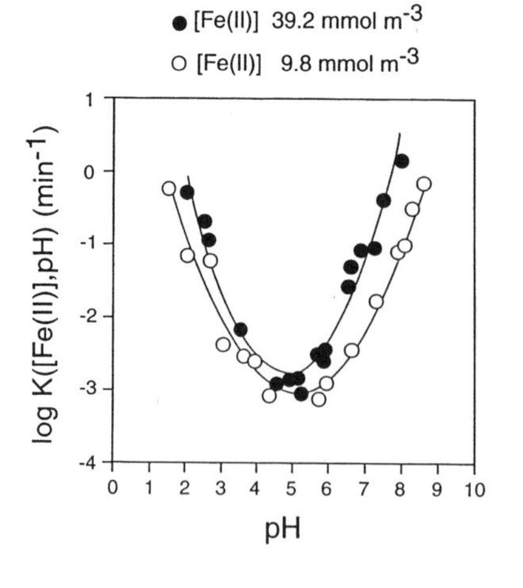

Fig. 3.2 Measured values of the pseudo first-order rate coefficient in eq. 3.4a at 10°C as a function of pH at two fixed Fe(II) concentrations in a 100 mol m^{-3} NaCl background.[6] The initial concentration of Cr(VI) (0.95 mmol m^{-3}) is well below that of Fe(II), so that eq. 3.4a applies. The two branches of each parabolic curve are nearly straight lines of unit negative or positive slope, irrespective of Fe(II) concentration.[6]

Equation 3.7 reflects each reduction half-reaction in eqs. 3.6, with the second-order rate coefficients k_i $(i = 0, 1, 2)$ being implicit functions of the speciation of Cr(VI) and the "env" variables other than pH. The pH dependence is assumed, therefore, to derive mainly from that of the Fe(II) hydrolytic species:

$$Fe^{2+} + H_2O = FeOH^+ + H^+ \qquad \log^* K_1 = -9.51 \tag{3.8a}$$

$$Fe^{2+} + 2H_2O = Fe(OH)_2^0 + 2H^+ \qquad \log^* \beta_2 = -20.6 \tag{3.8b}$$

with

$$[Fe(II)] \equiv [Fe^{2+}] + [FeOH^+] + [Fe(OH)_2^0] \tag{3.9}$$

Nonlinear least-squares fitting of data taken at 23°C on the pH dependence of K ([Fe(II)], pH) between pH 4.4 and 7.2 then leads to the parameter values:[7,10]

$$k_0 = 0.3 \pm 0.5 \text{ dm}^3 \text{ mol}^{-1} \text{ s}^{-1} \qquad k_1 = 1.4 \pm 0.3 \times 10^5 \text{dm}^3 \text{ mol}^{-1}\text{s}^{-1}$$
$$k_2 = 2.8 \pm 0.9 \times 10^9 \text{dm}^3\text{mol}^{-1}\text{s}^{-1}$$

The second-order rate coefficient k_0 is not necessarily different from zero, implying that the direct oxidation of Fe^{2+} by Cr(VI) may be a thermodynamically unfavorable process. Indeed, the pE value corresponding to the reduction half-reaction,

$$H_2CrO_4^0 + H^+ + e^- = H_3CrO_4^0 \qquad \log K_R = 9.3 \tag{3.10}$$

with pE = 9.3 – pH,[9] is always *below* that for the half-reaction in eq. 3.6a. Therefore, the redox reaction,

$$Fe^{2+} + H_2CrO_4^0 + H^+ \rightarrow Fe^{3+} + H_3CrO_4^0 \tag{3.11}$$

yields a *positive* $\Delta_r G$ value when (Fe^{3+}) $(H_3CrO_4^0)$ = (Fe^{2+}) $(H_2CrO_4^0)$.[9] In this circumstance, one expects $k_0 = 0$. On the other hand, $k_0 \neq 0$ is consistent with data from studies of eq. 3.11 at pH < 2, which extrapolate to give the value $k_0 \approx 0.2$ dm^3 mol^{-1} s^{-1} at pH 4.4. A nonzero k_0 also is predicted by the linear correlation equation,

$$\log k_i = 8.034 - 0.6271 \log K_{Ri} \qquad (i = 0, 1, 2; r^2 = 0.983) \tag{3.12}$$

that results from log–log least-squares fitting of the three k_i values given above to the inverse of the three corresponding K_R values given in eqs. (3.6). The correlation in eq. 3.12 is an example of a *linear free energy relationship* (LFER), which is a linear log–log relation between the rate coefficients for a suite of similar reactions and a set of corresponding equilibrium constants.[3,9] In the present case, the formation of a hydroxy complex donates electron density to Fe^{2+}, thereby facilitating electron transfer to Cr(VI).[11] This effect increases dramatically with the number of OH^- in the complex (i.e., $k_0 \ll k_1 \ll k_2$). For example, at pH 6, the relative contributions of the three terms on the right side of eq. 3.7 are, respectively, 1:144:23 for any $[Fe^{2+}]$, showing that the reaction in eq. 3.6b dominates in the pseudo first-order rate coefficient K($[Fe^{2+}]$, pH) at this pH value.

It is noteworthy that the species $FeOH^+$ accounts for only 0.3% of all the Fe(II) species (eq. 3.9) at pH 6, while Fe^{2+} accounts for 99.7%! This remarkable discrepancy illustrates the very important point that *the concentrations of kinetic species are unrelated to their role in a reaction mechanism if their formation is rapid on the time scale of the reaction.* That the fraction of Fe(II) in the form of $FeOH^+$ is tiny has no relevance to its dominance in the mechanism of Fe(II) oxidation, so long as the formation of this kinetic species through the hydrolysis reaction in eq. 3.8a is much faster than its loss through oxidation. Conversely, the fact that Fe^{2+} accounts for almost all of the *equilibrium* species of Fe(II) formed by hydrolysis at pH 6 is irrelevant to its importance as a *kinetic* species in the mechanism of Fe(II) oxidation.

The LFER exemplified in eq. 3.12, like the power-law expression for the rate of Cr(VI) reduction in eq. 3.5, is not a general result of kinetics theory, since there is no necessary connection between reaction rates and equilibria other than the constraint of positive-definiteness on the rate when $\Delta_r G < 0$.[5] In picturesque terms, the rate of a reaction depends on the reactants overcoming an "energy barrier" to form products, the magnitude of the barrier being inversely related to the likelihood of overcoming it. The magnitude of $\Delta_r G$, on the other hand, is related to the stability of the products of a reaction relative to the reactants, irrespective of the existence of an "energy barrier" along the pathway from one to the other. This stability is greater, the deeper is the "energy valley," whose depth is measured relative to the initial state for the reactants. In principle, the height of an "energy barrier" reactants must surmount is unrelated to the depth of the "energy valley" they tumble into as products. Thus, no general LFER between rate coefficients and standard Gibbs energy changes for reactions can be expected.[12]

The magnitude of the "energy barrier" for a reaction often is estimated on the basis of the observed temperature dependence of a rate coefficient, the rationale being that the likelihood of surmounting the barrier should be proportional to a Boltzmann probability factor.[3] One mathematical model realizing this picture is the *Arrhenius equation*,[3]

$$k = A \exp(-E_a/RT) \tag{3.13}$$

where k is a rate coefficient, A is termed the pre-exponential factor, R is the gas constant (8.3145 J mol^{-1} K^{-1}), T is absolute temperature, and E_a is termed the *apparent activation energy*, an empirical measure of the magnitude of the "energy barrier." Equation 3.13 has been applied to measurements of the pseudo first-order rate coefficient in eq. 3.7.[6,8] At pH 4, little or no variation with temperature is observed, the rate coefficient increasing by roughly a factor of two or less over the temperature range 5 to 40°C. At this pH value, and for 25°C, the first two terms on the right side of eq. 3.7 make approximately equal contributions to the pseudo first-order rate coefficient for any $[Fe^{2+}]$. Evidently the "energy barriers" implicit in these two terms are rather small relative to RT, the magnitude of an "energy barrier" that can be overcome by purely random thermal motions. At pH 6, however, a significant temperature effect on the rate coefficient is observed that is consistent with $E_a \approx 85$ kJ mol^{-1}.[6] This temperature effect must be dominated by the behavior of the second and third terms on the right side of eq. 3.7. Their change with temperature will be determined by the activation energies implicit in k_1 and k_2, as well as by those required for formation of the two hydrolytic species, which in turn, depend on $\Delta_r H^0$ for the hydrolysis reactions in eqs. 3.8[3,5] [$\Delta_r H^0 =$

55.8 kJ mol^{-1} ($*K_1$) and 103.3 kJ mol^{-1} ($*\beta_2$)]. That the observed value of E_a falls between these two $\Delta_r H^0$ values suggests again that the "energy barriers" implicit in k_1 and k_2 are small.[8]

3.2 Specific Adsorption Reactions

Specific adsorption was attributed to inner-sphere surface complexation in section 1.5. This mechanism is characterized by adsorption edges or envelopes that are robust against changes in ionic strength of a background electrolyte solution (or, equivalently, by adsorption isotherms that are robust against the introduction of competing adsorptives from addition of an indifferent electrolyte),[13] and by strong, positive correlations between adsorptive affinity for an hydroxylated adsorbent and affinity for OH^- (cations) or H^+ (anions). Thus, cations prone to form hydrolytic species at low pH values and anions remaining protonated at circumneutral pH values are likely to adsorb specifically. Intrinsic to the underlying mechanism for this process is the desolvation of the adsorptive ion necessary to ensure its direct contact with the adsorbent surface (section 1.2).

Studies of inner-sphere complexation by metals and ligands in aqueous solution have led to the elucidation of is mechanism in terms of a two-step process that may be termed the *Eigen-Wilkins-Werner mechanism.*[14] The elementary reactions that attend this mechanism are the formation of an outer-sphere surface complex by the metal and ligand, followed eventually by the desolvation of both to form an inner-sphere surface complex, with desolvation of the metal often being the rate-limiting step. This latter condition implies that the rate of inner-sphere complexation should be correlated positively with the rate of water exchange in the first solvation shell of the metal cation, at least to the extent that the replacement of a solvating water molecule by exchange with another ligand emulates water molecule self-replacement. This correlation is observed in many cases,[14] lending credibility to the mechanism. The rate coefficient for the water-exchange reaction,

$$M(H_2O)_n^{m+} + H_2O^* \rightarrow MH_2O^*(H_2O)_{n-1}^{m+} + H_2O \tag{3.14}$$

where M^{m+} is a metal cation, has been investigated extensively,[15] often by isotopic dilution with labeled solvent (*) or by ^{17}O NMR spectroscopy combined with temperature variation to expose the chemical shift of solvation-shell H_2O.[14] The second-order rate coefficient that follows naturally from a single rate law for the elementary reaction in eq. 3.14 usually is reported as the pseudo first-order rate coefficient that is obtained after multiplying by the molality of pure liquid H_2O (55.5 *m*). This latter rate coefficient, termed k_{wex}, is found to vary over about 16 orders of magnitude,[14,15] from just above 10^{-7} s^{-1} for Cr^{3+} to near 10^{10} s^{-1} for Cs^+. The data for most bivalent and trivalent metal cations can be described, within the spread of differing reported k_{wex} values for the same metal M in eq. 3.14, by the linear correlation equation ($r^2 = 0.872$):[16]

$$\log k_{wex} = 13.538 - 0.2365(Z/R) \tag{3.15}$$

where k_{wex} is in units of s^{-1}, Z is the valence and R is the ionic radius of the metal cation expressed in nm. For example, taking $Z = 2$ and $R = 0.1$ nm for Ca,[17] one estimates $k_{wex} = 6 \times 10^9\ s^{-1}$, as compared with the literature spread[15] of $6 - 9 \times 10^8\ s^{-1}$; or taking $Z = 3$ and $R = 0.054$ nm for Al,[17] one estimates $k_{wex} = 1.9\ s^{-1}$, as compared with the literature spread[15] of $0.2 - 16\ s^{-1}$. The physical basis of the correlation in eq. 3.15 is evident: solvation-shell water exchange is slower, the larger is the coulomb potential at the metal cation periphery. Strong cation-dipole interactions thus imply a sluggish replacement of solvation-shell water molecules by an incoming ligand.[14]

In the spirit of the Eigen-Wilkins-Werner mechanism, a two-step elementary reaction can be proposed for the specific adsorption of a metal cation by an hydroxylated adsorbent, taking alumina as an example:[18]

$$\begin{aligned} \equiv \mathrm{AlOH} + M^{m+} &\underset{k_{-1}}{\overset{k_1}{\Leftrightarrow}} \equiv \mathrm{AlOH} \cdots M^{m+} \\ &\underset{k_{-2}}{\overset{k_2}{\Leftrightarrow}} \equiv \mathrm{AlOM}^{(m-1)+} + \mathrm{H}^+ \end{aligned} \tag{3.16}$$

where the symbols are used in consonance with the adsorption reactions in eq. 2.1 and the k_i ($i = \pm 1, \pm 2$) are rate coefficients for the reactions in the forward (+) or backward (−) directions of the arrows. According to the Eigen-Wilkins-Werner mechanism, one expects the forward rate coefficient for the first step in eq. 3.16 to be much larger than that for the second step, and this condition can be used to simplify the modeling of a rate law describing the adsorption of the metal cation M^{m+}.

On the assumption that molecularity, stoichiometry, and reaction order are identical in the specific adsorption of the metal M^{m+} by an hydroxylated surface, the rate laws for the generic elementary reaction sequence,[19]

$$A + B \underset{k_b}{\overset{k_f}{\Leftrightarrow}} C \underset{k'_b}{\overset{k'_f}{\Leftrightarrow}} D + E \tag{3.17}$$

are applicable:

$$-\frac{dc_A}{dt} = -\frac{dc_B}{dt} = k_f c_A c_B - k_b c_C \tag{3.18a}$$

$$\frac{dc_D}{dt} = \frac{dc_E}{dt} = k'_f c_C - k'_b c_D c_E \tag{3.18b}$$

where c refers to an appropriate concentration variable (either aqueous solution molarity or the product of a surface excess with the solids concentration). Equations 3.18 constitute a pair of coupled bilinear rate laws for the consecutive reactions in eq. 3.17. Note that a rate law for the intermediate adduct C can be derived by subtracting eq. 3.18b from eq. 3.18a according to the stoichiometry of eq. 3.17. Because a common experimental method through which eq. 3.16 has been investigated involves the measurement of only small deviations of the concentrations in eqs. 3.18 from their equilibrium values,[18] it is useful to consider linearized versions of the right sides of the coupled rate laws.[20]

The linearization of eqs. 3.18 is initiated by setting $c_A \equiv c_A^{eq} + \Delta c_A$, $c_B \equiv c_B^{eq} + \Delta c_B$, and so on, where c^{eq} is an equilibrium concentration and Δc is a small deviation ($\Delta c/c \ll 1$). This produces the pair of equations:

$$-\frac{d\Delta c_A}{dt} = k_f\left(c_A^{eq}\Delta c_B + c_B^{eq}\Delta c_A\right) + k_f \Delta c_A \Delta c_B - k_B \Delta c_C \quad (3.19a)$$

$$\frac{d\Delta c_E}{dt} = k'_f \Delta c_C - k'_b\left(c_D^{eq}\Delta c_E + c_E^{eq}\Delta c_D\right) - k'_b \Delta c_D \Delta c_E \quad (3.19b)$$

where $dc_A^{eq}/dt = dc_E^{eq}/dt \equiv 0$ has been used to simplify the left side and terms containing only equilibrium concentrations have been deleted from the right side because they must sum algebraically to zero:

$$c_C^{eq}/c_A^{eq}c_B^{eq} = k_f/k_b \quad (3.20a)$$
$$c_D^{eq}c_E^{eq}/c_C^{eq} = k'_f/k'_b \quad (3.20b)$$

Thus far no approximation has been made. The linearization approximation involves deletion of the bilinear term $\Delta c_A \Delta c_B$ from eq. 3.19a and $\Delta c_D \Delta c_E$ from eq. 3.19b:

$$-\frac{d\Delta c_A}{dt} \approx k_f\left(c_A^{eq}\Delta c_B + c_B^{eq}\Delta c_A\right) - k_b \Delta c_C \quad (3.21a)$$

$$\frac{d\Delta c_E}{dt} \approx k'_f \Delta c_C - k'_b\left(c_D^{eq}\Delta c_E + c_E^{eq}\Delta c_D\right) \quad (3.21b)$$

A final simplification is made by incorporating the stoichiometry of the reactions in eq. 3.17,

$$\Delta c_A = \Delta c_B, \quad \Delta c_C = -\Delta c_A - \Delta c_D, \quad \Delta c_D = \Delta c_E \quad (3.22)$$

to obtain the coupled, linearized rate laws:

$$-\frac{d\Delta c_A}{dt} \approx \left[k_f\left(c_A^{eq} + c_B^{eq}\right) + k_b\right]\Delta c_A + k_b \Delta c_E \quad (3.23a)$$

$$\frac{d\Delta c_E}{dt} \approx -k'_f \Delta c_A - \left[k'_b\left(c_D^{eq} + c_E^{eq}\right) + k'_f\right]\Delta c_E \quad (3.23b)$$

Although the rate laws in eqs. 3.23 are first-order in all concentration variables, they are coupled as a relict of their connection to the intermediate species C (eqs. 3.22). If the second term on the right side of eq. 3.23a and the first term on the right side of eq. 3.23b were not present, two simple first-order rate laws would result whose solutions each exhibit the familiar decreasing exponential time dependence characterized by a time constant, τ.[3] Because of the linearity of the coupled rate laws, however, a theorem of algebra[21] has it that the solutions of eqs. 3.23 also exhibit this time dependence, the only complication being that they are linear combinations of Δc_A and Δc_D. Moreover, the theorem tells us that, irrespective of the explicit mathematical form these linear combinations may take, the two associated time constants are subject to the constraints:[20]

$$\frac{1}{\tau_1} + \frac{1}{\tau_2} = a_{11} + a_{22} \tag{3.24a}$$

$$\left(\frac{1}{\tau_1}\right)\left(\frac{1}{\tau_2}\right) = a_{11}a_{22} - a_{12}a_{21} \tag{3.24b}$$

where

$$a_{11} \equiv k_f(c_A^{\text{eq}} + c_B^{\text{eq}}) + k_b \tag{3.24c}$$

$$a_{12} \equiv k_b, \qquad a_{21} \equiv k_f' \tag{3.24d}$$

$$a_{22} \equiv k_b'(c_D^{\text{eq}} + c_E^{\text{eq}}) + k_f' \tag{3.24e}$$

in a convenient matrix notation. Equations 3.24 are sufficient to yield explicit mathematical expressions for the two time constants:

$$\frac{1}{\tau_1} = \frac{1}{2}\left\{(a_{11} + a_{22}) + \left[(a_{11} + a_{22})^2 - 4(a_{11}a_{22} - a_{12}a_{21})\right]^{\frac{1}{2}}\right\} \tag{3.25a}$$

$$\frac{1}{\tau_2} = \frac{1}{2}\left\{(a_{11} + a_{22}) - \left[(a_{11} + a_{22})^2 - 4(a_{11}a_{22} - a_{12}a_{21})\right]^{\frac{1}{2}}\right\} \tag{3.25b}$$

with the a_{ij} $(i, j = 1, 2)$ given by eqs. 3.24c – 3.24e. Two special cases of eqs. 3.25 are of interest. First, if there were no coupling, $a_{12} = a_{21} \equiv 0$ and eqs. 3.25 reduce to the simple expressions:

$$\frac{1}{\tau_1} = a_{11}, \quad \frac{1}{\tau_2} = a_{22} \tag{3.25c}$$

which also follows directly from eqs. 3.23 and 3.24c – 3.24e. Second, if the condition $a_{11} \gg a_{21}$ or a_{22} is met, i.e., the first reaction step in eq. 3.17 occurs on a much shorter time scale than the second step, eqs. 3.25 reduce to the approximate expressions:

$$\frac{1}{\tau_1} \approx a_{11}, \quad \frac{1}{\tau_2} \approx a_{22} - \left(\frac{a_{12}a_{21}}{a_{11}}\right) \tag{3.25d}$$

In this special case, the smaller time constant (τ_1) is decoupled from the other one because the reaction step it represents has equilibrated long before the other step is significantly underway.

Returning to eq. 3.16, one can make the associations:

$$A \leftrightarrow \equiv \text{AlOH}, B \leftrightarrow M^{m+}, C \leftrightarrow \equiv \text{AlOH} \cdots M^{m+}, D \leftrightarrow \equiv \text{AlOM}^{(m^{-1})+}, E \leftrightarrow H^+ \tag{3.26}$$

which then lead to an explicit equation for the larger time constant, τ_2, under the condition that the second step in eq. 3.16 is rate determining (eq. 3.25d):

$$\begin{aligned}\frac{1}{\tau_2} \approx & k_{-2}\left(\left[\equiv \text{AlOM}^{(m-1)+}\right]_{\text{eq}} + \left[H^+\right]_{\text{eq}}\right) + k_2 \\ & - k_2\left\{1 + K_{\text{OS}}\left([\equiv \text{AlOH}]_{\text{eq}} + \left[M^{m+}\right]_{\text{eq}}\right)\right\}^{-1}\end{aligned} \tag{3.27}$$

where

$$K_{OS} \equiv \frac{[\equiv \text{AlOH} \cdots M^{m+}]_{eq}}{[\equiv \text{AlOH}]_{eq}[M^{m+}]_{eq}} = \frac{k_1}{k_{-1}} \tag{3.28}$$

by analogy with eq. 3.20a, is the conditional equilibrium constant for the formation of the outer-sphere surface complex in eq. 3.16, with [ ] defined for surface species as the product of surface excess and solids concentration (cf. eq. 1.6). The subscript "eq" implies only equilibration of the first step in eq. 3.16 prior to the second step. The time constant on the left side of eq. 3.27 can be measured by a pressure-jump method,[18,20,22] in which a submillisecond pulse of high pressure is applied to an equilibrated aqueous suspension containing all of the chemical species indicated in eq. 3.27, after which the electrolytic conductivity of the suspension is monitored as the system restores itself to equilibrium. On the assumption that the pressure pulse has impulsively desorbed M^{m+} to increase its aqueous solution concentration, and given the fundamental hypothesis that the system relaxes by the same mechanisms as it would had the small concentration increase occurred as a spontaneous fluctuation, the time-decay of the electrolytic conductivity can be analyzed mathematically to extract a value for τ_2. (Companion experiments are necessary to demonstrate that no relaxation occurs in suspensions containing only the adsorbent and background electrolyte solution or in aqueous solutions of the adsorptive.) Measurements of τ_2 as pH and $[M^{m+}]_{eq}$ are varied then lead to a determination of the kinetics parameters in eq. 3.27.

Figures 3.3 and 3.4 show the dependence of $1/\tau_2$ on pH and $[M^{m+}]_{eq}$ for the adsorption of five bivalent metal cations by γ-Al_2O_3 (p.z.n.p.c. $\approx$ 8.3, $a_s \approx$ 100 $m^2\ g^{-1}$) suspended in 7.5 mol m^{-3} $NaNO_3$ ($c_s = 30$ kg m^{-3}) at 25°C.[23] The data indicate millisecond time scales for the adsorption process. In addition, τ_2 decreases sharply with increasing pH and, more gradually, with increasing metal cation concentration over the ranges of these two variables that were observed. These trends imply that the adsorption process is faster as pH and metal cation concentration are increased. The effect of concentration is consistent with eq. 3.27 if the second term in curly brackets on the right side is much smaller than 1.0, thus permitting the approximation,

$$\{1 + K_{OS}\ ([\equiv \text{AlOH}]_{eq} + [M^{m+}]_{eq}\}^{-1} \approx 1 - K_{OS}\ ([\equiv \text{AlOH}]_{eq} + [M^{m=}]_{eq}) \tag{3.29}$$

such that eq. 3.27 takes the form:

$$\frac{1}{\tau_2} \approx k_{-2}\Big\{\Big(\big[\equiv \text{AlOM}^{(m-1)+}\big]_{eq} + \big[H^+\big]_{eq}\Big) + (k_2 K_{OS}/k_{-2})\Big([\equiv \text{AlOH}]_{eq} + \big[M^{m+}\big]_{eq}\Big)\Big\} \tag{3.30}$$

A plot of $1/\tau_2$ versus the quantity inside curly brackets in eq. 3.30 is shown in fig. 3.5.[23] Straight lines through the origin can be calculated by linear regression analysis, leading to estimates of the rate coefficient k_{-2} (as the slope parameter). The parameter $k_2 K_{OS}/k_{-2}$ in eq. 3.30 is equal to the product of the equilibrium expressions in eqs. 3.20:

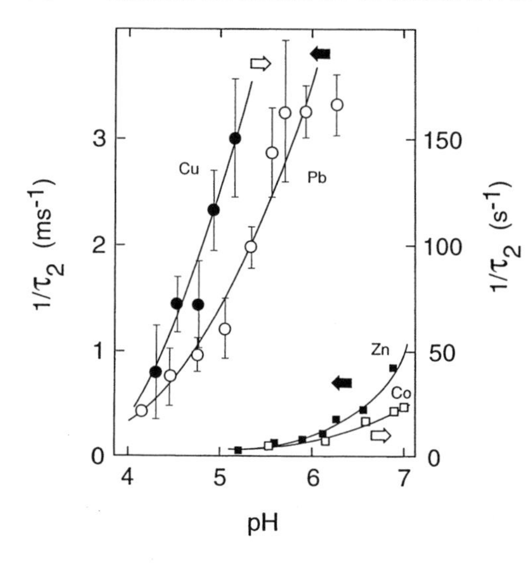

Fig. 3.3 Measured values of $1/\tau_2$ (eq. 3.27) in a suspension of γ-Al_2O_3 for four trace metal cations using the pressure-jump technique with varying pH (Hachiya et al.[18]. The curves through the data points were calculated with eq. 3.30.

$$k_2K_{OS}/k_{-2} = k_2k_1/k_{-1}k_{-2} = \frac{[\equiv \text{AlOM}^+]_{\text{eq}}[H^+]_{\text{eq}}}{[\equiv \text{AlOH}]_{\text{eq}}[M^{2+}]_{\text{eq}}} \tag{3.31}$$

where eq. 3.28 has been incorporated in the first step. This equilibrium parameter can be measured independently of the kinetics experiments,[24] then used to calculate the product k_2K_{OS} with the associated value of k_{-2}. Some results of this computational scheme are shown in the following table:[23]

Metal Cation	k_{-2} $dm^3\ mol^{-1}\ s^{-1}$	k_2K_{OS} $dm^3\ mol^{-1}\ s^{-1}$	k_{wex} s^{-1}
Pb^{2+}	$4.1 \pm 1.0 \times 10^6$	$6.4 \pm 1.6 \times 10^4$	7×10^9
Cu^{2+}	$3.1 \pm 0.9 \times 10^5$	$7.4 \pm 2.0 \times 10^3$	1×10^9
Zn^{2+}	$1.3 \pm 0.2 \times 10^5$	$5.1 \pm 0.8 \times 10^2$	7×10^7
Mn^{2+}	$1.8 \pm 0.3 \times 10^6$	$3.2 \pm 0.5 \times 10^1$	3×10^7
Co^{2+}	$6.9 \pm 1.9 \times 10^4$	$1.5 \pm 0.4 \times 10^1$	2×10^6

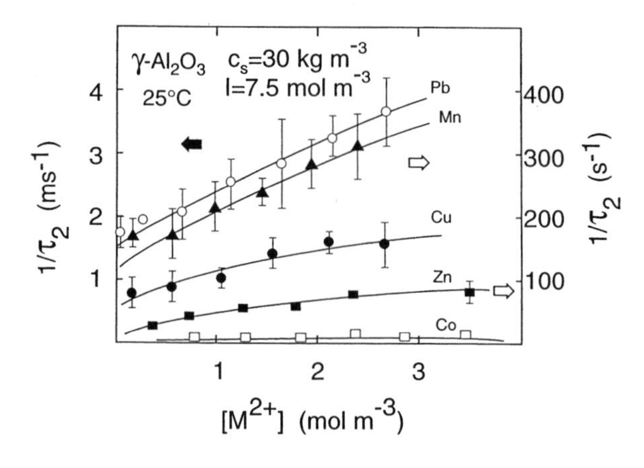

Fig. 3.4 Dependence of $1/\tau_2$ on metal cation concentration in suspensions of γ-Al_2O_3 (Hachiya et al.[18]). The curves through the data points were calculated with eq. 3.30.

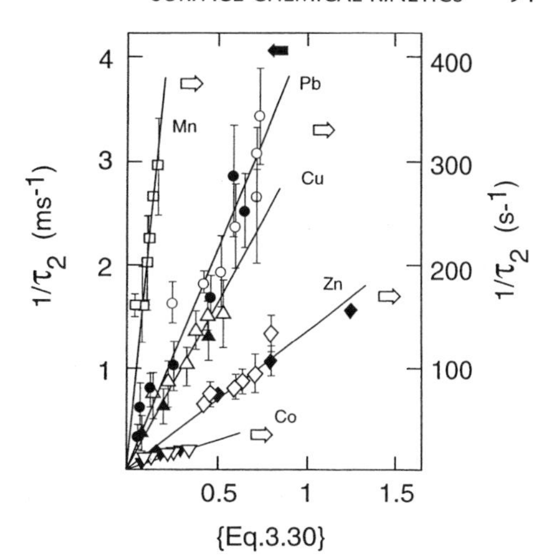

Fig. 3.5 Graph of the $1/\tau_2$ data in fig. 3.3 (filled symbols) and 3.4 (open symbols) versus the quantity inside curly brackets in eq. 3.30 expressed in mol m^{-3} (Hachiya et al.[18]). The slope of each line is equal to the rate coefficient k_{-2} in eq. 3.30.

The chemical significance of k_2K_{OS} can be deduced by returning to eq. 3.18b, keeping in mind the associations in eq. 3.26:

$$\frac{dc_D}{dt} = k'_f c_C - k'_b c_D c_E = k'_f\left(\frac{k_f}{k_b}\right)c_A c_B - k'_b c_D c_E \tag{3.18b}$$

and, therefore,

$$\frac{d}{dt}[\equiv \text{AlOM}^+] = k_2K_{OS}[\equiv \text{AlOH}][M^{2+}] - k_{-2}[\equiv \text{AlOM}^+][\text{H}^+] \tag{3.32}$$

where the rapid equilibration of species C with species A and B has been assumed so as to insert eq. 3.20a into the rate law for species D. (Note that "eq" in eq. 3.20a applies to species A and B only in relation to species C, *not* to species D or E.) The parameter k_2K_{OS} is the second-order rate coefficient for the specific adsorption of M^{2+} by the alumina adsorbent. It is this parameter that, by analogy with the Eigen-Wilkins-Werner mechanism for inner-sphere complexes in aqueous solution,[13,14] should correlate positively with k_{wex} for the metal cation. The values of k_{wex} do indeed correlate positively with k_2K_{OS}, yielding the linear regression equation ($r^2 = 0.999$):

$$\log k_2K_{OS} = -5.272 + 1.02 \log k_{wex} \tag{3.33}$$

This proportionality between k_2K_{OS} and k_{wex}, which also has been observed for bivalent metal cation adsorption on goethite,[25] lends strong support to eq. 3.16.

An elementary reaction sequence analogous to eq. 3.16 for specific anion adsorption by a ligand-exchange mechanism to form an inner-sphere surface complex can be exemplified by iron oxyhydroxide as the adsorbent:

$$\equiv \text{FeOH}_2^+ + L^{\ell-} \underset{k_{-1}}{\overset{k_1}{\Leftrightarrow}} \equiv \text{FeOH}_2{}^+ \cdots L^{\ell-} \underset{k_{-2}}{\overset{k_2}{\Leftrightarrow}} \equiv \text{FeL}^{(1-\ell)} + \text{H}_2\text{O} \tag{3.34}$$

where $L^{\ell-}$ is the adsorptive anion and pH < p.z.n.p.c. has been assumed. This reaction sequence can be identified with that in eq. 3.17 after making the associations:

$$A \leftrightarrow \equiv \mathrm{FeOH}_2^+ \quad B \leftrightarrow L^{\ell-} \quad C \leftrightarrow \equiv \mathrm{FeOH_2}^+ \cdots L^{\ell-} \\ D \leftrightarrow \equiv \mathrm{Fe}L^{(1-\ell)} \quad E \leftrightarrow \mathrm{H_2O} \tag{3.35}$$

In this application, however, any concentration deviation of H_2O ("species E") will be quite negligible and, therefore, $c_E = c_E^{\mathrm{eq}}$ in the steps leading from eqs. 3.18 to 3.19. Moreover, the third stoichiometric condition in eq. 3.22 is not applicable and eq. 3.21b must be replaced by the rate law:

$$\frac{d\Delta c_D}{dt} = k_f' \; \Delta c_C - k_b' \; c_E^{\mathrm{eq}} \; \Delta c_D \tag{3.19c}$$

The coupled, linear rate laws in eqs. 3.23 then become:

$$-\frac{d\Delta c_A}{dt} = \left[k_f\left(c_A^{\mathrm{eq}} + c_B^{\mathrm{eq}}\right) + k_b\right]\Delta c_A + k_b \Delta c_D \tag{3.36a}$$

$$\frac{d\Delta c_A}{dt} = -k_f' \Delta c_A - \left(k_f' + k_b' c_E\right)\Delta c_D \tag{3.36b}$$

The time constants that characterize these rate laws are given by eqs. 3.25, but with

$$a_{22} \equiv k_b' c_E + k_f' \equiv K_b' + k_f' \tag{3.24f}$$

instead of the definition in eq. 3.24e. In the likely special case that $a_{11} \gg a_{12}$ or a_{22}, corresponding to a very rapid first step in eq. 3.34, eq. 3.25d applies, and

$$\frac{1}{\tau_1} \approx k_1\left(\left[\equiv \mathrm{FeOH}_2^+\right]_{\mathrm{eq}} + \left[L^{\ell-}\right]_{\mathrm{eq}}\right) + k_{-1} \tag{3.37a}$$

$$\frac{1}{\tau_2} \approx k_2\left\{k_1\left(\left[\equiv \mathrm{FeOH}_2^+\right]_{\mathrm{eq}} + \left[L^{\ell-}\right]_{\mathrm{eq}}\right)\tau_1\right\} + K_{-2} \tag{3.37b}$$

where $K_{-2} \equiv K_{-2}\ [\mathrm{H_2O}]_{\mathrm{eq}}$. Equations 3.37 have been applied to pressure-jump kinetics data on the adsorption of molybdate ($L^{\ell-} \equiv \mathrm{MoO}_4^{2-}$) by goethite.[26] The resulting values of the rate coefficients, obtained from linear regression of measured values of $\frac{1}{\tau_1}$ on the term in parentheses in eq. 3.37a and of $\frac{1}{\tau_2}$ on the term in curly brackets in eq. 3.37b, were:

$$k_1 = 4 \times 10^3\ \mathrm{dm^3\ mol^{-1}\ s^{-1}} \qquad k_{-1} = 3.9 \times 10^2\ \mathrm{s^{-1}}$$
$$k_2 = 1.9\ \mathrm{dm^3\ mol^{-1}\ s^{-1}} \qquad k_{-2} = 42\ \mathrm{s^{-1}}$$

from which it follows that $K_{\mathrm{OS}} = 10\ \mathrm{dm^3\ mol^{-1}}$ (eq. 3.28), a value comparable to what is found for outer-sphere complexes in aqueous solution.[13] As was observed in fig. 3.3 for specific bivalent metal adsorption, the time scales for τ_1 and τ_2 were millisecond and both time constants decreased as pH increased.[26]

It is important to bear in mind that the expressions for characteristic time constants in eqs. 3.27, 3.30, and 3.37 are not unique. The elementary reactions in eqs. 3.16 and 3.34 are *hypotheses* whose validation cannot be accomplished by goodness-of-fit of kinetics models to rate data any more than can those of surface complexation models (section 2.1). Even if the premise that the Eigen-Wilkins-Werner mechanism applies to inner-sphere surface complexation is sound, more than one way exists to realize this mechanism through elementary reactions.[23] In the broadest terms, the relaxation of an adsorptive-adsorbate system as observed in pressure-jump experiments typically can be deconvoluted into a pair of exponentially decaying processes whose characteristic time scales are significantly different from one another.[18,22–26] The time constants that characterize these exponential decays usually display monotonic dependence on the initial pH and adsorptive concentration, and it is this variation that a kinetics model interprets with one or two adjustable parameters. Disagreement with the observed pH or concentration dependence of the time constants can invalidate a proposed kinetics model, but consistency with the observations cannot logically validate it.

For example, eq. 3.34 describes an elementary reaction between an anion and a single Lewis acid site on the surface of an iron oxide, but it is well established that oxyanions often react with a *pair* of adjacent Lewis acid sites to form a binuclear inner-sphere surface complex (fig. 2.7).[27] Therefore, a two-step elementary reaction alternative to that eq. 3.34 could be of the form:

$$\equiv \begin{matrix} \mathrm{FeOH_2^+} \\ \mathrm{FeOH_2^+} \end{matrix} + L^{\ell-} \underset{k_{-1}}{\overset{k_1}{\Leftrightarrow}} \equiv \begin{matrix} \mathrm{FeL}^{(1-\ell)} \\ \mathrm{FeOH_2^+} \end{matrix} + \mathrm{H_2O} \qquad (3.38)$$

$$\underset{k_{-2}}{\overset{k_2}{\Leftrightarrow}} \equiv \begin{matrix} \mathrm{Fe} \\ \mathrm{Fe} \end{matrix} L^{(2-\ell)} + \mathrm{H_2O}$$

Because in both steps water is a product whose concentration is not perturbed by the pressure jump, this reaction sequence is a special case of the generic two-step process:

$$A + B \underset{k_b}{\overset{k_f}{\Leftrightarrow}} C \underset{k'_b}{\overset{k'_f}{\Leftrightarrow}} D \qquad (3.39)$$

for which the coupled, linearized rate laws are:[28]

$$-\frac{d\Delta c_A}{dt} = \left[k_f\left(c_A^{\mathrm{eq}} + c_B^{\mathrm{eq}}\right) + k_b\right]\Delta c_A + k_b \Delta c_D \qquad (3.40a)$$

$$\frac{d\Delta c_D}{dt} = -k'_f \Delta c_A - \left(k'_f + k'_b\right)\Delta c_D \qquad (3.40b)$$

in exact analogy with eqs. 3.36. Equations 3.40 can be decoupled to derive eqs. 3.24 and 3.25 for the two characteristic time constants, but with the a_{ij} $(i, j = 1, 2)$ defined by:

$$a_{11} \equiv k_f(c_A^{\mathrm{eq}} + c_B^{\mathrm{eq}}) + k_b \qquad (3.41a)$$

$$a_{12} \equiv k_b, \qquad a_{21} \equiv k'_f \qquad (3.41b)$$

$$a_{22} \equiv k'_f + k'_b \qquad (3.41c)$$

In applications,[28] Eqs. 3.24 are often the most useful:

$$\left(\frac{1}{\tau_1}\right) + \left(\frac{1}{\tau_2}\right) = k_f(c_A^{eq} + c_B^{eq}) + k_b + k_f' + k_b' \tag{3.42a}$$

$$\left(\frac{1}{\tau_1}\right)\left(\frac{1}{\tau_2}\right) = k_f(k_f' + k_b')(c_A^{eq} + c_B^{eq}) + k_b k_b' \tag{3.42b}$$

Graphs of the sum and product of the measured values of τ_1 and τ_2 then yield the rate coefficients from linear regression analysis:

$$k_f = \text{slope of eq. 3.42a} \tag{3.43a}$$

$$k_b = y\text{-intercept of eq. 3.42a} - (\text{slope of eq. 3.42b} \div k_f) \tag{3.43b}$$

$$k_b' = (y\text{-intercept of eq. 3.42b} \div k_b) \tag{3.43c}$$

$$k_f' = y\text{-intercept of eq. 3.42a} = k_b - k_b' \tag{3.43d}$$

Note that eqs. 3.42 are exact and do not depend on any assumption about the relative magnitudes of τ_1 and τ_2.

Equations 3.42 have been applied to pressure-jump kinetics data for arsenate and chromate on goethite.[29] However, the reactant adsorbent species was assumed to be $\equiv$ FeOH (despite the fact that pH $<$ p.z.n.p.c. in the experiments), with each surface OH group reacting independently (despite the binuclear nature of the surface complex assumed). Time constants with millisecond time scales were observed, and rate coefficients roughly comparable to those tabulated above for molybdate adsorption were reported. This example serves to illustrate the inherent ambiguity in attempting to deduce adsorption mechanisms solely from adsorption kinetics modeling.[30]

3.3 Surface Oxidation-Reduction Reactions

Surface oxidation-reduction (or *redox*) reactions are electron-transfer processes in which the oxidant and reductant interact as adsorbate species. *The adsorbent does not participate directly in the redox reaction.* Surface redox reactions are ubiquitous and important agents of transformation in soils, sediments, and aquatic systems.[31] Their usual pathway follows a sequence initiated by inner-sphere surface complexation of either the oxidant or the reductant. Then a complex forms between the adsorbed species and its counterpart reactant as a precursor to an electron-transfer step, after which this ternary surface complex becomes destabilized by the production of newly reduced and oxidized species.[32] If the complex formed between oxidant and reductant is outer-sphere, the electron-transfer step is termed a *Marcus process*, whereas if it is inner-sphere, the electron-transfer step is termed a *Taube process*.[33] Electron transfer is likely to be the rate-limiting step in surface redox reactions proceeding by the Marcus process.[32,33]

A prototypical surface redox reaction based on the two-step sequence outlined above can be expressed as a special case of the generic reaction in eq. 3.39:

$$\equiv A + B \underset{k_b}{\overset{k_f}{\Leftrightarrow}} \equiv A - B \overset{k_{\mathrm{ET}}}{\rightarrow} \equiv P \tag{3.44}$$

where $\equiv A$ represents an adsorbate species that may be either the oxidant or the reductant, B represents its counterpart reactant in the vicinal aqueous solution, $\equiv A - B$ is the ternary surface complex they form, either outer-sphere or inner-sphere, and $\equiv P$ represents the oxidized/reduced products formed after electron transfer has taken place. These products may then desorb (for example, CO_2 produced from the oxidation of an organic reductant may desorb) or remain in the adsorbate, eventually becoming any one of the surface species introduced in fig. 1.4.[31–33] The rate laws describing the kinetics of this two-step process may be expressed by a set of coupled ordinary differential equations (cf. eqs. 3.18):

$$-\frac{dc_A}{dt} = -\frac{dc_B}{dt} = k_f c_A c_B - k_b c_{AB} \tag{3.45a}$$

$$\frac{dc_{AB}}{dt} = k_f c_A c_B - (k_b + k_{\mathrm{ET}}) c_{AB} \tag{3.45b}$$

$$\frac{dc_p}{dt} = k_{\mathrm{ET}} c_{AB} \tag{3.45c}$$

For the important special case that the concentration of reactant B is held constant (e.g., by maintaining $c_B \gg c_A$) while the kinetics of the reaction in eq. 3.44 are monitored, eqs. 3.45a and 3.45b can be linearized through the introduction of the pseudo first-order rate coefficient, $K_f \equiv k_f c_B$:

$$-\frac{dc_A}{dt} = K_f c_A - k_b c_{AB} \tag{3.46a}$$

$$\frac{dc_{AB}}{dt} = K_f c_A - (k_b + k_{\mathrm{ET}}) c_{AB} \tag{3.46b}$$

This pair of coupled linear differential equations can be solved exactly:[34]

$$c_A(t) = c_A(0)\left[\frac{K_f(k - k_{\mathrm{ET}})}{k(k - k')}\exp(-kt) - \frac{K_f(k' - k_{\mathrm{ET}})}{k'(k - k')}\exp(-k't)\right] \tag{3.47a}$$

$$c_{AB}(t) = \frac{K_f c_A(0)}{k - k'}\left[\exp(-k't) - \exp(-kt)\right] \tag{3.47b}$$

where $(k > k')$:

$$k \equiv \frac{1}{2}\left\{K_f + k_b + k_{\mathrm{ET}} + \left[(K_f + k_b + k_{\mathrm{ET}})^2 - 4K_f k_{\mathrm{ET}}\right]^{\frac{1}{2}}\right\} \tag{3.48a}$$

$$k' \equiv \frac{1}{2}\left\{K_f + k_b + k_{\mathrm{ET}} - \left[(K_f + k_b + k_{\mathrm{ET}})^2 - 4K_f k_{\mathrm{ET}}\right]^{\frac{1}{2}}\right\} \tag{3.48b}$$

are composite first-order rate coefficients. Equation 3.47b then can be introduced into eq. 3.45c to produce the solution:

$$c_P(t) = c_A(0)\left\{\frac{K_f k_{\mathrm{ET}}}{k'(k-k')}[1-\exp(-k't)] - \frac{K_f k_{\mathrm{ET}}}{k(k-k')}[1-\exp(-kt)]\right\} \qquad (3.47c)$$

Note that the exponential decay of the concentrations of the species $\equiv A$, $\equiv A - B$, and $\equiv P$ is governed by the two composite rate coefficients, k and k', not by any one of the three rate coefficients in eqs. 3.46. This behavior is a direct result of the coupling between eqs. 3.46a and 3.46b.

Equations 3.47 can be applied to describe the reduction of adsorbed Cr(VI) on TiO_2 by organic reductants in aqueous solution,[35] a heterogeneous redox process analogous to the homogeneous process in eq. 3.1, but with the organic reductant replacing Fe and $\equiv$ Cr(VI) replacing dissolved chromate. This heterogeneous process can be initiated by rapid ($<$ 12 h) adsorption of Cr(VI) on TiO_2 (p.z.n.p.c. = 6.5) in the presence of excess organic reductant at pH 4.7 [20 mmol m^{-3} Cr(VI) and 200 mmol m^{-3} organic reductant initially added to 1 kg m^{-3} TiO_2 suspended in 100 mol m^{-3} $NaClO_4$ solution]. Thus, A $\leftrightarrow$ Cr(VI) and B $\leftrightarrow$ organic reductant in eqs. 3.44 to 3.47, given that the initial excess of organic reductant [i.e.,10:1 molar ratio with initial Cr(VI)] is sufficient to validate the condition of constant concentration during the redox reaction. The empirical rate law that has been used to describe this reaction,[36]

$$-\frac{d}{dt}\mathrm{Cr(VI)_T} = K[\equiv \mathrm{Cr(VI)}] \qquad (3.49)$$

where

$$\mathrm{Cr(VI)_T} = [\mathrm{Cr(VI)}] + [\equiv \mathrm{Cr(VI)}] \qquad (3.50)$$

and K is a pseudo first-order rate constant, is based on the hypothesis that the rate of total Cr(VI) decline is governed principally by the interaction of the organic reductant with adsorbed Cr(VI). [Independent experiments have demonstrated that chromate reduction in aqueous solution alone (i.e., the homogeneous redox reaction) does not occur on the time scale of the heterogeneous process (about 200 h).[35,36]] Therefore, this rate law should be consistent with eqs. 3.45 to 3.47.

The initial concentration of Cr(VI) was selected low enough to justify the use of a distribution ratio (eq. 1.6) to relate the two Cr(VI) concentrations on the right side of eq. 3.50:[36]

$$[\equiv \mathrm{Cr(VI)}] = D_{\mathrm{Cr}}[\mathrm{Cr(VI)}] \qquad (3.51)$$

where $D_{\mathrm{Cr}} = 0.43$ at pH 4.7.[35] Given that this relationship is established on a time scale much shorter than that over which eq. 3.49 applies, the empirical rate law can be expressed entirely in terms of [$\equiv$Cr(VI)]:

$$-\frac{d}{dt}[\equiv \mathrm{Cr(VI)}] = \frac{K\, D_{\mathrm{Cr}}}{1+D_{\mathrm{Cr}}}[\equiv \mathrm{Cr(VI)}] \qquad (3.52)$$

This rate law can be compared to the one in eq. 3.46a, since A $\leftrightarrow \equiv$ Cr(VI). It is evident that eq. 3.52 does not contain the concentration of the intermediate species, $\equiv A - B \leftrightarrow \equiv$ Cr(VI)-organic reductant. Consistency between the two rate laws is possible either if the back-reaction in eq. 3.44 is negligible over the time scale of observation ($k_b \ll K_f$) or if the intermediate species concentration is at steady-state ($dc_{AB}/dt = 0$).[37] In this latter case, eq. 3.46a takes the form:

$$-\frac{dc_A}{dt} = \frac{K_f k_{ET}}{k_b + k_{ET}} c_A \tag{3.53}$$

with the composite pseudo first-order rate coefficient on the right side also replacing k_{ET} in eq. 3.45c. Thus two alternative chemical interpretations,

$$\frac{K\ D_{Cr}}{1 + D_{Cr}} = K_f \tag{3.54a}$$

$$\frac{K\ D_{Cr}}{1 + D_{Cr}} = \frac{K_f k_{ET}}{k_b + k_{ET}} \tag{3.54b}$$

are possible for the empirical rate coefficient, K. We note in passing that eq. 3.49 can also be expressed in terms of [Cr(VI)] instead of [$\equiv$ Cr(VI)] by introducing eq. 3.51.[35] The pseudo first-order rate coefficient linking the reaction rate to [Cr(VI)], termed k_{obs},[35,36] is then equal to the left side of eqs. 3.54. This approach is useful because [Cr(VI)] is much less difficult to measure than [$\equiv$ Cr(VI)], but leads to the same kinetics data.

Values of the empirical rate coefficients, k_{obs} and K, for several organic reductants are listed in the table below:[35]

Organic Reductant	R-Substituent[a]	$k_{obs}(10^{-3}\ h^{-1})$	$K(10^{-2} h^{-1})$[b]
Glycolic acid	H	4.2 ± 0.7	2.0 ± 0.3
Lactic acid	CH_3	6.5 ± 0.9	2.6 ± 0.3
Mandelic acid	C_6H_5	43 ± 7	16 ± 3
Tartaric acid	CH(OH)COOH	4.8 ± 0.7	3.9 ± 0.5
Methyl glycolate	H	4.8 ± 0.7	1.4 ± 0.2
Methyl lactate	CH_3	3.3 ± 0.5	1.0 ± 0.1
Methyl mandelate	C_6H_5	7.8 ± 0.3	2.5 ± 0.8
Glyoxylic acid	H	6.0 ± 0.9	4.1 ± 0.6
Oxalic acid	OH	2.1 ± 0.2	2.4 ± 0.2

[a] Chemical formulas for the three categories of reductant are: R-CH(OH)COOH; R-CH(OH)$COOCH_3$; and R-COCOOH. Thus, the second group of reductants is the methyl ester of the first group. The pK_a values are < 4 at 25°C [35] for the carboxyls in the first and third groups of reductants.

[b] Calculated as $K = (1 + D_{Cr})k_{obs}/D_{Cr}$, where D_{Cr} is the distribution ratio measured in the presence of the reductant at pH 4.7. In general, $0.095 < D_{Cr} < 0.54$,[35] the difference from 0.43 (the value measured in the absence of reductant) being attributed to reductant competition with Cr(VI) for adsorption by TiO_2.

The rate coefficient K (fourth column in the table above) shows important trends with the substituents that distinguish among the reductants in a given category (column 2 in the table) and with the substituents on either carbon atom in the "backbone" of the reductant molecule. For example, the value of

K is increased significantly if the substituent on the α-hydroxy group [R-CH(OH)-] is a benzene ring (C_6H_5), but it is decreased by methylation of the carboxyl bound to the α-hydroxy group (e.g., mandelic acid versus methyl mandelate). Replacement of the α-hydroxy group by a carbonyl (glycolic acid versus glyoxylic acid) also decreases K. These trends are associated with differing oxidation pathways of the organic reductants,[35] which implies that K reflects electron-transfer rates and, therefore, that eq. 3.54b is the more likely chemical interpretation of this empirical rate coefficient. The dependence of K for mandelic acid on pH and on the type of hydroxylated adsorbent adds to this interpretation[35]. The value of K decreases with increasing pH and with increasing p.z.n.p.c. of the adsorbent, suggesting that strong adsorbent Brønsted acidity enhances the rate of the redox reaction between adsorbed Cr(VI) and a dissolved organic reductant.[35] This property, in turn, derives from surface OH groups coordinated to a single metal ion in the adsorbent, with stronger Brønsted acidity favored by a larger bond valence of the metal ion (section 1.2 and problem 3 in chapter 1). If chromate adsorbs according to the inner-sphere surface complexation mechanism in eq. 3.34, the metal cation plays a role similar to that of a proton (section 1.5), but the metal cation does so more effectively because of its much larger valence. *The result of this interaction is a shift in electron density away from Cr(VI) to the adsorbent metal cation, which then facilitates electron transfer to Cr(VI) from an organic reductant.* Moreover, as with protonation, coordination of chromate to a metal cation makes the oxygen ions in the anion easier to replace and thereby facilitates inner-sphere complex formation with the reductant.[33]

The reaction in eq. 3.44 and its kinetics model in eqs. 3.45 and 3.47 can also be applied to a variant of eq. 3.1 in which the Fe(II) reductant is adsorbed while Cr(VI) remains an aqueous solution species.[39] In this case, by analogy with eqs. 3.5 and 3.7, the empirical rate law,

$$-\frac{d}{dt}[\mathrm{Cr(VI)}] = k_1\left[\mathrm{FeOH^+}\right][\mathrm{Cr(VI)}] + k_1^{\mathrm{surf}}[\equiv \mathrm{Fe(II)}][\mathrm{Cr(VI)}] \tag{3.55}$$

can be used to describe the kinetics of Cr(VI) reduction in the presence of dissolved and adsorbed Fe(II), where $k_1 = 1.4 \pm 0.3 \times 10^5\ \mathrm{dm^3\ mol^{-1}\ s^{-1}}$ for the homogeneous reaction,[7] which cannot be vitiated in an experiment performed with an aqueous suspension of the adsorbent. [The choice of $FeOH^+$ as the reactant species is appropriate for Cr(VI) reduction at circumneutral pH values, as is evident from examining the contribution of each term to the right side of eq. 3.7 in this pH range.] If Fe(II) adsorption is rapid and is described accurately by a distribution ratio, then the second term in eq. 3.55 [a special case of eq. 3.45a with the identifications $A \leftrightarrow$ Fe(II), $B \leftrightarrow$ Cr(VI)] is proportional to [Fe(II)] throughout the redox reaction. If pH is constant, the same is true for the first term. Under these circumstances, eq. 3.55 can be recast into the convenient form:[39]

$$-\frac{d}{dt}[\mathrm{Cr(VI)}] = k_{\mathrm{obs}}\ \mathrm{Fe(II)_T}[\mathrm{Cr(VI)}] \tag{3.56}$$

where $Fe(II)_T$ is defined analogously to eq. 3.50 and

$$k_{obs} = k_1 \frac{[FeOH^+]}{Fe(II)_T} + k_1^{surf} \frac{[\equiv Fe]}{Fe(II)_T} \tag{3.57}$$

is a composite, second-order rate coefficient for Cr(VI) reduction by Fe(II).

Values of k_{obs} measured in suspensions of amorphous SiO_2 (i.e.p. = 2.3) containing 60 mmol m^{-3} Fe(II) and 20 mmol m^{-3} Cr(VI) initially in 10 mol m^{-3} KCl at pH 4.9 are listed in the table below for several solids concentrations:[39]

c_s (kg m^{-3})	1.0	2.0	5.0	10.0
k_{obs} (dm^3 mol^{-1} s^{-1})	4.6	9.6	19.8	34.7
k_1^{surf} (m^3 mol^{-1} s^{-1})	2.9 ± 1.0	7.4 ± 1.9	7.9 ± 2.0	7.5 ± 1.9

The low point of zero charge for SiO_2(am) ensures the positive adsorption of Fe^{2+} and the negative adsorption of chromate. It is evident that k_{obs} increases in approximate proportion to the solids concentration, thus indicating the importance of adsorbed Fe(II) in the reduction reaction. At pH 4.9, its first term (eq. 3.57) has the value 3.4 ± 0.3 dm^3 mol^{-1} s^{-1}, according to eq. 3.8a and the value of k_1 given below it. That this term is always smaller than k_{obs} is also evidence supporting the participation of $\equiv$ Fe(II) as a reductant. Given the observed distribution coefficient for Fe^{2+} adsorption on SiO_2(am),[39] $K_{dFe} = 0.42 \pm 0.11$ L kg^{-1} and eq. 1.6, eq. 3.57 leads to the values of k_1^{surf} given in the third row of the table above. The mean of the three clustered values is $7.6 \pm 2.0 \times 10^3$ dm^3 mol^{-1} s^{-1}. This value is smaller than k_1 ($= 1.4 \pm 0.3 \times 10^5$ dm^3 mol^{-1} s^{-1}),[7] indicating that adsorption by SiO_2(am) is not as effective as formation of the hydrolytic species $FeOH^+$ is at enhancing the rate of Cr(VI) reduction by Fe(II). On the other hand, $k_1{}^{surf} \gg k_0$ ($= 0.3 \pm 0.5$ dm^3 mol^{-1} s^{-1}),[7] demonstrating the catalytic effect of adsorption relative to Cr(VI) reduction solely by the aqueous species Fe^{2+}. This catalytic effect can be understood in the same way as for hydrolytic species: *formation of surface complexes donates electron density to Fe^{2+}, thereby facilitating its electron transfer to Cr(VI).*[32]

Given this mechanistic conclusion, it is reasonable to ask whether the LFER in eq. 3.12 can also relate $k_1{}^{surf}$ to the corresponding equilibrium constant for the reduction half-reaction,

$$\equiv Fe^{3+} + e^- = \equiv Fe^{2+} \tag{3.58}$$

No directly measured value of log K_R for this redox couple is available, but an estimate, log K_R = 6.1,[40] has been made on the basis of a LFER analysis of the oxidation of Fe(II) by $O_2(g)$. This estimate and the value of $k_1^{surf} \approx 7.6 \times 10^3$ dm^3 mol^{-1} s^{-1} can be compared with a rather complete LFER for the second-order rate coefficient characterizing Cr(VI) reduction by Fe(II) in a variety of complexed forms:[41]

$$\log k = 8.129 - 0.6021 \log K_R \qquad (r^2 = 0.946) \tag{3.59}$$

which is very nearly the same as eq. 3.12. As can be seen in fig. 3.6, the data for Cr(VI) reduction by surface-complexed Fe(II) are very compatible with eq. 3.59.

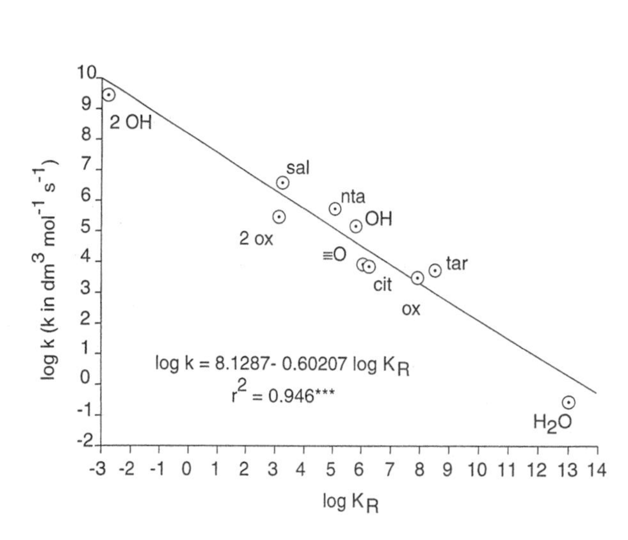

Fig. 3.6 Linear free energy relationship for the rate coefficient k in the rate law: $-d/dt[\mathrm{Cr(VI)}] = k\,[\mathrm{FeL}_n]\,[\mathrm{Cr(VI)}]$ where FeL_n is a soluble complex of Fe(II). Plotted on the ordinate are values of log k at pH 5, while on the abscissa are values of log K_R for the reduction half-reaction: $\mathrm{Fe(III)L}_n + e^- = \mathrm{Fe(II)L}_n$ $(n = 1, 2)$. The ligand L is: citrate (cit), hydroxide (OH), nitrilotriacetate (nta), oxalate (ox), salicylate (sal), surface O (≡O), tartrate (tar), or water (H_2O).

This strong correlation over 16 orders of magnitude is satisfying evidence for the electron-density donation interpretation of the role of Fe complexation in the kinetics of the reaction in eq. 3.1.[11] We note in passing that the clustering of data points around that for $FeOH^+$ suggests intense competition for Cr(VI) will occur among hydrolyzed, organic, and adsorbed forms of Fe(II) in natural waters.[41]

These examples of surface redox reactions illustrate several important general features of their kinetics that can be understood in a unified fashion with reference to eqs. 3.44, 3.45, and 3.47. In most experimental studies, the chemical termed A in eq. 3.44 is redox-active both as an adsorptive and as an adsorbate, such that the rate law for its transformation takes the form:

$$-\frac{d}{dt}A_T = k_1[A][B] + k_1^{\mathrm{surf}}[\equiv A][B] \tag{3.60}$$

where

$$A_T = [A] + [\equiv A] \tag{3.61}$$

is its total concentration in both chemical forms. If experimental conditions are such that adsorption is rapid on the time scale of eq. 3.60 and can be described by a constant distribution ratio (eq. 1.6),

$$D_A = \frac{[\equiv A]}{A} = \frac{x_{\mathrm{Aads}}}{x_{\mathrm{Asoln}}} \tag{3.62}$$

then eq. 3.60 can be expressed in two equivalent forms:

$$-\frac{d}{dt}[\equiv A] = \left(x_{\mathrm{Asoln}}\,k_1 + x_{\mathrm{Aads}}\,k_1^{\mathrm{surf}}\right)[\equiv A][B] \tag{3.63a}$$

$$-\frac{d}{dt}[A] = \left(x_{\mathrm{Asoln}}\,k_1 + x_{\mathrm{Aads}}\,k_1^{\mathrm{surf}}\right)[A][B] \tag{3.63b}$$

Equation 3.63a can be compared to the second-order term in eq. 3.45a, showing that a mixed homogeneous/heterogeneous redox reaction leads to a rate law that is similar to that for the heterogeneous reaction alone. Equation 3.63b demonstrates that the rate coefficient can be measured by monitoring the time dependence of A as the aqueous species, a fact of great practical importance.

In the special case, $k_1 \ll k_1^{\text{surf}}$, eq. 3.63a reduces to a rate law like eq. 3.52 ($K \equiv k_1^{\text{surf}}\,[B]$). This case is expected when the donation of electron density to an adsorbed reductant or the removal of electron density from an adsorbed oxidant by a surface site greatly exceeds that which occurs for the reductant or oxidant as an aqueous solution species. In the example of $\equiv$Cr(VI) reduction by carboxylic acids, this case was observed (the aqueous species $H{-}O{-}CrO_3^-$ versus the surface species $\equiv Ti{-}O{-}CrO_3^-$), whereas in the example of Cr(VI) reduction by $\equiv$Fe(II), it did not (the hydrolytic species $HOFe^+$ versus the surface species $\equiv SiOFe^+$).

Measured values of the composite rate coefficient in eqs. 3.63 can be decomposed into homogeneous and heterogeneous parts by subtracting the value of $x_{\text{Asoln}}k_1$ obtained in separate kinetics and adsorption experiments. If the pH dependence of the composite rate coefficient, of k_1, and of the distribution ratio D_A is known, that of k_1^{surf} also can be determined. A relation between k_1^{surf} and K_f in eq. 3.46a then can be developed under several different conditions. The simplest is $k_b \ll K_f$, which is characteristic of very strong intermediate complex formation, in which case

$$x_{\text{Aads}}k_1^{\text{surf}} = K_f \tag{3.64a}$$

and no information about the electron-transfer process is per se contained in k_1^{surf}. Alternatively, the intermediate species may be at steady state, in which case the left side of eq. 3.45b is zero, and

$$x_{\text{Aads}}k_1^{\text{surf}} = \frac{K_f k_{\text{ET}}}{k_b + k_{\text{ET}}} \tag{3.64b}$$

Since c_B is maintained constant, both c_A and c_P will show an exponential time dependence governed by k_1^{surf}, which in turn depends on all three rate coefficients in eqs. 3.46. It is under constant c_B also that the most general time dependence of c_A is given by the two-term exponential series in eq. 3.47a. The absence of this general time dependence should be verified by a careful statistical analysis of the observed behavior of $c_A(t)$ before invoking an approximate rate law such as that in eq. 3.60, whose common use begs the question of the applicability of the exact result in eq. 3.47a.

3.4 Proton-Promoted Mineral Dissolution Reactions

Mineral dissolution reactions are termed *surface-controlled* if a surface complex is involved as an intermediate kinetic species. Otherwise, these reactions are considered to be *transport-controlled* and are described by rate laws whose parameters reflect ion diffusion and advection processes, not surface chemistry. Minerals that comprise metal carbonates, oxyhydroxides, and silicates typically exhibit surface-controlled dissolution phenomena.[42]

A surface-controlled dissolution reaction can occur in the presence of liquid water alone according to the schematic two-step sequence:[43]

$$\equiv \mathrm{SR} + \mathrm{H}^+ \Leftrightarrow \equiv \mathrm{SRH}^+ \rightarrow \mathrm{P} \tag{3.65a}$$

$$\equiv \mathrm{SR} + \mathrm{H_2O} \Leftrightarrow \equiv \mathrm{SRHOH} \rightarrow \mathrm{P} \tag{3.65b}$$

$$\equiv \mathrm{SR} + \mathrm{OH}^- \Leftrightarrow \equiv \mathrm{SROH}^- \rightarrow \mathrm{P} \tag{3.65c}$$

where $\equiv$ SR is a reactive functional group exposed at the surface of a mineral (e.g., $\equiv$ AlOH or $\equiv MgCO_3H$) and P represents the products of the dissolution reaction [e.g., Al(III) or Mg(II) species in aqueous solution]. Equations 3.65 illustrate proton-, solvation-, and hydroxide-promoted dissolution, respectively, in terms of an adsorption step followed by a rearrangement/detachment step at the mineral surface. The adsorption step occurs on a time scale that is orders of magnitude smaller than that of the subsequent step, which usually encompasses days to months.[44]

A rate law for d[P]/dt evidently can be formulated in terms of the concentrations of the surface complexes that appear as intermediate species in eqs. 3.65, under the assumption of additivity for the three parallel reactions.[3] In many studies of mineral dissolution, however, the overall rate law for conditions far from equilibrium is expressed in terms of the concentrations or activities of the three adsorptives in eqs. 3.65, with the concentration of H_2O merely subsumed into a pseudo zero-order rate coefficient:

$$\frac{d[\mathrm{P}]}{dt} = k_{\mathrm{H}}[\mathrm{H}^+]^{n_{\mathrm{H}}} + \mathrm{K}_{\mathrm{H_2O}} + k_{\mathrm{OH}}[\mathrm{OH}^-]^{n_{\mathrm{OH}}} \tag{3.66}$$

where n_H and n_{OH} are partial reaction orders whose observed values often lie in the interval (0,1).[44] Fractional reaction orders for the adsorptives can be expected—even if the rate law is first-order with respect to the concentration of the intermediate adsorbed species—on the basis of rapid equilibration of the first step in eqs. 3.65 and a nonlinear adsorption isotherm (such as that in eq. 1.13) relating the concentration of the adsorbate species to that of the corresponding adsorptive. If the mineral surface is heterogeneous with respect to the affinity of the surface group $\equiv$ SR for adsorptive H^+ or OH^-, then a power-law adsorption isotherm (van Bemmelen-Freundlich model[45]) is known to result, leading naturally to power-law terms in eq. 3.66. We note in passing that a dissolution rate law based on Eqs. 3.65a or 3.65c will be greater than first-order in the concentration of the intermediate adsorbed species if there are multiple adsorption-desorption steps before the final rearrangement/detachment step.[46]

Mineral dissolution experiments typically involve measurements of the evolving aqueous solution concentration of one or more of the cationic constituents found in a mineral in order to quantify the left side of eq. 3.66.[44] Given the absence of [P] on the right side of the equation, a constant value of its left side should be observed at a given pH, with either [P] or the equivalent extent of reaction parameter, ξ (section 3.1) showing linear dependence on the elapsed time of reaction.[3] For example, a 1.3 g sample of the olivine, forsterite (Mg_2SiO_4), dissolves in water congruently at 25°C to release Mg and Si at the constant rate,[47] $d\xi/dt = 5.7 \pm 0.8 \times 10^{-11}$ mol s^{-1} at pH 5, where $d\xi/dt$ is either the rate of Si release or one-half the rate of Mg release ([Mg]/[Si] = 1.8 ± 0.2 experimentally[47]). A time scale for this proton-promoted dissolution reaction can be calculated with the expression:[48]

$$\tau_{\text{dis}} \equiv [M_r \, d(\xi/m)/dt]^{-1} \tag{3.67}$$

where M_r is the relative molecular mass of a dissolving mineral whose *initial* mass (in grams) is m. In the present example, $M_r = 140.7$ Da, yielding $\tau_{\text{dis}} \approx 4$ years for an initial 1 g (or 7.1 mmol) of forsterite to dissolve. Dissolution time scales ranging from decades to millennia are typical for minerals in soils and sediments.[48]

It is almost universally assumed that the dissolution rate of a mineral is proportional to its specific surface area, with the resulting practice of reporting the rate in units of mol m^{-2} s^{-1} [division of $d(\xi/m)/dt$ by the specific surface area a_s or of $d(\xi/V)/dt$ by $a_s c_s$, where c_s is the solids concentration of the dissolving mineral].[44] This convention raises the important issue of whether measurements of a_s (by gas adsorption or microscopy methods)[49] *prior* to a dissolution experiment are sufficient.[50] Recent experimental studies of the dependence of $d(\xi/m)/dt$ on a_s suggest that the relationship between the two variables is proportionality, irrespective of whether specific surface area is measured before or after a dissolution reaction. It is important to bear in mind, however, that, like reaction order, the dependence of a dissolution rate on specific surface area is an *empirical* issue, *not* a theoretical one. It may be noted in passing that, if $d(\xi/m)/dt$ is modeled with a rate law which is first-order in the adsorbed intermediate species concentration (eqs. 3.65a,c),[51] then the area-normalized rate, $d(\xi/a_s c_s V)/dt$ will be proportional to the surface excess of the intermediate species. With the latter expressed in units of mol m^{-2}, as is often done,[48] normalization by a_s is quite redundant, since it appears in the denominator of both the area-normalized rate and the area-normalized surface excess to which the rate is assumed proportional.

Figure 3.7 shows the pH dependence of the mass-normalized rate of dissolution [$d(\xi/m)/dt$] of the clay mineral, kaolinite [$Si_4Al_4O_{10}(OH)_8$], at 25°C.[52] Zero-order kinetics were observed after 25 days of reaction, with a congruent release of Si and Al, except in the range of pH between 5 and 9, within which secondary gibbsite [γ-$Al(OH)_3$] precipitation was likely.[52] The reaction rate in fig. 3.7 has been normalized by the initial sample mass (80 mg) on the assumption that the release of Si and Al by the mineral will scale with its mass. The data in the log-log plot can be fit to the rate law in eq. 3.66 to yield the expression:

$$\frac{d(\xi/m)}{dt} = 10^{-8.28}(\text{H}^+)^{0.55} + 10^{-10.45} + 10^{-6.80}(\text{OH}^-)^{0.75} \tag{3.68}$$

where ( ) is thermodynamic activity for an aqueous species and the reaction rate is in units of mol kg^{-1} s^{-1}. The fractional reaction orders, $n_{\text{H}} = 0.55$ and $n_{\text{OH}} = 0.75$, lie within the typical interval (0, 1) observed for layer silicate minerals.[53] A

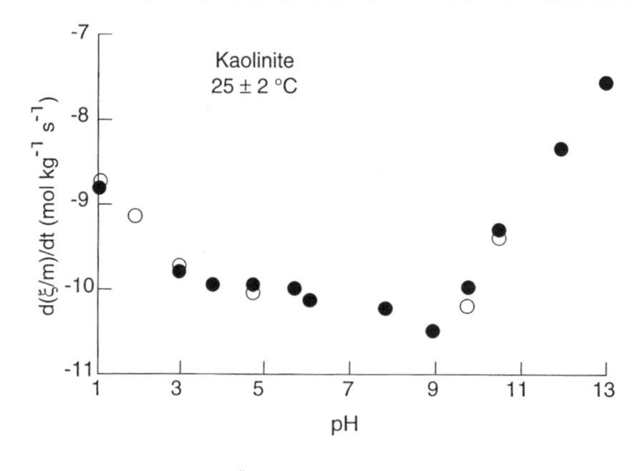

Fig. 3.7 Log-log plot of the rate of kaolinite dissolution (a_s = 8.16 m^2 g^{-1}, c_s = 2 kg m^{-3}) versus proton concentration in 1 mol dm^{-3} NaCl solution: (○) rate based on Al release, (•) rate based on Si release.[52]

dissolution rate described by eq. 3.66 has a minimum value at $pH_m = (n_H + n_{OH})^{-1} \log (n_H k_H / n_{OH} k_{OH} K_w)$. In the present example, with $K_w = 10^{-14}$, $pH_m = 9.5$ is predicted by eq. 3.68, in good agreement with the observed value, $pH_m = 9.7$.[52] Note that this latter pH value can be used to reduce the number of adjustable parameters in eq. 3.66 to four while introducing the useful constraint, $n_H k_H < n_{OH} k_{OH}$, given the typical range of values for n_H, n_{OH}. The dissolution time scale at pH 5 that follows from eq. 3.68 is approximately 1,370 years for an initial 1 g (1.9 mmol, M_r = 516.3 Da). On the basis of an equal initial number of moles, this time scale is three orders of magnitude larger than that found for olivine, illustrating the much more refractory nature of kaolinite.[54]

The temperature dependence of mineral dissolution rates has been investigated often in connection with climatic effects on weathering and biogeochemical cycling.[55] In a typical application, the dissolution rate normalized by specific surface area [$d(\xi/a_s m)/dt$ or $d([P]/a_s c_s)/dt$] is modeled with the Arrhenius equation (eq. 3.13) to extract values of the pre-exponential factor and the apparent activation energy. Values of this latter parameter usually fall in the range 20 to 90 kJ mol^{-1}, irrespective of mineral structure and composition, or of pH. As a typical example,[56] the proton-promoted dissolution rate of quartz (α-SiO_2) is described over the temperature range 25 to 300°C (fig. 3.8) by the Arrhenius equation:

$$d([\mathrm{Si}]/a_s c_s)/dt = 24 \exp[-(87.7 \pm 4.7 \times 10^3/RT)] \tag{3.69}$$

where the prefactor is in units of mol m^{-2} s^{-1}, the exponential numerical factor is in units of J mol^{-1}, and a_s is a specific surface area measured by gas adsorption techniques. In some studies, the temperature dependence of one or more terms on the right side of the rate law in eq. 3.66 is measured to obtain estimates of A and E_a in eq. 3.13. For example, the temperature dependence of the low-pH dissolution rate of forsterite[47] can be described by the multiple-regression equation:

$$k_H (\mathrm{H}^+)^{n_H} = (0.124 \pm 0.040) \exp\left[-\left(42.6 \pm 0.8 \times 10^3/RT\right)\right]\left(\mathrm{H}^+\right)^{0.5} \tag{3.70}$$

where the prefactor on the right side is now in units of mol g^{-1} s^{-1} and the exponential numerical factor is in units of J mol^{-1}. The reaction order in this

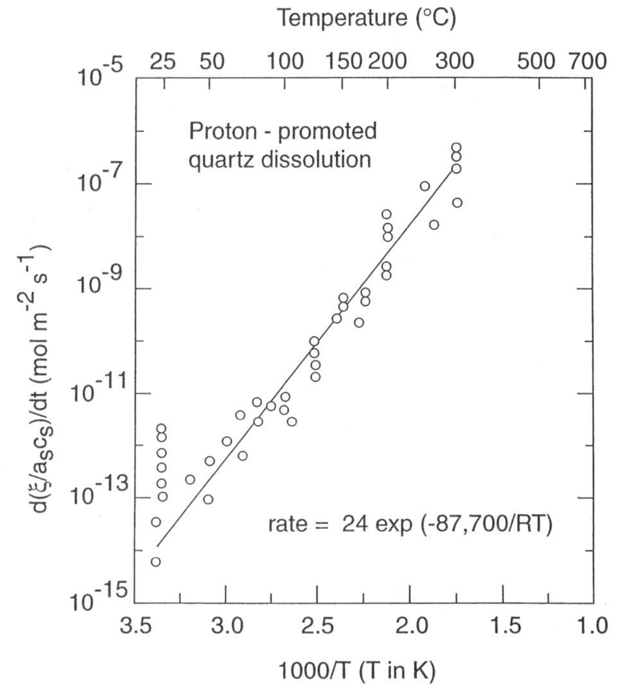

Fig. 3.8 Arrhenius plot of quartz dissolution data taken under varying pH and background electrolyte conditions. (Tester et al.[56]) The straight line through the data points represents eq. (3.69).

example is independent of temperature over the range 25 to 45°C ($1.7 < pH < 4.0$).[47] In other studies, the reaction order is found to depend on temperature and, therefore, the apparent activation energy varies with pH,[57] evidently reflecting a temperature dependence of the proton adsorption process.[58]

Scrutiny of the database on the pre-exponential factor, A, and the apparent activation energy, E_a, has led to the discovery of a strong correlation between the two parameters, such that A increases when E_a increases. This kind of positive correlation is termed a *compensation effect* and the correlation equation linking A with E_a is termed a *compensation law*.[59] In general terms, compensation laws can arise if the energetics of a rate-limiting process (E_a) are coupled to configurational changes of its participating reactant species in such a way that a large "energy barrier" for the reactants to overcome is compensated by a correspondingly large configurational change in the reactants which facilitates the process.[59] These configurational changes are manifest in the rate coefficient through the pre-exponential factor, A. The mathematical expression of this compensation effect usually takes the form:

$$\ln A = a + bE_a \tag{3.71}$$

where a and $b > 0$ are adjustable parameters. Equation 3.71 has been applied successfully to a variety of data on the temperature dependence of proton-promoted mineral dissolution rate coefficients normalized by specific surface area.[59]

Equations 3.13 and 3.71 provide constraints on the values of the Arrhenius parameters, A and E_a. If the compensation law parameters, a and b are known, for example, then the Arrhenius parameters can be calculated with a *single* measured rate coefficient for a chosen temperature:

$$\ln A = \ln k(T) + \frac{\ln[k(T)/a]}{bRT - 1} \tag{3.72a}$$

$$E_a = RT\,\frac{\ln[k(T)/a]}{bRT - 1} \tag{3.72b}$$

as follows from solving eqs. 3.13 and 3.71 simultaneously for the two Arrhenius parameters. Equations 3.72 represent a powerful ramification of the compensation effect. Another important result of the compensation law is the existence of an *isokinetic temperature*, T_{iso}, at which all rate coefficients whose Arrhenius parameters are related through eq. 3.71 have the *same* value, k_{iso}:

$$T_{iso} \equiv 1/bR \qquad k(T_{iso}) \equiv k_{iso} = \exp(a) \tag{3.73}$$

which follows after introducing the compensation law into the Arrhenius equation and setting $T = 1/bR$.[59] Note that the compensation law parameters, a and b, can be determined, according to eq. 3.73, simply by plotting several sets of rate coefficient data as a function of temperature (say, in Arrhenius plots like that in fig. 3.8) and extracting the values of k_{iso} and T_{iso} at their observed common point of intersection.

As a numerical example of an application of the compensation law, proton-promoted dissolution rates for andalusite (Al_2SiO_5), forsterite, kaolinite, and quartz[60] yield the regression equation:

$$\ln A = -19.146 + 0.387E_a \qquad (r^2 = 0.887) \tag{3.74}$$

with A in units of mol kg^{-1} s^{-1} (mass-normalized dissolution rate) and E_a in units of kJ mol^{-1}. For the typical range of E_a values (20 to 90 k J mol^{-1}),[48] eq. 3.74 predicts A values in the broad range 10^{-5} to 10^7 mol kg^{-1} s^{-1}. The isokinetic temperature is about 310 K (37°C), with $k_{iso} \approx 5 \times 10^{-9}$ mol kg^{-1} s^{-1}. The order-of magnitude quality of this prediction can be appreciated by noting that, at 37°C, quartz and forsterite have dissolution rates equal approximately to 10^{-10} and 10^{-8} mol kg^{-1} s^{-1}, respectively.[61]

3.5 Ligand-Promoted Mineral Dissolution Reactions

Ligand-promoted mineral dissolution reactions are surface-controlled processes in which a ligand adsorptive forms a surface complex with a reactive constituent of a mineral as an intermediate species in a pathway ultimately leading to detachment of the constituent into aqueous solution.[43] Similarly, *metal-promoted* dissolution reactions involve surface complex formation between an adsorptive metal cation and a detachable mineral constituent,[62] whereas *complex-promoted* dissolution reactions have ternary mineral constituent-metal-ligand surface complexes as intermediate species prior to the detachment of the mineral constituent into aqueous solution.[63] Examples of these three processes include gibbsite dissolution promoted by sulfate ligands;[64] quartz dissolution promoted by alkali metal cations;[62] and goethite dissolution promoted by transition metal-EDTA (ethylenedinitrilotetraacetate) complexes.[63] In each type of dissolution reaction, prior formation of a surface complex is critical to the destabilization of a mineral structure unit that constrains the detachable constituent (i.e., Al^{3+}, Si^{4+}, and Fe^{3+} in the three examples cited).[65]

Ligand-promoted dissolution reactions have been studied extensively because of their evident connection to mineral weathering processes in biologically active zones below the land surface.[66] The consensus of these studies is that organic ligands can enhance the rate of mineral dissolution relative to the proton-promoted rate, although a strong pH modulation of this enhancement still exists and has the same character as the data trend in fig. 3.7.[67] A useful illustrative example is provided by oxalate-promoted dissolution rates far from equilibrium ($H_2C_2O_4$, $pK_{a1} = 1.25$, $pK_{a2} = 4.27$). The table below lists mass-normalized steady-state rates at pH 5 in the presence of 1 mol m^{-3} oxalate (total concentration) at laboratory temperature.[68]

Mineral	$d(\xi/m)/dt$ (mol kg^{-1} s^{-1})
Alumina	4.3×10^{-7}
Andesine plagioclase	2.2×10^{-8}
Bytownite (plagioclase)	5.8×10^{-8}
Goethite	4.1×10^{-8}
Hematite	1.1×10^{-7}
Kaolinite	1.0×10^{-8}
Goethite[a]	1.5×10^{-6}
Manganite[a]	5.1×10^{-3}
Pyrolusite[a]	3.6×10^{-5}

[a]Reductive dissolution by oxalate at pH 5.

Oxalate is a ubiquitous organic ligand in terrestrial weathering zones, particularly those under humid climate and forest canopy. Its concentration in pore waters ranges up to 1 mol m^{-3} in bulk soil, but larger concentrations can occur in localized zones with high microbial populations.[69] Notable in the table of oxalate-promoted dissolution rates above is the clustering of the first half-dozen values around 10^{-7} mol kg^{-1} s^{-1}, irrespective of mineral composition and structure, suggesting that the key steps in the mechanism of ligand-promoted dissolution are more sensitive to the local coordination environment of the detachable metal center (Al^{3+} or Fe^{3+}) than to the structure of the unit cell in which it resides. Another general trend is the marked decrease in dissolution time scale relative to that for proton-promoted kinetics at the same pH value. In the case of goethite, for example, the proton-promoted dissolution time scale is about 240 years for an initial 1 g (or 11.3 mmol, $M_r = 88.9$ Da) to dissolve at pH 5 [$d(\xi/m)/dt = 1.5 \times 10^{-9}$ mol kg^{-1} s^{1}],[70] whereas $\tau_{dis} \approx 9$ years for oxalate-promoted dissolution under the same initial condition, according to the table above. Similarly, the time scale for the proton-promoted dissolution of 1.9 mmol kaolinite at pH 5, calculated in section 3.4 as 1,370 years, may be compared to $\tau_{dis} \approx 6$ years for the oxalate-promoted dissolution rate under the same initial conditions. These large enhancements of the dissolution rate at pH 5 illustrate the typical effectiveness of millimolar concentrations of bidentate carboxylate ligands over protons in the destabilization of surface structures in metal oxide and aluminosilicate minerals.

Figure 3.9 is a schematic cartoon of the proton-promoted (1) and ligand-promoted (2) pathways of mineral dissolution, as exemplified by gibbsite [γ-$Al(OH)_3$] and fluoride ligands at pH 4. The inner-sphere surface complex between F^- and Al^{3+} on a gibbsite edge surface is facilitated by the protonation of OH^- groups there which, according to Pauling Rule 2 (Section 1.2), should be bound to a pair of Al^{3+}, each contributing a bond valence of about 0.5, but are instead bound only to one peripheral Al^{3+}. Protonation thus produces a Lewis acid site $\equiv Al{-}OH_2{}^+$ that is very reactive, the water molecule being easily replaced by F^- to form the surface complex.

The effect of coordination of F^- to Al^{3+} is to donate electron density to the metal cation, thereby making its bonds to OH^- in the mineral structure more labile.[71] This effect can be appreciated by comparing the values of k_{wex} for $Al(H_2O)_6{}^{3+}$ and $AlF(H_2O)_5{}^{2+}$ in dilute aqueous solutions:[72]

$$Al(H_2O)_6^{3+} : k_{wex} = 1.3 - 2.0\ s^{-1} \qquad AlF(H_2O)_5^{2+} : k_{wex} = 1.1 \pm 0.2 \times 10^2\ s^{-1}$$

These data show that inner-sphere complex formation between Al^{3+} and F^- increases the rate of water exchange by two orders of magnitude. Similarly, inner-sphere surface complex formation between the metal and ligand should enhance the lability of the remaining Al–O bonds and, therefore, increase the likelihood of their breaking, with replacement of structural OH by water molecules and release of the AlF^{2+} complex into aqueous solution.[73] This tendency to lability in Al–O bonds should increase with the Lewis basicity of the ligand coordinating to Al^{3+}, since high basicity means high ability to donate electron

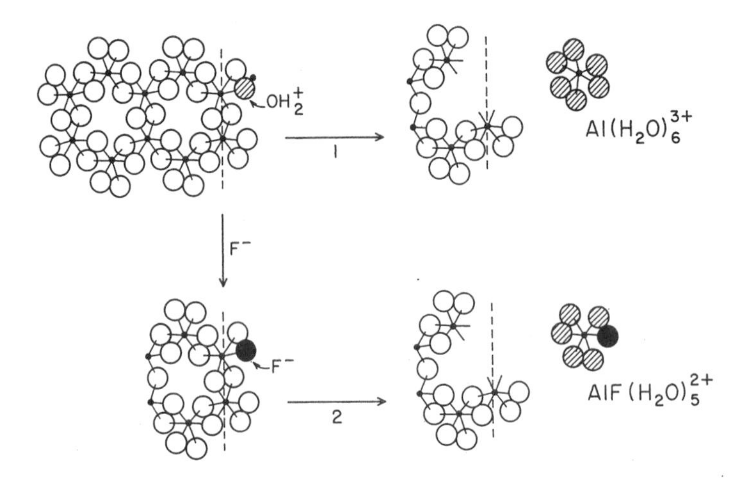

Fig. 3.9 Schematic drawing of the proton-promoted (1) and ligand-promoted (2) pathways of gibbsite dissolution.[42]

density in bond formation. Lewis basicity, in turn, can be quantified conveniently by log K for the protonation of a ligand, with high values of log K associated with high Lewis basicity. Thus a LFER is expected between the common logarithm of k_{wex} for $AlL(H_2O)_n^{3-\ell}$ and log K for the equilibrium protonation of the coordinating ligand, $L^{-\ell}$. This kind of correlation is exemplified in the table below:[71–75]

Ligand (ℓ, n)	k_{wex} (s^{-1})	log $\bar{K}_G^a$	$\log K^b$
H_2O (0, 6)	1.3 ± 0.6	0	0
Oxalate (2, 4)	$1.09 \pm 0.13 \times 10^2$	2.7	5.97
Fluoride (1, 5)	$1.11 \pm 0.14 \times 10^2$	3.17	7.11
Methylmalonate (2, 4)	$6.60 \pm 0.20 \times 10^2$	4.39	5.65
Sulfosalicylate (2, 4)	$3.00 \pm 0.08 \times 10^3$	7.69	11.71
Salicylate (2, 4)	$4.90 \pm 0.07 \times 10^3$	8.34	12.77
Hydroxide (1, 5)	$3.10 \pm 0.25 \times 10^4$	14.0	8.48

[a] $\log \bar{K}_G \equiv \left[\prod_{i=1}^{\ell} K_i\right]^{\frac{1}{\ell}}$ is the geometric mean value of the ℓ equilibrium constants for protonation of the ligand, where $\ell > 1$.

[b] K is the stability constant for a 1:1 Al-ligand complex (I = 0.6 *m*).

The LFER implicit in this set of data can be epitomized by the linear regression equation:

$$\log k_{wex} = 1.007 + 0.2883 \log \bar{K}_G \qquad (r^2 = 0.869) \tag{3.75}$$

where $\bar{K}_G$ is the geometric mean value of the protonation constants for a multiprotic ligand, serving as an aggregated measure of its overall basicity. It is noteworthy that log $\bar{K}_G$ spans the entire range of pH in this linear regression equation, which predicts log k_{wex} to within about 0.4 log units. In section 1.5, the strong positive correlation between log K_G and the affinity of an adsorptive ligand for a metal oxide surface was mentioned. Equation 3.75 adds to this concept the further insight that strong ligand adsorption tends to destabilize the metal-oxygen bonds that restrain the metal-ligand surface complex from detachment into aqueous solution. Completing the circle of this argument, one can hypothesize that the stability constant for a metal-ligand complex in aqueous solution should correlate with k_{wex} for the complex in the same manner as does log $\bar{K}_G$ in eq. 3.75.[75] Indeed, the LFER that emerges from the data in the second and fourth columns of the table above is:

$$\log k_{wex} = 0.5826 + 0.2826 \log K \qquad (r^2 = 0.702) \tag{3.76}$$

with the same slope as in eq. 3.75, illustrating a close connection between ligand protonation and ligand-metal complexation.[73]

If strong ligand adsorption enhances mineral dissolution, then strong metal cation adsorption may *inhibit* mineral dissolution. As pointed out in section 1.2, inner-sphere surface complexation of a metal cation by a mineral adsorbent is qualitatively akin to a crystal growth process, particularly if the metal adsorptive is a constituent of the dissolving mineral. Moreover, if the metal adsorbate is chelated by a pair of adjacent sites on a mineral surface (fig. 1.10), inhibition of dissolution should be facile. Metal-inhibited dissolution can be pictured as a result of sufficient electron density being donated to the metal cation adsorbate by a surface site to preclude the site from adsorbing a proton, a reaction that would destabilize its bonding to metal centers in the mineral structure. In this sense, metal inhibition of proton-promoted dissolution is the result of simply "blocking" pronatable sites on a mineral surface with a more recalcitrant cation.[76] These concepts of ligand promotion and metal inhibition of mineral dissolution reactions can be captured by considering the inner-sphere surface complexation reactions exemplified in eqs. 3.16 and 3.34 as *parallel* pathways to adsorption, but with the two adsorption reactions followed by a metal-center detachment step:[77]

$$\equiv \mathrm{SOH} + M^{m+} \overset{\mathrm{ads}}{\rightarrow} \equiv \mathrm{SOM}^{(m-1)+} + \mathrm{H}^+ \overset{\mathrm{det}}{\rightarrow} P'' \tag{3.77a}$$

$$\equiv \mathrm{SOH} + \mathrm{H}^+ \overset{\mathrm{ads}}{\Leftrightarrow} \equiv \mathrm{SOH}_2^+ \overset{\mathrm{det}}{\rightarrow} P \tag{3.77b}$$

$$\equiv \mathrm{SOH}_2^+ + L^{\ell-} \overset{\mathrm{ads}}{\rightarrow} \equiv SL^{(1-\ell)} + \mathrm{H_2O} \overset{\mathrm{det}}{\rightarrow} P' \tag{3.77c}$$

where all symbols are used in consonance with eqs. 3.16, 3.34, and 3.65 (R = OH in eq. 3.65a), the primes denoting the possibility of a dissolution product that can differ from the one induced by the proton-promoted pathway in eq. 3.77b. The corresponding rate laws for dissolution far from equilibrium can be expressed in terms of the concentrations of either the surface species (e.g., $\equiv SOH_2{}^+$) or the aqueous solution adsorptives (e.g., H^+) under the assumption that the adsorption steps in eqs. 3.77 occur on a time scale much smaller than that for the detachment steps.[77]

Metal cation inhibition of proton-promoted dissolution can be modeled kinetically by assuming that eqs. 3.77a and 3.77b contribute to the overall rate law:

$$\frac{d[P]}{dt} + \frac{d[P'']}{dt} = \phi([H^+], [M^{m+}], \mathrm{env}) \tag{3.78}$$

where ϕ ( ) is a positive-valued function of its arguments and "env" refers to all variables other than the concentrations of the two adsorptive cations (see eq. 3.3). If M^{m+} inhibits proton-promoted dissolution, the detachment step in eq. 3.77a must occur on a time scale that is much larger than that for the detachment step in eq. 3.77b and, therefore, the second term on the left side of eq. 3.78 can be neglected. The right side of eq. 3.78 then must depend on the concentrations of the two adsorptives through those of the reactants on the left side of eq. 3.77b, given that the adsorption step equilibrates rapidly. The concentration of the reactant $\equiv$SOH indeed does depend on both $[H^+]$ and $[M^{m+}]$ because of the mass balance condition:

$$\mathrm{SOH_T} = [\equiv \mathrm{SOH}] + [\equiv \mathrm{SOH}_2^+] + [\equiv \mathrm{SOM}^{(m-1)+}] \tag{3.79}$$

The explicit dependence on $[H^+]$ and $[M^{m+}]$ is determined by the order of the rate law with respect to [$\equiv$SOH] and by the relations between the two adsorbate concentrations on the right side of eq. 3.79 and their corresponding adsorptive concentrations (i.e., their adsorption isotherms). Figure 3.10 shows an application of eq. 3.78 to proton-promoted dissolution data for three aluminosilicates, with

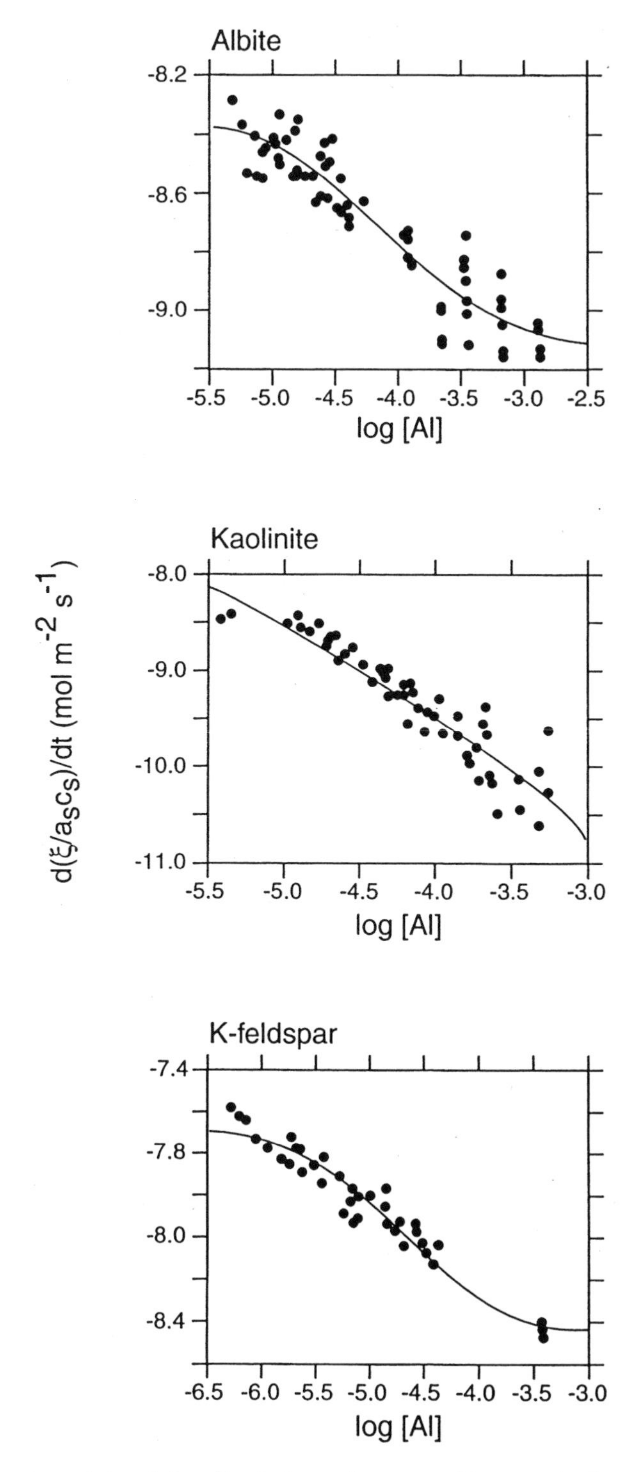

Fig. 3.10 Log-log plots of the dissolution rate versus Al concentration (m) for three aluminosilicate minerals.[78] The curve through the data points represents eq. 3.80.

Al^{3+} as the inhibiting metal cation.[78] The functional form of $\phi([H^+], [M^{m+}], \text{env})$ selected was:

$$\phi([\mathrm{H}^+], \mathrm{Al_{TS}}) = k\left(1 - \frac{F\,K\,\mathrm{Al_{TS}}}{1 + K\,\mathrm{Al_{TS}}}\right)(\mathrm{H}^+) \tag{3.80}$$

where Al_{TS} is the total soluble Al concentration and with (OH^-) substituted for (H^+) if pH > 7. Equation 3.80 may be considered as a generalization of eq. 3.66 to permit k_H (or k_{OH}) to depend on the total molal concentration of the inhibiting metal cation, Al_{TS}. [In this context, $n_H = 1$ (or $n_{OH} = 1$) in eq. 3.66.] Equation 3.80 features a Langmuir adsorption isotherm for $[\equiv SOAl^{2+}]$ (eq. 1.13), with "affinity parameter" K and "capacity parameter" F (evidently equal to the ratio of the maximum value of $[\equiv SOAl^{2+}]$ to $SOH_T - [\equiv SOH_2^{\ +}]$), and an area-normalized, pseudo first-order rate coefficient k. Because the area-normalized dissolution data were obtained at fixed pH values, a pH dependence of F and K is not manifest. The dotted curves through the rather scattered data points represent eq. 3.80 with the parameter values ($F \approx 1$):[78]

Mineral	pH	k (mol m^{-2} s^{-1})	K (kg mol^{-1})
Albite	9[a]	9.5×10^{-6}	1.7×10^5
Kaolinite	2	1.52×10^{-6}	4×10^5
K-feldspar	9[a]	1.8×10^{-6}	3.2×10^4

[a] (OH^-) appears in the rate law at pH 9 (eq. 3.80).

Equations (3.77a) and (3.77b) not only describe the enhancement or inhibition of mineral dissolution resulting from the donation of electron density to or from a surface site by an adsorbate species, but also can describe these two effects on mineral dissolution when they involve complete electron transfer to or from a surface site by an adsorbate species. (The distinction here is between Lewis acid-base reactions and oxidation-reduction reactions.) These latter processes are termed *reductive* or *oxidative dissolution reactions.*[79] They differ from surface redox reactions (section 3.3) because the adsorbent participates *directly* in the electron transfer process and is transformed into a different chemical species because of this participation. Like surface redox reactions, reductive or oxidative dissolution reactions are common in soils, sediments, and aquatic systems and offer major pathways for the biogeochemical cycling of both nutrient and toxicant chemical species.[80]

Examples of *reductive* dissolution reactions include those of oxalate (and many other natural and synthetic organic ligands) with Fe(III), Mn(IV), and Mn(III) oxides or oxyhydroxides, some rate coefficients for which are listed near the bottom of the tabulation of oxalate-promoted dissolution kinetics given above;[68,79,80] those of Fe(II) and Co(II) with Mn(IV, III) oxides;[81] those of NH_4^+ with Mn(IV) oxides;[82] and those of pollutant elements, such as Cr(III), As(III), or Se(IV), with Fe(III) and Mn(IV) oxides.[83] Examples of *oxidative* dissolution reactions include those of NO_3^-, NO_2^-, $HCrO_4^-$, and $HSeO_4^-$ with elemental iron and green rust ["fougérite," $Fe(III)_x\ Fe(II)_{1-x}\ (OH)_2\ (L_x \cdot nH_2O)_{1/\ell}$ where $L^{\ell-} = Cl^-$, CO_3^{2-}, or SO_4^{2-} and $x = 1/4$ or $1/3$[84]];[85] and those of Fe(III), Cu(II), V(V), and Cr(VI) with Fe(II)-bearing micas, ilmenite, goethite, and magnetite (Section 1.2, *Metal Oxides* inset).[86] Thus, a broad variety of important oxidant and reductant species adsorb on and destabilize solid-phase adsorbents containing Fe or Mn metal centers.

This kind of dissolution reaction can be exemplified by the reductive dissolution of hydrated Na-birnessite ("Na-buserite") by Cr(III).[87] This form of birnessite (Section 1.2, *Metal Oxides* inset) has a layer spacing of 1.0 nm, with both Mn(III) and Mn(IV) in the octahedral sheet. The principal interlayer cation is (solvated) Na^+, the unit cell formula being $Na_{0.3}Mn(III, IV)O_2 \cdot 0.9H_2O$, although interlayer Mn^{2+} and OH also exist in the mineral.[88] The overall redox reaction between Mn(IV) and Cr(III) can be expressed schematically as it was in eq. 3.1:

$$\text{Cr(III)} + \equiv \text{Mn(IV)} \rightarrow \text{Cr(IV)} + \text{(III)} \tag{3.81}$$

where the Mn reactant is an exposed metal center at the periphery of the mineral and a one-electron process is assumed to be rate-limiting in the three-electron transfer which ultimately produces Cr(VI) with a 3:1 stoichiometric ratio between Mn and Cr.[89]

A rate law for the redox reaction in eq. 3.81 can be formulated analogously to that in eq. 3.3:

$$-\frac{d[\text{Cr(III)}]}{dt} = \phi([\text{Cr(III)}], [\equiv \text{Mn(IV)}], \text{env}) = \frac{d[\text{Cr(VI)}]}{dt} \tag{3.82}$$

where $\phi(\ )$ is a positive-valued function of its arguments. At pH 4, with a fixed [≡Mn(IV)], measurements[87] of the loss of Cr(III) and the production of Cr(VI) show that both sides of eq. 3.82 are equal and proportional to [≡Mn(IV)] if $[\equiv \text{Mn(IV)}] < 10\ \text{mol m}^{-3}$, although the left side of eq. 3.82 is strictly proportional to [≡Mn(IV)] up to $60\ \text{mol m}^{-3}$. Moreover, the loss of Cr(III) follows first-order kinetics as in eq. 3.4a for the analogous case of Cr(VI) loss.[87] Therefore, the rate law for eq. 3.81 can be expressed:

$$-\frac{d[\text{Cr(III)}]}{dt} = k[\text{Cr(III)}]\,[\equiv \text{Mn(IV)}] = \frac{d[\text{Cr(VI)}]}{dt} \tag{3.83}$$

where k is a second-order rate coefficient. Figure 3.11 shows an application of eq. 3.83 with $k = 0.6\ \text{dm}^3\ \text{mol}^{-1}\ \text{s}^{-1}$ based on curve-fitting.[87] It is noteworthy that, as indicated by the rate law, there is no lag in the production of Cr(VI), substantiating the hypothesis that electron-transfer steps subsequent to that in eq. 3.81 are rapid. There is, however, a difference between the total concentration loss of Cr(III) and the total concentration gain of Cr(VI) that increases with the initial concentration of ≡Mn(IV), as shown by the open squares plotted in fig. 3.11. This discrepancy is likely caused by adsorbed Cr intermediates that figure in the steps following the initial one-electron transfer.[89]

In eq. 3.81, Cr(III) is the reductant, playing a role analogous to that of Fe(II) in eq. 3.1. It is therefore reasonable to consider eq. 3.83 in the pseudo first-order form,

$$-\frac{d[\text{Cr(III)}]}{dt} = K([\text{Cr(III)}], \text{pH})\,[\equiv \text{Mn(IV)}] \tag{3.84}$$

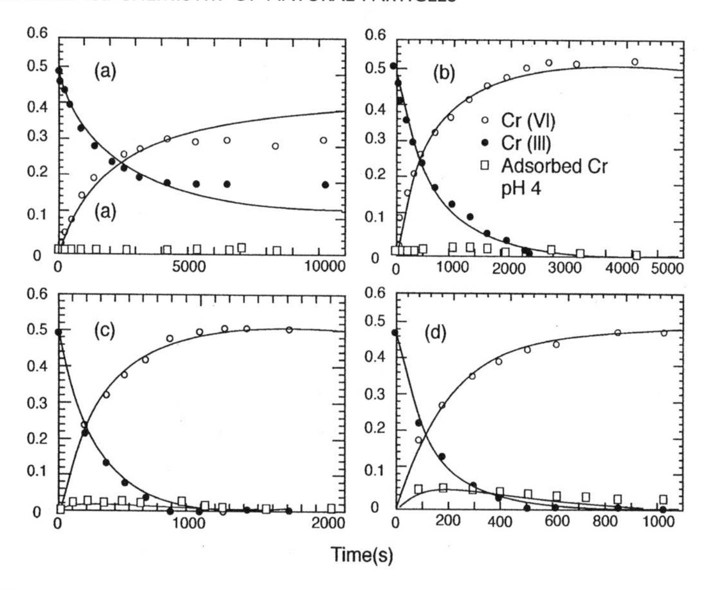

Fig. 3.11 Time dependence of the concentration of Cr(III) (•), Cr(VI) (○), and adsorbed Cr (□) during the reaction in eq. 3.81 at pH 4, with an initial Cr(III) concentration of 0.5 mol m^{-3} and the initial solids concentration of ≡Mn(IV) varied: (a) 0.75 mol m^{-3}, (b) 2.9 mol m^{-3}, (c) 7.4 mol m^{-3}, and (d) 15 mol m^{-3}. Note the increase in adsorbed Cr with increasing solids concentration. The curves through the soluble Cr concentration data represent eq. 3.83, while the adsorbed Cr data represent the difference between Cr_T and the sum of soluble Cr concentrations (after Manceau et al.[87]).

where, by analogy with eq. 3.7,

$$K([\text{Cr(III)}, \text{pH}) = k_0[\text{Cr}^{3+}] + k_1[\text{CrOH}^{2+}] + k_2[\text{Cr(OH)}_2^+] \quad (3.85)$$

provides a model for the pH dependence of the rate coefficient k in eq. 3.83. The half-reactions analogous to those in eqs. 3.6 are:[87]

$$\text{Cr}^{4+} + e^- = \text{Cr}^{3+} \qquad \log K_R = 35.5 \quad (3.86a)$$

$$\text{CrOH}^{3+} + e^- = \text{CrOH}^{2+} \qquad \log K_R = 31.8 \quad (3.86b)$$

$$\text{Cr(OH)}_2^{2+} + e^- = \text{Cr(OH)}_2^+ \qquad \log K_R = 25.5 \quad (3.86c)$$

And the corresponding hydrolysis reactions are:

$$\text{Cr}^{3+} + \text{H}_2\text{O} = \text{CrOH}^{2+} + \text{H}^+ \qquad \log^* K_1 = -4.0 \quad (3.87a)$$

$$\text{Cr}^{3+} + 2\text{H}_2\text{O} = \text{Cr(OH)}_2^+ + 2\text{H}^+ \qquad \log^* \beta_2 = -10.7 \quad (3.87b)$$

Figure 3.12 shows the fit of eq. 3.85 to the observed pH dependence of k (eq. 3.83) based on the parameter values:[87]

$$k_0 = 0.19 \text{ dm}^3 \text{ mol}^{-1}\text{s}^{-1}, k_1 = 0.775 \text{ dm}^3 \text{ mol}^{-1} \text{ s}^{-1}, k_2 = 1.95 \text{ dm}^3 \text{ mol}^{-1}\text{s}^{-1}$$

In this example, as the closeness of the log K_R values in eqs. 3.86 also suggest, the second-order rate coefficients in eq. 3.85 do not vary over as broad a range as did

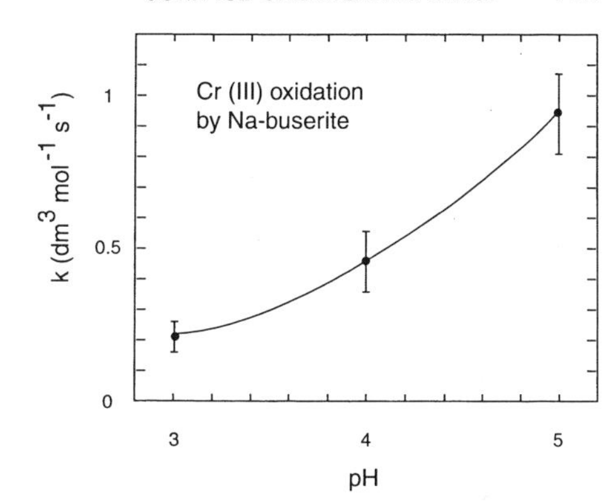

Fig. 3.12 Plot of the pH dependence of the second-order rate coefficient in eq. 3.83. The curve through the data points represents eq. 3.85 divided by the total Cr(III) concentration (After Manceau et al.[87]).

those for the Fe(III)/Fe(II) couples in eq. 3.7. The LFER analogous to that in eq. 3.12 reflects this behavior in its small slope parameter:

$$\log k_i = 2.832 - 0.09787 \log K_{\mathrm{R}i} \qquad (i = 0, 1, 2; r^2 = 0.953) \tag{3.88}$$

Thus, donation of electron density to Cr^{3+} through its complexation by OH^- has less impact on Cr^{3+} electron transfer to Mn(IV) than it did on Fe^{2+} electron transfer to Cr(VI).

Notes

1. These techniques and apparatus are discussed in exemplary fashion by M.C. Amacher in chapter 2 of D.L. Sparks and D.L. Suarez (Eds.), *Rates of Soil Chemical Processes*, Soil Science Society of America, Madison, WI, 1991.
2. Many examples are discussed in D.L. Sparks and D.L. Suarez, ibid.
3. For a review of basic concepts in chemical kinetics, see, e.g., K.J. Laidler, *Chemical Kinetics*, Harper and Row, New York, 1987.
4. For a summary of these and other kinetics terms, see section 2.12 in I. Mills, T. Cvitaš, K Homann, N. Kallay, and K. Kuchitsu, *Quantities, Units and Symbols in Physical Chemistry*, Blackwell Scientific Publications, Oxford, 1993.
5. This critically important point is explained cogently in chapter 5 of K. Denbigh, *The Principles of Chemical Equilibrium*, Cambridge Univ. Press, Cambridge.
6. M. Pettine, L. D'Ottone, L. Campanella, F.J. Millero, and R. Passino, *Geochim. Cosmochim. Acta* **62**:1509 (1998). This article also provides a review of previous research on the kinetics of Cr(VI) reduction by Fe(II) in aqueous solutions.
7. I.J. Buerge and S.J. Hug, *Environ. Sci. Technol.* **31**:1426 (1997).
8. D.L. Sedlak and P.G. Chan, *Geochim. Cosmochim. Acta* **61**:2185 (1997).
9. The difference in pE values for the reduction half-reactions of two redox couples that are combined in a redox reaction like eq. 3.1 is proportional to $\Delta_r G$ for the redox reaction $\Delta_r G = \mathrm{RT} \ln 10\,(\mathrm{pE_{Fe}} - \mathrm{pE_{Cr}})$, where the subscripts refer to the couples Fe(III)/Fe(II) and Cr(VI)/Cr(V), respectively. For a useful discussion of this relationship, see chapter 7 in F.M.M. Morel and J.G. Hering, *Principles and Applications of Aquatic Chemistry*, John Wiley, New York, 1993.
10. Values of these rate coefficients reported by other investigators can differ from those given here by up to an order of magnitude. However, the relative order, $k_0 \ll k_1 \ll k_2$, is found uniformly, and this ordering is the essential point in the present discussion. For a comparison of the rate coefficients, see M. Pettine et al., *Geochim. Cosmochim. Acta* **62**:1509 (1998).

11. For a discussion of the LFER in eq. 3.12 as applied to eq. 3.1 with a variety of complexant compounds reacting with Fe to alter its oxidation kinetics, see R.P. Schwarzenbach, W. Augst, C. Holliger, S.J. Hug, and J. Klausen, *Chimia* **51**:908 (1997). Technical aspects of the effect hydrolysis has on Fe(II) oxidation are discussed by G.W. Luther in chapter 6 of W. Stumm (Ed.), *Aquatic Chemical Kinetics*, Wiley-Interscience, New York, 1990.
12. An excellent counterexample to the positive correlation in eq. 3.12 is provided by the specific adsorption of Al^{3+} on goethite, for which the rate-limiting step is adsorbate desolvation, a very slow process for a trivalent cation (eq. 3.15), whereas the equilibrium constant for specific adsorption is very large because of the strong attraction between a hydrolyzing cation and an hydroxylated surface (section 1.5). In this case, the rate of adsorption is low, but the strength of adsorption is high. In picturesque terms, the height of the "energy barrier" and the depth of the "energy valley" are *both* large. For Fe(II) oxidation, on the other hand, the "energy barrier" *decreases* in height as the "energy valley" *increases* in depth.
13. Evidence that this criterion is not *necessary* for inner-sphere surface complexation is presented for the case of selenate (SeO_4^{2-}) by A. Manceau and L. Charlet, *J. Colloid Interface Sci.* **168**:87 (1994) and by C. Su and D.L. Suarez, *Soil Sci. Soc. Am. J.* **64**:101 (2000).
14. For a useful introductory discussion, see chapter 10 in J. Burgess, *Ions in Solution*, John Wiley, New York, 1988 and, at a more technical level, chapter 1 in D.T. Richens, *The Chemistry of Aqua Ions*, John Wiley, New York, 1997. The desolvation step is not always the only critical one to consider. For some metal cations, the binding ligand also influences the rate of inner-sphere complexation. A detailed technical discussion is given in chapter 4 of R.G. Wilkins, *Kinetics and Mechanism of Reactions of Transition Metal Complexes*, VCH, Weinheim, Germany, 1991.
15. The classic reference on this topic is D.W. Margerum, G.R. Cayley, D.C. Weatherburn, and G.K. Pagenkopf, pp. 1–220 in *Coordination Chemistry*, vol. 2, A.E. Martell (Ed.), American Chemical Society, Washington, DC, 1978. For a brief introduction to solvent-exchange studies, see chapter 11 in J. Burgess, *Metal Ions in Solution*, John Wiley, New York, 1978. A comprehensive discussion is given by D.T. Richens, *The Chemistry of Aqua Ions*, John Wiley, New York, 1997.
16. A.L. Crumbliss and J.M. Garrison, *Comments Inorg. Chem.* **8**:1 (1988). See also chapter 6 in F.M.M. Morel and J. G. Hering, *Principles and Applications of Aquatic Chemistry* John Wiley, New York, 1993.
17. R.D. Shannon, *Acta Cryst.* **A32**:751 (1976).
18. The reaction in eq. 3.16 is similar to that proposed by K. Hachiya, M. Sasaki, T. Ikeda, N. Mikami, and T. Yasunaga, *J. Phys. Chem.* **88**:27 (1984), which, in turn, was inspired in part by an ESR study by M.B. McBride, *Soil Sci. Soc. Am. J.* **42**:27 (1978). A review of the kinetics approach used to evaluate the rate coefficients is given by T. Yasunaga and T. Ikeda, pp. 230–253 in J.A. Davis and K.F. Hayes (Eds.), *Geochemical Processes at Mineral Surfaces*, American Chemical Society, Washington, DC, 1986. See also chapter 3 in D.L. Sparks and D.L. Suarez (Eds.), *Rates of Soil Chemical Processes*, Soil Science Society of America, Madison, WI, 1991; and chapter 13 in W. Stumm and J.J. Morgan, *Aquatic Chemistry*, John Wiley, New York, 1996.
19. See, e.g., section 1.5 in G. Sposito, *Chemical Equilibria and Kinetics in Soils*, Oxford University Press, New York, 1994, for a discussion of coupled rate laws.
20. The mathematical description of coupled rate laws under linearization is developed fully in C.F. Bernasconi, *Relaxation Kinetics*, Academic Press, New York, 1976.
21. See, e.g., chapter 2 in G. Goertzel and N. Tralli, *Some Mathematical Methods of Physics*, McGraw-Hill, New York, 1960. Equations 3.24 c–3.24e permit eqs. 3.23

to be cast into a matrix notation. Their solution then involves finding a linear transformation from the coupled independent variables (Δc_A, Δc_D) to a new pair of independent variables, such that the matrix elements a_{12} and a_{21} both vanish when the rates are expressed in the new variables. Equations 3.24 reflect the fact that, irrespective of the explicit form of the linear transformation required, the sum of the principal diagonal elements in the matrix of the a_{ij} $(i, j = 1, 2)$—termed its *trace*—and the determinant of this matrix are invariant in value. Thus, the left sides of eqs. 3.24a and 3.24b are equal to the trace and determinant of the matrix in terms of the new independent variables, while the right sides are the same quantities expressed in terms of the original independent variables. This invariance of the trace and the determinant is a result of the linearity of the rate laws and their symmetry under exchange of the labels "dependent" and "independent" for the rate and concentration variables.

22. K. Hachiya, M. Ashida, M. Sasaki, T. Inoue, and T. Yasunaga, *J. Phys. Chem.* **83**:1866 (1979).
23. K. Hachiya et al., *J. Phys. Chem.* **88**:27 (1994). The model equation used by Hachiya et al. has the same appearance as eq. 3.30, but the kinetics parameters in it do not have the same interpretation as k_{-2} and k_2 in terms of the mechanism in eq. 3.16. Hachiya et al. assume that proton desorption is the rate-limiting process during the formation of an inner-sphere surface complex, whereas eq. 3.30 is based on the assumption that metal cation desolvation or proton desorption may be rate-limiting. Moreover, in the present chapter, K_{OS} is assumed to be small in a sense that justifies the approximation in eq. 3.29 (i.e., $K_{OS} \approx 1 - 10$ $dm^3\ mol^{-1}$, as is true for outer-sphere aqueous solution complexation; see note 14).
24. K. Hachiya, M. Sasaki, Y. Saruta, N. Mikami, and T. Yasunaga, *J. Phys. Chem.* **88**:23 (1984).
25. P.R. Grossl, D.L. Sparks, and C.C. Ainsworth, *Environ. Sci. Technol.* **28**:1422 (1994). The data in this paper also can be interpreted with eq. 3.30 to evaluate k_{-2} and k_2K_{OS}. C. Liu and P.M. Huang [*Geoderma* **102**:1 (2001)] have applied eq. 3.30 to Pb^{2+} adsorption by $Fe(OH)_3$ (*am*), finding a value of k_2K_{OS} close to that predicted with the correlation equation reported by Grossl et al.
26. P.C. Zhang and D.L. Sparks, *Soil Sci. Soc. Am. J.* **53**:1028 (1989), **54**:624 (1990).
27. A. Manceau, L. Charlet, M.C. Boisset, B. Didier, and L. Spadini, *Appl. Clay Sci.* **7**:201 (1992).
28. The kinetics modeling of the reaction sequence in eq. 3.39 is discussed fully in chapter 3 of C.F. Bernasconi, *Relaxation Kinetics*, Academic Press, New York, 1976.
29. P.R. Grossl, M. Eick, D.L. Sparks, S. Goldberg, and C.C. Ainsworth, *Environ. Sci. Technol.* **31**:321 (1997).
30. Both XAS and FTIR spectroscopy have been applied to determine the molecular structure of adsorbed arsenate and related oxyanions. [See X. Sun and H. Doner, *Soil Sci.* **161**:865 (1996); S. Fendorf, M.J. Eick, P. Grossl, and D.L. Sparks, *Environ. Sci. Technol.* **31**:315 (1997); B.A. Manning, S.E. Fendorf, and S. Goldberg, *Environ. Sci. Technol.* **32**:2383 (1998).] This kind of molecular-scale research must precede the development of an adsorption kinetics model, just as it must precede the development of an adsorption equilibrium model, as discussed in section 2.1.
31. See R.J. Bartlett and B.R. James, *Advan. Agron.* **50**:151 (1993) and W. Stumm, *Croatica Chem. Acta* **70**:71 (1997) for useful reviews.
32. B. Wehrli, B. Sulzberger, and W. Stumm, *Chem. Geol.* **78**:167 (1989). See also G.W. Luther, chap. 6 in W. Stumm (Ed.), *Aquatic Chemical Kinetics*, Wiley-Interscience, New York, 1990.
33. A.T. Stone, K.L. Godtfredsen, and B. Deng, pp. 337–374 in *Chemistry of Aquatic Systems*, G. Bidoglio and W. Stumm (Eds.), Kluwer Academic, Boston, 1994. A

fine overview of outer-sphere electron-transfer reaction mechanisms is given by R.A. Marcus, *Pure Appl. Chem.* **69**:13 (1997).

34. See section 1.V in chapter 2 of C.H. Bamford and C.F.H. Tipper, *Comprehensive Chemical Kinetics*, vol. 2, *The Theory of Kinetics*, Elsevier, New York, 1969.
35. B. Deng and A.T. Stone, *Environ. Sci. Technol.* **30**:2484 (1996). See also Stone et al., pp. 337–374 in *Chemistry of Aquatic Systems*, G. Bidologlio and W. Stumm (Eds.), Kluwer Academic, Boston, 1994.
36. B. Deng and A.T. Stone, *Environ. Sci. Technol.* **30**:463 (1996).
37. Yet another possibility, considered in section 1.6.4 of R.G. Wilkins, *Kinetics and Mechanism of Reactions of Transition Metal Complexes*, VCH, Wenheim, Germany, 1991, is the rapid equilibration of the intermediate species ($\equiv A - B$ with the reactants, such that $c_{AB} = K_{eq}\ c_A c_B$ on the time scale of the electron-transfer step and the composite rate coefficient in eq. 3.53 is replaced by $K_f - k_b$ $K_{eq} c_B$, where K_{eq} describes the rapid equilibration. In this case, all three concentration variables in eqs. 3.45 will exhibit single-exponential decay.
38. The empirical rate coefficient, k_{rxn}, defined by Deng and Stone, *Environ. Sci. Technol.* **30**:2484 (1996), is the same as K in eq. 3.49. The empirical rate coefficient, k_{surf}, defined by Deng and Stone, is equal to $(1 + D_{Cr})K$. Note that the dependence of $k_{obs} = KD_{Cr}/(1 + D_{Cr})$ on reductant concentration is expected to be complex because of the dependence of D_{Cr} on reductant concentration, which, in turn, implies that $K_f/(k_b + k_{ET})$, the conditional equilibrium constant for the formation of the intermediate species ($\equiv A - B$) from adsorbed Cr(VI) and dissolved reductant, depends in a complex way on the concentration of the reductant as well (eq. 3.54b).
39. I.J. Buerge and S.J. Hug, *Environ. Sci. Technol.* **33**:4285 (1999). An evident extension of eq. 3.55 is to include *adsorbed* species of the oxidant. This case was also studied by Buerge and Hug, and by E. Liger, L. Charlet, and P. van Cappellen, *Geochim. Cosmochim. Acta* **63**:2939 (1999), with U(VI) as the oxidant, using a rate law like eq. 3.55, but with $k_1 = 0$.
40. This estimate of log K_R for the reduction half-reaction, $\equiv Fe^{3+} + e^- = \equiv Fe^{2+}$ $\log K_R = 6.1$ was made by B. Wehrli in chapter 11 of W. Stumm, *Aquatic Chemical Kinetics*,[11] on the basis of a LFER like that in eq. 3.12 for the oxidation of Fe(II) by $O_2(g)$ along with a measured value of k_1^{surf} for this oxidation reaction involving Fe^{2+} adsorbed on goethite. Note that the value of log K_R given above is rather close to that for the reduction of $FeOH^{2+}$ (eq. 3.6b and fig. 3.6).
41. I.J. Buerge and S.J. Hug, *Environ. Sci. Technol.* **32**:2092 (1998). See also R.P. Schwarzenbach et al., *Chimia* 51:908 (1997). Figure 3.6 differs slightly from the LFERs presented in these articles because of differences among the thermodynamic data selected.
42. Basic concepts of mineral dissolution reactions are discussed in chapter 5 of G. Sposito, *The Chemistry of Soils*, Oxford University Press, New York. Criteria for rate laws expressing transport control are outlined through a simple model in section 4.5 of G. Sposito, *Chemical Equilibria and Kinetics in Soils*, Oxford University Press, New York, 1994. V. Metz and J. Ganor [*Geochim. Cosmochim. Acta* **65**:3475 (2001)] have investigated the influence of stirring—usually employed as the criterion for determining whether transport control is operative—on kaolinite dissolution kinetics at pH ≤ 4 over the temperature range from 25 to 70°C. They observed an increase in the rate of dissolution with increased stirring rate, the effect being greater at lower temperature and higher pH. This observation was not connected with transport control, however, but instead was attributed to the abrasion of the stirred particles and subsequent dissolution of the very small colloids released by the abrasion process.
43. G. Furrer and W. Stumm, *Geochim. Cosmochim. Acta* **50**:1847 (1986). See also W. Stumm, B. Wehrli, and E. Wieland, *Croatica Chem. Acta* **60**:429 (1987).
44. A comprehensive introduction to mineral dissolution reactions is given in A.F. White and S.L. Brantley (Eds.), *Chemical Weathering Rates of Silicate Minerals*,

Mineralogical Society of America, Washington, DC, 1995. See also S.M. Colman and D.P. Dethier (Eds.), *Rates of Chemical Weathering of Rocks and Minerals*, Academic Press, San Diego, CA, 1986. S.D. Samson and C.M. Eggleston [*Environ. Sci. Technol.* **32**:2871 (1998), *Geochim. Cosmochim. Acta* **64**:3675 (2000)] have observed conditions under which the regeneration of $\equiv$SR, a step necessary to the formation of the intermediate species in Eqs. 3.65, may be relatively slow.

45. A derivation of the van Bemmelen-Freundlich isotherm equation, $n_i^{(w)} = bKc_i^{\beta}$, $0 < \beta \leq 1$ (notation the same as in eq. 1.13) based on heterogeneous adsorption-desorption kinetics is given in section 4.2 of G. Sposito, *Chemical Equilibria and Kinetic in Soils*, Oxford University Press, 1994. If the rate law for $d[P]/dt$ based on eqs. 3.65 is first-order in $\equiv SRH^+$ or $\equiv SROH^-$, then the exponent β in the isotherm equation above can be identified with the partial reaction order, n_H or n_{OH}.
46. See, e.g., E. Wieland, B. Wehrli, and W. Stumm, *Geochim. Cosmochim. Acta* **52**:1969 (1988); W.H. Casey and C. Ludwig, *Nature* **381**:506 (1966) for descriptions of mineral dissolution mechanisms leading to rate laws that are *n*-order ($n > 1$) with respect to $\equiv SRH^+$. In the Casey-Ludwig model, *n* is determined by the difference between the number of H^+ consumed in the detachment of a product species and the number released by restoration of the protonated surface through reaction with H_2O. By contrast with the Furrer-Stumm model (see note 43) the Casey-Ludwig *n* is not directly related to the number of proton adsorption steps prior to the detachment step in eq. 3.65a.
47. J.J. Rosso and J.D. Rimstidt, *Geochim. Cosmochim. Acta* **64**:797 (2000); O.S. Pokrovsky and J. Schott, *Geochim. Cosmochim. Acta* **64**:3313 (2000).
48. The dissolution time scale is the time required for complete dissolution of m/M_r initial moles of a mineral. See section 1.11 in A.C. Lasaga, *Kinetic Theory in the Earth Sciences*, Princeton University Press, Princeton, NJ, 1998. Equation 1.164 in this latter book can be rewritten in the form:

$$t_{\text{lifetime}} = \frac{3(V/V_m)}{d\xi/dt} = 3\tau_{\text{dis}}$$

where V is the initial volume of a mineral sample and V_m is its molar volume. The tabulated values of t_{lifetime} given by Lasaga (table 1.5), however, reflect an *arbitrary* choice of the initial volume $V(4.19 \times 10^{-9}\ m^3)$ and the initial surface area ($1.26 \times 10^{-5}\ m^2$) to apply to all minerals, whereas the values of m/M_r ($= V/V_m$) and specific surface area used in the present chapter are taken from the experiment in which $d(\xi/m)/dt$ was measured.

49. A review of these methods is given in section 1.4 of G. Sposito, *The Surface Chemistry of Soils*, Oxford University Press, New York, 1984.
50. Useful working discussions of this unresolved issue are given by C. Anbeek, *Geochim. Cosmochim. Acta* **56**:1461 (1992), **57**:4963 (1993); H. Zhang, P.R. Bloom, and E.A. Nater, *Geochim. Cosmochim. Acta* **57**:1681 (1993); J.W. Tester, W.G. Worley, B.A. Robinson, C.O. Grigsby, and J.L. Feerer, *Geochim. Cosmochim. Acta* **58**:2407 (1994); S.L. Brantley and Y. Chen, *Rev. Mineral.* **31**:119 (1995); A.F. White, A.E. Blum, M.S. Schulz, T.D. Bullen, J.W. Harden, and M.L. Peterson, *Geochim. Cosmochim. Acta* **60**:2533 (1996); M.E. Hudson, *Geochim. Cosmochim. Acta* **62**:3429 (1999); J. Ganor, J.L. Mogollón, and A.C. Lasaga, *Geochim. Cosmochim. Acta* **63**:1635 (1999); L.R. Kump, S.L. Brantley, and M.A. Arthur, *Annu. Rev. Earth Planet. Sci.* **28**:611 (2000); S.L. Brantley and N.P. Mellott, *Am. Mineral.* **85**:1767 (2000); and J.-M. Gaultier, E.H. Oelkers, and J. Schott, *Geochim. Cosmochim. Acta* **65**:1059 (2001).
51. The proton-promoted dissolution rate of a number of minerals has been observed to be first-order in the intermediate adsorbed species [S.A. Carrol-Webb and J.V. Walther, *Geochim. Cosmochim. Acta* **52**:2609 (1988); A.E. Blum and A.C. Lasaga,

Geochim. Cosmochim. Acta **55**:2193 (1991); P.V. Brady and J.V. Walther, *Am. J. Sci.* **292**:639 (1992); P.M. Dove, *Am. J. Sci.* **294**:665 (1994); F.J. Huertas, L. Chou, and R. Wollast, *Geochim. Cosmochim. Acta* **63**:3261 (1999)], but not universally so [R. Oxburgh, J.I. Drever, and Y.-T. Sun, *Geochim. Cosmochim. Acta* **58**:661 (1994); O.S. Pokrovsky and J. Schott, *Geochim. Cosmochim. Acta* **63**:881 (1999)].

52. F.J. Huertas et al., *Geochim. Cosmochim. Acta* **63**:3261 (1999).
53. K.L. Nagy, *Rev. Mineral.* **31**:173 (1995).
54. See also table 1 in A.C. Lasaga, J.M. Soler, J. Ganor, T.E. Burch, and K.L. Nagy, *Geochim. Cosmochim. Acta* **58**:2361 (1994).
55. See, e.g., A.F. White and A.E. Blum, *Geochim. Cosmochim. Acta* **59**:1729 (1995) and L.R. Kump et al., *Annu. Rev. Earth Planet. Sci* **28**:611 (2000) for a discussion of the role of the Arrhenius equation in weathering processes.
56. J.W. Tester et al., *Geochim. Cosmochim. Acta* **58**:2407 (1994).
57. This issue is reviewed by L.R. Kump et al., *Annu. Rev. Earth Planet. Sci* **28**:611 (2000).
58. See, e.g., W.H. Casey and G. Sposito, *Cosmochim. Geochim. Acta* **56**:3825 (1992); P.V. Brady and J.V. Walther, *Am. J. Sci* **292**:639 (1992); ;[51] and W.H. Casey and M.A. Cheney, *Aquat. Sci.* **55**:304 (1993). If the rate of proton-promoted dissolution is proportional to the concentration of $\equiv SRH^+$ (eq. 3.65a), then its apparent activation energy must reflect the standard enthalpy change associated with proton adsorption equilibria, which is known to depend on pH (M.L. Machesky, pp. 282–292 in D.C. Melchior and R.L. Bassett, *Chemical Modeling of Aqueous Systems II*, American Chemical Society, Washington, DC, 1990). This issue is reviewed briefly by A.C. Lasaga et al., *Geochim. Cosmochim. Acta* **58**:2361 (1995). It is important to bear in mind also that pH- and temperature-dependence of a dissolution rate will also be convoluted if conditions far from solubility equilibrium are not maintained [J. Cama, C. Ayora, and A.C. Lasaga, *Geochim. Cosmochim. Acta* **63**:2481 (1999)].
59. See section 1.7.2 in A.C. Lasaga, Kinetic Theory in the Earth Sciences, Princeton University Press, Princeton, NJ, 1998; and W.H. Casey, M.F. Hochella, and H.R. Westrich, *Geochim. Cosmochim. Acta* **57**:785 (1993). A theoretical model for the compensation effect can be developed in the context of the transition state theory of chemical reactions by postulating a linear relationship between the entropy and enthalpy of activation. But a compensation effect also can be *induced* in a given set of rate data if the observable variation in a rate coefficient is significantly less than the observable variability in A and E_a (i.e., $a \approx k$ in eq. 3.71, with ln A rising or falling in sympathy with E_a solely to "compensate" for the narrow range of observable variation in k). An induced compensation effect also can occur if k depends on a number of chemical parameters with compensating temperature dependencies.
60. The data were compiled from W.H. Casey and G. Sposito, *Geochim. Cosmochim. Acta* **56**:3825 (1992) (andalusite and kaolinite), J.J. Rosso and J.D. Rimstidt, *Geochim. Cosmochim. Acta* **64**:2481 (1999) (forsterite), and J.W. Tester et al., *Geochim. Cosmochim. Acta* **58**:2407 (1994) (quartz).
61. Nonetheless, these kinds of simplifications take on a special significance in the practice of scaling laboratory-based mineral dissolution rates for differences in temperature, specific surface area, pH, and water content in order to predict field-scale weathering. See, e.g., M.E. Malström, G. Destouni, S.A. Banwart, and B.H.E. Strömberg, *Environ. Sci. Technol.* **34**:1375 (2000).
62. G. Berger, E.Cadore, J.Schott, and P.M. Dove, *Geochim. Cosmochim. Acta* **58**:541 (1994); P.M. Dove, *Am. J. Sci.* **294**:665 (1994); P.M. Dove and C.J. Nix, *Geochim. Cosmochim. Acta* **61**:3329 (1997); P.M. Dove, *Geochim. Cosmochim. Acta* **63**:3715 (1999).
63. See B. Nowack and L. Sigg, *Geochim. Cosmochim. Acta* **61**:951 (1997) and references cited therein.

64. J.L. Mogollón, A. Pérez-Diaz, and S. Lo Monaco, *Geochim. Cosmochim. Acta* **64**:781 (2000).
65. It is important to note that the surface complexes formed may be either inner-sphere or outer-sphere. J.R. Bargar, P.Persson, and G.E. Brown [*Geochim. Cosmochim. Acta* **63**:2957 (2000)] suggest the term, *hydration-sphere complex*, for surface complexes mediated by hydrogen bonding between surface functional groups and an adsorbate, such that the solvation shell of the former is disrupted but not necessarily that of the latter. This kind of surface complex has been proposed for metal-EDTA complexes adsorbed on goethite and for oxalate adsorbed on silica.
66. A compendium of useful reviews can be found in E.D. Pittman and M.D. Lewan, *Organic Acids in Geological Processes*, Springer, New York, 1994.
67. J.I. Drever, *Geochim. Cosmochim. Acta* **58**:2325 (1994).
68. Data compiled from G. Furrer and W. Stumm, *Geochim. Cosmochim. Acta* **50**:1847 (1986) [alumina, δ-Al_2O_3]; S.R. Poulson, *Environ. Sci. Technol.* **32**:2856 (1998) [andesine plagioclase, 46 mol % anorthite]; S.A. Welch and W.J. Ullman, *Geochim. Cosmochim. Acta* **60**:2939 (1996) [bytownite, $K_{0.01}Na_{0.23}Ca_{0.77}Al_{1.75}Si_{2.25}O_8$]; B. Zinder, G. Furrer, and W. Stumm, *Geochim. Cosmochim. Acta* **50**:1861 (1986) [goethite, hematite]; E. Wieland and W. Stumm, *Geochim. Cosmochim. Acta* **56**:3339 (1992) [kaolinite]; A.G. Xyla, B. Sulzberger, G.W. Luther, J.G. Hering, P. van Cappellen, and W. Stumm, *Langmuir* **8**:95 (1992) [manganite (γ-MnOOH), pyrolusite (β-MnO_2)].
69. See, e.g., chapter 16 in F.J. Stevenson, *Humus Chemistry*, John Wiley, New York, 1994.
70. B. Zinder et al., *Geochim. Cosmochim. Acta* **50**:1861 (1986).
71. B.L. Phillips, J.A. Tossell, and W.H. Casey, *Environ. Sci. Technol.* **32**:2865 (1998).
72. B.L. Phillips, W.H. Casey, and S. Neugebauer Crawford, *Geochim. Cosmochim. Acta* **61**:3041 (1997); J.P. Nordin, D.J. Sullivan, B.L. Phillips, and W.H. Casey, *Inorg. Chem.* **37**:4760 (1998).
73. These concepts are discussed comprehensively by W.H. Casey, B.L. Phillips, and J. Nordin, pp. 244–264 in D.L. Sparks and T.J. Grundl (Eds.), *Mineral-Water Interfacial Reactions,* American Chemical Society, Washington, DC, 1998.
74. D.J. Sullivan, J.P. Nordin, B.L. Phillips, and W.H. Casey, *Geochim. Cosmochim. Acta*, **63**:1471 (1999).
75. B.L. Phillips, S. Neugebauer Crawford, and W.H. Casey, *Geochim. Cosmochim. Acta* **61**:4965 (1997).
76. M.V. Biber, M. Dos Santos Afonso, and W. Stumm, *Geochim. Cosmochim. Acta* **58**:1999 (1994).
77. Equations 3.77 describe metal-, proton-, and ligand-promoted dissolution reactions, respectively. Enhancement of the rate of metal- or ligand-promoted dissolution relative to the proton-promoted rate occurs if the rate coefficient for the adsorption and detachment steps is larger than that for the corresponding proton steps, whereas inhibition occurs if the rate coefficient is smaller, given that the metal or ligand adsorbate occurs at significant concentration.
78. J. Ganor and A.C. Lasaga, *Geochim. Cosmochim. Acta* **62**:1295 (1998).
79. An introductory survey of this complex topic is given by W. Stumm, B. Sulzberger, and J. Sinniger, *Croat. Chem. Acta* **63**:277 (1990). See also A.T. Stone et al., pp. 337–374 in Chemistry of Aquatic Systems, G. Bidologlio and W. Stumm (Eds.), Kluwer Academic, Boston, 1994; and D. Suter, S. Banwart, and W. Stumm, *Langmuir* **7**:809 (1991).
80. D.R. Lovley, *Annu. Rev. Microbiol.* **47**:263 (1993); Y. Deng and W. Stumm, *Appl. Geochem.* **9**:23 (1994); K.L. Godtfredsen and A.T. Stone, *Environ. Sci. Technol.* **28**:1450 (1994); S.A. Banwart, *Geochim. Cosmochim. Acta* **63**:2919 (1999); E.G. Majcher, J. Chorover, J.-M. Bollag, and P.M. Huang, *Soil Sci. Soc. Am. J.* **64**:157 (2000).

81. J.M. Zachara, P.L. Gassman, S.C. Smith, and D. Taylor, *Geochim. Cosmochim. Acta* **59**:4449 (1995) [Co(II)]; A. Manceau, V.A. Drits, E. Silvester, C. Bartoli, and B. Lanson, *Am. Mineral.* **82**:1150 (1997) [Co(II)]; D. Postma and C.A.J. Appelo, *Geochim. Cosmochim. Acta* **64**:1237 (2000) [Fe(II)].
82. G.W. Luther, B. Sundby, B.L. Lewis, P.J. Brendel, and N. Silverberg, *Geochim. Cosmochim. Acta* **61**:4043 (1997), **62**:2218 (1998); S. Hulth, R.C. Aller, and F. Gilbert, *Geochim. Cosmochim. Acta* **63**:49 (1999); P. Anschutz, B. Sundby, L. Lefrançois, G.W. Luther, and A. Mucci, *Geochim. Cosmochim. Acta* **64**:2751 (2000).
83. S.E. Fendorf, *Geoderma* **67**:55 (1995) [Cr(III)]; M.J. Scott and J.J. Morgan, *Environ. Sci. Technol.* **29**:1898 (1995) [As(III)], **30**:1990 (1996) [Se(IV)]; X. Sun and H.E. Doner, *Soil Sci.* **163**:278 (1998) [As(III)]; H.W. Nesbitt, G.W. Canning, and G.M. Bancroft, *Geochim. Cosmochim. Acta* **62**:2097 (1998) [As(III)]; V.Q. Chiu and J.G. Hering, *Environ. Sci. Technol.* **34**:2029 (2000) [As(III)]; B.A. Manning and D.L. Suarez, *Soil Sci. Soc. Am. J.* **64**:128 (2000) [As(III)].
84. F. Trolard, J.-M.R. Génin, M. Abdelmoula, G. Bourrié, B. Humbert, and A. Herbillon, *Geochim. Cosmochim. Acta* **61**:1107 (1997); J.-M.R. Génin, G. Bourrié, F. Trolard, M. Abdelmoula, A. Jaffrezic, P. Refait, V. Maître, B. Humbert, and A. Herbillon, *Environ. Sci. Technol.* **32**:1058 (1998).
85. H. Chr. B. Hansen [N(V)]; H. Chr. B. Hansen, O.K. Borggaard, and J. Sørensen, *Geochim. Cosmochim. Acta* **58**:2599 (1994) [N(III)]; G. Bidoglio, P.N. Gibson, M. O'Gorman, and K.J. Roberts, *Geochim. Cosmochim. Acta* **57**:2389 (1993); H. Chr. B. Hansen, C.B. Koch, H. Nancke-Krogh, O.K. Broggaard, and J. Sørensen, *Environ. Sci. Technol.* **30**:2053 (1996); A.R. Pratt, D.W. Blowes, and C.J. Ptacek, *Environ. Sci. Technol.* **31**:2492 (1997); S. Loyaux-Lawniczak, P. Refait, J.-J. Ehrhardt, P. Lecomte; J.-M.R. Génin, *Environ. Sci. Technol.* **34**:438 (2000); S.M. Ponder, J.G. Darab, and T.E. Mallouk, *Environ. Sci. Technol.* **34**:2564 (2000); and E.S. Ilton, C.O. Moses, and D.R. Veblen, *Geochim. Cosmochim. Acta* **64**:1437 (2000) [Cr(VI)]; S.C.B. Myneni, T.K. Tokunada, and G.E. Brown, *Science* **278**:1106 (1997); and P. Refait, L. Simon, and J.-M.R. Génin, *Environ. Sci. Technol.* **34**:819 (2000) [Se(VI)].
86. A.F. White and M.L. Peterson, *Geochim. Cosmochim. Acta* **60**:3799 (1996).
87. E. Silvester, L. Charlet, and A. Manceau, *J. Phys. Chem.* **99**:16, 662 (1995). See also J.-B. Chung, R.J. Zasoski, and S.-U. Lim, *Agric. Chem. Biotech.* **37**:414 (1994).
88. V.A. Drits, E. Silvester, A.I. Gorshkov, and A. Manceau, *Am. Mineral.* **82**:946 (1997); E. Silvester, A. Manceau, and V.A. Drits, *Am. Mineral.* **82**:962 (1997); D. Banerjee and H.W. Nesbitt, *Geochim. Cosmochim. Acta* **63**:1671 (1999); B. Lanson, V.A. Drits, E. Silvester, and A. Manceau, *Am. Mineral.* **85**:826 (2000). For a useful overview of Mn oxide structures, see J.E. Post, *Proc. Natl. Acad. Sci. USA* **96**:3447 (1999).
89. Although adsorption of Cr(IV) as an intermediate step is indicated clearly by the available data on Mn(IV) reduction by Cr(III), questions remain as to whether the lifetime of the adsorbed Cr intermediate is related to Cr(IV) desorption kinetics and coordination number change [6 to 4, required for the formation of Cr(V)], or to slow formation of Cr(V) directly from adsorbed Cr(III) in parallel with the reaction in eq. 3.81. See E. Silvester et al., *J. Phys. Chem.* **99**:16, 662 (1995) and D. Banerjee and H.W. Nesbitt, *Geochim. Cosmochim. Acta* **63**:1671 (1999) for additional discussion of the kinetic steps that may be involved. It is salient to note that the reductive dissolution of birnessite by Cr(III) is actually slower than the proton-promoted dissolution of this mineral. The inhibiting effect of Cr(III) on Mn(IV) reduction is discussed by S.E. Fendorf, M. Fendorf, D.L. Sparks, and R. Gronsky, *J. Colloid Interface Sci.* **153**:37 (1992). The importance of Mn(III) in enhancing the rate of Mn(IV) reduction by Cr(III) is discussed by P.S. Nico and R.J. Zasoski, *Environ. Sci. Technol.* **34**:3363 (2000).

For Further Reading

P. L. Brezonik, *Chemical Kinetics and Process Dynamics in Aquatic Systems,* Lewis Publishers: Boca Raton, FL, 1999. An encyclopedic, advanced introduction to chemical kinetics for application to aqueous systems.

A. C. Lasaga, *Kinetic Theory in the Earth Sciences*, Princeton University Press: Princeton, 1998. This advanced textbook offers a comprehensive introduction to the concepts and practice of chemical kinetics in the context of geochemistry. Chapters 1, 2, and 7 are particularly relevant to the present chapter.

D. L. Sparks and T. J. Grundl, Eds., *Mineral–Water Interfacial Reactions: Kinetics and Mechanisms*, American Chemical Society: Washington, DC, 1998. This edited volume describes recent instrumental, computational, and conceptual innovations in geochemical kinetics and equilibria.

D. L. Sparks and D .L. Suarez, Eds., *Rates of Soil Chemical Processes*, Soil Science Society of America: Madison, WI, 1991. A most useful compendium of kinetics methodologies for use in the study of the surface chemistry of natural particles.

A. F. White and S. L. Brantley, Eds., *Chemical Weathering Rates of Silicate Minerals,* Mineralogical Society of America: Washington, DC, 1995. This edited volume contains outstanding, comprehensive reviews on the dissolution reactions of every important class of naturally occurring silicate mineral.

Research Matters

1. The reaction sequences in eqs. 3.17, 3.39, 3.44, 3.65, and 3.77 are similar in structure, with the last four being special cases, for the most part, of the first one. The rate laws corresponding to these sequences can be solved exactly (as ordinary differential equations) after linearization by either (i) assuming constant values of the concentrations of one reactant and one product in eq. 3.17 or (ii) assuming small deviations from the equilibrium values of the concentrations of the five species in eq. 3.17, then neglecting all terms that are bilinear in the concentration deviations. The first approach is described in section 3.3, whereas the second approach appears in section 3.2.
   (a) Consider again the coupled, linear rate laws in eqs. 3.21 and add to them a linearized rate law for the intermediate species C in eq. 3.17 to produce three coupled rate laws with three dependent variables. Under what conditions are eqs. 3.47 the *exact* solutions of these three rate laws? Make a table of correspondence for the concentration variables and constant parameters in eqs. 3.21 and 3.47 to confirm your conclusions.
   (b) What is the relationship among eqs. 3.24a,b; 3.25a,b; and 3.48?
   (c) Answer the same questions posed in (a) and (b) for the kinetics analysis of the reaction sequence in eq. 3.34.
   (d) Show that eqs. 3.47a,c are, in fact, formally *exact* solutions of eqs. 3.23, 3.36, and 3.40 by substituting eqs. 3.47a,c into the three sets of coupled differential equations to develop a new set of correspondences for the parameters K_f, k_{ET}, k, and k'. (N.B. Equation 3.47a will require your adding a constant term on its right side to reflect the nonzero value of the second-order rate coefficient, $k_{b'}$ in eqs. 3.17 and 3.39.)
2. B. Wehrli, S. Ibric, and W. Stumm [*Colloids Surf.* **51**:77 (1990)] have determined a rate law for VO^{2+} adsorption on δ-Al_2O_3 (p.z.n.p.c. = 8.7) in aqueous suspension [2.1 mol $\equiv$AlOH m^{-3}, 0.1 mol VO^{2+} m^{-3} initial concentrations]. Their data were interpreted in the context of eq. 3.32 ($M^{2+} = VO^{2+}$) under the assumption of negligible desorption.
   (a) Establish a relationship between the second-order rate coefficient k_2K_{OS} in eq. 3.32 and $[H^+]$ based on the data presented in fig. 3(a) of the article by Wehrli et al. Use this relationship and the line of reasoning in eqs. 3.7 and 3.8

to develop a new second-order rate law that features [$VOOH^+$] instead of [VO^{2+}]. N.B. The hydrolysis reaction analogous to eq. 3.8a for vanadyl ion is: $VO^{2+} + H_2O = VOOH^+ + H^+ \qquad \log{}^*K_1 = -5.67$.

(b) Calculate the pH-independent, second-order rate coefficient that is analogous to k_1 in eq. 3.7 (and equivalent to k_2K_{OS} in eq. 3.32 if $M^{2+} \leftrightarrow VOOH^+$). Compare your result with the value of k_2K_{OS} predicted by eq. 3.33 using $k_{wex} = 3 \times 10^4\ s^{-1}$ for $VOOH^+$. Given that $k_{wex} = 300\ s^{-1}$ for VO^{2+}, explain why the rate law for vanadyl adsorption does not feature [VO^{2+}] with a pH-independent rate coefficient analogous to k_0 in eq. 3.37.

(c) Estimate the fraction of the total vanadyl ion concentration that is in the form of $VOOH^+$ over the pH range of the adsorption kinetics data in fig. 2(a) of the article by Wehrli et al. How can the very small values of this fraction be reconciled with the explicit appearance of [$VOOH^+$] in the rate law for vanadyl adsorption (eq. 4 in the article)?

3. S. H. R. Davies and J. J. Morgan [*J. Colloid Interface Sci.* **129**:63 (1989), *Geochim. Cosmochim. Acta* **62**:361 (1998)] have studied the kinetics of Mn(II) oxidation by O_2 gas in aqueous solution and in the presence of the iron oxides, goethite (α-FeOOH, p.z.n.p.c. = 7.6) and lepidocrocite (γ-FeOOH, p.z.n.p.c. = 7.1 ± 0.1). Equation 3.60 provides a rate law that can be adapted to describe their data.

(a) The data in table IV of the 1989 article can be used to calculate the rate coefficient k_1 (eq. 3.60) given $K_H = 1.26 \times 10^{-3}$ mol dm^{-3} atm^{-1} for O_2(g). Determine the pH dependence of this rate coefficient in the form, $k_1 = K_1$ [OH^-]^n, taking advantage of insight provided by fig. 4 in P. J. von Langen, K. S. Johnson, K. H. Coale, and V. A. Elrod, *Geochim. Cosmochim. Acta* **61**:4945 (1997). Use linear regression analysis to establish confidence intervals for the parameters, K_1 and n.

(b) Calculate D_{Mn} (eq. 1.6) for goethite and lepidocrocite using the adsorption-edge data in fig. 2 of the 1989 article. Then calculate x_{Mnads}.

(c) Measured values of the composite second-order rate coefficient in eq. 3.63b can be reconstructed from the data in table V of the 1989 article after division of the pseudo first-order rate coefficient (third column) by the corresponding concentration of O_2(aq), followed by addition of the value of k_1 as determined in (a) for the appropriate pH value. Tabulate the composite second-order rate coefficient for each iron oxide, then use the results obtained in (a) and (b) to calculate k_1^{surf} in eq. 3.63b. Compare k_1^{surf} to k_1 at the same pH value and interpret their difference. (N.B. Comment on the relation of figs. 1 and 4 and table V in the 1989 article, as well as figs. 1a and 1b in the 1998 article, to the basic assumptions underlying the rate law in eq. 3.60.)

4. Y. Chen and S. L. Brantley [*Chem. Geol.* 135:275 (1997)] have reported measurements of the rate of proton-promoted dissolution of albite ($NaAlSi_3O_8$) under conditions of varying pH, temperature, and Al(III) concentration.

(a) Use the data in tables 2 and 3 of the article by Chen and Brantley to calculate the dissolution time scale parameter, τ_{dis} as a function of pH and temperature. Compare your results to that cited for forsterite below eq. 3.67. What are the trends in τ_{dis} with pH and temperature?

(b) Apply eq. 3.66 to the rate data [$d(\xi/m)/dt$] used in (a) under the assumption that $n_H = 0.5$ (see table 4 in the article by Chen and Brantley). Calculate k_H in mol kg^{-1} s^{-1} for each $d(\xi/m)/dt$ value and determine the apparent activation energy (E_a) governing its temperature dependence.

(c) Use the value of E_a obtained in (b) to predict the Arrhenius parameter, ln A according to the compensation law in eq. 3.74. Compare your result with the value of ln A obtained as the y-intercept of the Arrhenius plot developed in (b). How does the value of k_{iso} based on eq. 3.74 compare with the estimated value of k_H at 37°C? (N.B. The compensation law in eq. 3.74 is averaged over pH.)

(d) Discuss the relationship between eq. 3.80 and eq. 18 in the article by Chen and Brantley. Show that both model equations have the generic mathematical form: $\phi([H^+], [Al^{3+}]) = k([H^+]/1 + K[Al^{3+}])^n$ after the observed numerical values of the parameters F and K_H are considered carefully.

5. C. A. Johnson and A. G. Xyla [*Geochim. Cosmochim. Acta* 55:2861 (1991)] have investigated the kinetics of reductive dissolution of manganite (γ-MnOOH, p.z.n.p.c. = 6.2) by Cr(III) in aqueous suspension [7.27 mmol ≡Mn(III) m^{-3}, 0.5 mmol Cr(III) m^{-3} initial concentrations].
   (a) Use the data in figs. 2 to 5 of the article by Johnson and Xyla to establish the concentration dependence of the rate of Cr(III) oxidation (following the example in eq. 3.82), as well as its overall Mn:Cr stoichiometry.
   (b) The data in fig. 1 of the article show the existence of an adsorbed Cr intermediate species (> MnCr) whose concentration achieves a maximum value of about 0.07 mmol m^{-3} after 20 s of reaction, when the concentration of Cr(III) has declined to 0.25 mmol m^{-3}. [The concentration of ≡Mn(III) is effectively constant because of its large value relative to that of Cr(III).] Adapt the rate laws in eqs. 3.46 to describe the kinetics data in fig. 1, with $K_f = k_f$ [≡Mn(III)] = 0.01 s^{-1}, according to the data in table 1 of the article, and $k_b = 0$ (no Cr desorption). Show that eqs. 3.47 are consistent with the data in fig. 1 of the article for [Cr(III)], [> MnCr], and [Cr(VI)], respectively, with the value of k_{ET} determined by applying eq. 3.46b to the maximum in [> MnCr] at $t = 20$ s.
   (c) Use the data in table 1 of the article by Johnson and Xyla to calculate the apparent activation energy at pH 4.5 for the rate coefficient K_f as defined in (b) above.

4

Modeling Ion Adsorption

4.1 Modeling the Diffuse Ion Swarm

A net particle surface charge different from zero signals the existence of adsorbed ions that are not bound in surface complexes. These ions constitute a *diffuse swarm* (or diffuse layer), the contribution of which to surface charge balance is epitomized in eq. 1.43. Diffuse swarm species are relatively free to move about in the aqueous solution contacting a charged particle, although they are expected to respond to surface charge by distributing themselves in order to screen it effectively, either by accumulating near an adsorbent (*counterions*) or by depleting themselves from the interfacial region (*coions*). Screened particle charge, in turn, mediates interparticle interactions in colloidal suspensions, with important consequences for suspension structure and stability (see section 5.3).

Description of the diffuse ion swarm in molecular terms begins with consideration of an aqueous electrolyte solution as influenced significantly by a charged solid surface, excluding, however, any chemical bonding that would imply strong ion adsorption. This straightforward concept is, nonetheless, made complicated by ion-ion and ion-solvent interactions in an aqueous solution (including soluble complex formation) and by the vagaries of the topographic and charge-producing properties that are found typically on examination of the surface of a solid adsorbent over molecular spatial scales. Development of a comprehensive molecular description of the diffuse ion swarm thus remains an important and difficult item on the agenda of natural particle surface chemistry.[1]

Given the absence of chemical bonding to the adsorbent surface, the principal features of a diffuse ion swarm should be explicable in terms of adsorptive and adsorbent charge and size, much in the spirit of a Schindler diagram (see section 1.5). The fundamental question to be posed, then, is how a swarm of aqueous ions of given valence and radius distributes itself near a particle of given surface charge and radius according to the dictates of the coulomb force. This basic question can be answered in perhaps the simplest way by means of a purely physical model, the electrostatic theory of an ion swarm that is immersed in a continuum dielectric medium. With its mathematical description being taken to the level of approximation known as *modified Gouy-Chapman theory*,[2] a quantitative description of the diffuse ion swarm results that is tractable enough to be used in software applied routinely to compute the surface speciation of natural particles.[2,3]

The principal objective of the modified Gouy-Chapman model is to calculate the distribution of counterions and coions in the vicinity of a uniformly charged, molecularly smooth surface. This distribution is calculated by combining the *Poisson equation*, a differential equation that relates the mean electrostatic potential in a swarm of ions to the net charge density in the swarm, with the *potential of mean force*, which is the average energy required to move an ion from a reference point, defined to be at zero mean electrostatic potential, to a point somewhere in the diffuse swarm.[4] Implicit in the use of the Poisson equation are two key postulates:

- The charged surface is uniform and smooth, characterized by a charge density σ and by high symmetry, such that the diffuse-swarm electrostatic potential varies spatially only along a coordinate axis perpendicular to the charged surface.
- The liquid phase is a uniform continuum characterized solely by its dielectric permittivity.

These two postulates have the effect of smoothing the molecular structure of both the adsorbent and the solvent in the aqueous phase. (The first postulate above can be weakened a bit, however, to admit a spatially variable surface charge density or an electrostatic potential that varies in three dimensions because of irregular surface topography.[5]) Only the ions in the diffuse swarm now retain a discrete molecular character; but this, too, is highly constrained by a third postulate concerning the potential of mean force:

- The potential of mean force, $W_i(x)$, is proportional to $\psi(x)$, the mean electrostatic potential, as measured into the aqueous solution phase along an axis perpendicular to the charged surface.

The potential of mean force, by definition,[4] includes interactions other than the coulomb force which creates a mean electrostatic potential in the diffuse swarm. For example, an ion must endure short-range repulsive interactions with other ions, irrespective of their charge, along its path from the reference point, where $\psi = 0$, to its place in the swarm, these interactions coming as a direct effect of finite ion size. Moreover, the clustering of counterions against a charged surface is likely, in turn, to attract nearby coions, because the effect of like surface charge has been mitigated by counterion screening. These kinds of short-range spatial correlations are not reflected in the mean electrostatic potential, which depends only on the averaged coulomb interaction. Thus, in modified Gouy-Chapman theory, *there are no short-range correlations except those which may follow from long-range correlations.*[6] The impact of this dictum is to leave only a "distance of closest ion approach" to the charged surface, $d_0/2$, as a relict of any direct manifestation of the molecular character of the electrolyte ions.[7]

The Poisson equation for the mean electrostatic potential in a diffuse ion swarm has the generic mathematical form:[2]

$$\frac{d^2\psi}{dx^2} + \frac{(1-2\nu)}{x}\frac{d\psi}{dx} = -\frac{1}{\varepsilon_0 D}\sum_i Z_i F c_i(x) \tag{4.1}$$

where D is the dielectric constant of water (78.3 at 298 K), ε_0 is the permittivity of vacuum (8.85419×10^{-12} C V^{-1} m^{-1}) and F is the Faraday constant (96, 485 C mol^{-1}). The coordinate x in eq. 4.1 is measured along an axis normal to the particle surface and extending from the center of the particle into the aqueous solution phase. The particle surface may be planar ($\nu = \frac{1}{2}$); cylindrical ($\nu = 0$); or spherical ($\nu = -\frac{1}{2}$). Naturally occurring examples of these three surface geometries include nanoparticles formed by 2:1 clay minerals ($\nu = \frac{1}{2}$), goethite or halloysite ($\nu = 0$), and ferrihydrite ($\nu = -\frac{1}{2}$), respectively. Particle aggregates can have any of the three shapes.

The sum on the right side of eq. 4.1 includes each different type of counterion or coion in the diffuse swarm. The quantity $c_i(x)$ is the concentration of ions of type i at a point x. It is subject to the theoretical constraint:[4]

$$RT\frac{d\ln c_i}{dx} = -\frac{dW_i}{dx} \tag{4.2}$$

and to its boundary value, c_{i0}, the concentration at the reference point, where W_i vanishes. Equation 4.2 is the defining relationship between gradients of $c_i(x)$ and the mean force acting on an ion of type i that produces them [hence the name, "potential of mean force," for $W_i(x)$]. The premise in modified Gouy-Chapman theory is that the right side of eq. 4.2 can be approximated as proportional to the coulomb force, that is, to minus the gradient of the mean electrostatic potential:

$$RT\frac{d\ln c_i}{dx} = -Z_i F\frac{d\psi}{dx} \tag{4.3}$$

It follows that, in the modified Gouy-Chapman model, $c_i(x)$ has the explicit mathematical form which is obtained by integration of eq. 4.3:

$$c_i(x) = c_{i0}\exp[-Z_i F\psi(x)/RT] \tag{4.4}$$

subject to the boundary condition at the reference point. Therefore, eq. 4.1 now can be rewritten as a nonlinear differential equation solely in the dependent variable, $\psi(x)$:

$$\frac{d^2\psi}{dx^2} + \frac{(1-2\nu)}{x}\frac{d\psi}{dx} = -\frac{1}{\varepsilon_0 D}\sum_i Z_i F c_{i0}\exp\left(\frac{-Z_i F\psi}{RT}\right) \tag{4.5}$$

This nonlinear differential equation is termed the *Poisson-Boltzmann equation.*[4,8]

The chemical significance of the Poisson-Boltzmann equation, whose solution is ultimately to be substituted into eq. 4.4 to calculate the spatial distribution of each ion species in a diffuse swarm, can be understood through an examination of the intrinsic scales of electrostatic potential and length (along the x-axis) that are implicit in it. (Note that eq. 4.5 is not invariant in form under multiplication of either ψ or x by an arbitrary scale factor, which means that the modified Gouy-Chapman model must exhibit intrinsic scales of both electrostatic potential and length.) The intrinsic scale of electrostatic potential is RT/F (= 25.693 mV at 298 K), which appears in the exponent on the right side of eq. 4.5. On scaling both sides of the equation with this factor to create the dimensionless dependent variable, $y = F\psi/RT$, the quantity

$$\begin{aligned} \beta &\equiv 2F^2/\varepsilon_0 \mathrm{DRT} \\ &= 1.08335 \times 10^{16} \text{ m mol}^{-1} \text{ (298 K)} \end{aligned} \tag{4.6}$$

appears naturally as a coefficient of the summation on the right side, the factor 2 being introduced ad hoc to avoid its appearance in several important special cases of the equation to be considered. The product βc_{i0} has the dimensions of inverse length-squared, thus leading to the definition of an intrinsic length scale as the inverse square-root of the parameter,

$$\kappa_i^2 \equiv \beta c_{i0} \tag{4.7}$$

The value of the length scale κ_i^{-1} ranges from 1 to 30 nm at 298 K as c_{i0} ranges from 0.1 to 100 mol m^{-3}, typical of natural freshwaters ($\kappa^{-1} \approx 0.36$ nm for seawater). Thus, *nanometer length scales characterize the spatial variability of $\psi(x)$, whereas millivolt scales characterize the variation in its magnitude.*

Upon consideration of the full boundary-value problem for $\psi(x)$, two other length scales are implicit in the Poisson-Boltzmann equation. One is an extrinsic length scale determined by the distance of closest approach of an ion to the particle surface, which is measured from an origin placed arbitrarily at the center of the particle, considered in eq. 4.5 as either a rhombohedron (or a disc), a cylinder, or a sphere. This distance, which ranges typically from tens of nanometers to tens of micrometers, may be denoted a and set equal to the sum of the distance from the center of a particle to its surface (R) plus the distance $d_0/2$ defined above: $a \equiv R + d_0/2$. This length scale, a, which has no intrinsic significance for eq. 4.5, is nonetheless a convenient scale factor for the independent variable x, such that $x' \equiv x/a$ becomes a new dimensionless independent variable, always defined on the same semiopen interval $[1,\infty)$ for systems described by eq. 4.5. The correspondent scaling for κ_i is then $\varepsilon_i \equiv \kappa_i a$.

A second intrinsic length scale (i.e., besides that defined in eq. 4.7) appears through the boundary condition on $\psi(x)$. It involves the surface charge density, σ_p:

$$\begin{aligned} \ell &\equiv \varepsilon_0 \mathrm{DRT/F}|\sigma_p| \\ &= 2F/\beta|\sigma_p| \\ &= 0.111/\tilde{\sigma} \qquad (\ell \text{ in nm}, \ T = 298\ K) \end{aligned} \tag{4.8}$$

where $\tilde{\sigma} \equiv N_A|\sigma_p|/F$ is the absolute value of the net total particle surface charge density expressed in units of protonic charges per square nanometer, which is convenient for use in applications, and N_A is the Avogadro constant (6.02214 $\times$ 10^{23} mol^{-1}). Given the typical range of $|\sigma_0|$ for smectite and vermiculite clay minerals, $0.5 < \tilde{\sigma} < 2.0\ nm^{-2}$ (see section 1.3), and the maximal value, $|\tilde{\sigma}_H| \approx 2\ nm^{-2}$ for oxide minerals,[9] the minimal range of ℓ is between 0.05 and 0.25 nm. Thus the minimal ℓ lies much below the typical values for κ_i^{-1} cited above. On the other hand, if charge neutralization by surface complexes reduces $|\tilde{\sigma}|$ to values near zero, the length scale ℓ will grow correspondingly large.

The physical significance of the parameter ℓ is seen after dividing the conventional boundary condition on eq. 4.5,[8,10]

$$\left(\frac{d\psi}{dx}\right)_{x=a} = -\frac{\sigma_p}{\varepsilon_0 D} \tag{4.9}$$

by the electrostatic potential scale factor, RT/F, in order to derive the result:

$$\left(\frac{dy}{dx}\right)_{x=a} = -\frac{\text{sgn}(\sigma_p)}{\ell} \tag{4.10}$$

where sgn(·) is the signum function,

$$\text{sgn}(\sigma) = \begin{cases} +1 & \sigma_p > 0 \\ -1 & \sigma_p < 0 \end{cases} \tag{4.11}$$

Equation 4.10 shows that ℓ defines an intrinsic length scale for *spatial variation of $\psi(x)$ at the distance of closest approach of an ion to a particle surface* (as measured from the center of the particle). Thus, small values of ℓ translate to large spatial variation of $\psi(x)$ outward from the particle surface.

The three spatial scales attendant to the Poisson-Boltzmann equation are summarized in the table below. Note that the corresponding dimensionless extrinsic/intrinsic length ratios, $\kappa_i a = \varepsilon_i$ and a/ℓ, give measures of particle size in units of the length scales over which $\psi(x)$ varies either in the diffuse swarm or at its boundary (the particle surface), respectively. Thus, charged particles are "small" if the aqueous solution phase is very dilute or if the particle surface charge density is very low.

κ^{-1}	spatial variability of $\psi(x)$ in the diffuse ion swarm
ℓ	spatial variability of $\psi(x)$ at the particle surface
a	lower boundary for x, an extrinsic measure of particle size

Equations 4.5 and 4.10, along with the stipulation that $\psi(x)$ is to vanish at some reference point, constitute a well-defined boundary-value problem for the mean electrostatic potential in the diffuse swarm.[8,10] Without yet solving this boundary-value problem, one can prove some rather general statements about its solution that transcend any particular results that may be obtained by integrating eq. 4.10.[11] These statements, known as *PBE Theorems*, also are useful to developing a chemical interpretation of the modified Gouy-Chapman model that is uncluttered by the complexities of detailed numerical calculations. The PBE

Theorems apply to the mathematical problem posed by the scaled nonlinear differential equation,

$$\frac{d^2y}{dx'^2} + \frac{(1-2\nu)}{x'}\frac{dy}{dx'} = -\frac{1}{2}\sum_i \varepsilon_i^2 Z_i \exp(-Z_i y) \tag{4.12}$$

subject to the scaled boundary conditions,

$$\left(\frac{dy}{dx'}\right)_{x'=1} = -\mathrm{sgn}(\sigma_p)\frac{a}{\ell} \tag{4.13}$$

with $y(x') \equiv 0$ at the reference point, where $c_i(x') = c_{i0}$. It is important to note that eqs. 4.12 and 4.13 are invariant in mathematical form under the transformation of variables defined by:

$$x^* = x', \quad y = -y, \quad Z_i^* = -Z_i, \quad \sigma_p* = -\sigma_p \tag{4.14}$$

This means that any particular solution of eq. 4.12, $y(x')$, can be transformed immediately to another solution $y^*(x^*)$, corresponding to opposite ion and surface charge. It also means that the PBE Theorems need be demonstrated solely for the case $\sigma_p > 0$, with no loss of generality.

The PBE Theorems are facilitated by consideration of eq. 4.12 after it is rearranged to have the form:

$$\frac{d^2y}{dx'^2} = -\frac{1}{2}\sum_i \varepsilon_i^2 Z_i \exp(-Z_i y) - \frac{(1-2\nu)}{x'}\frac{dy}{dx'} \tag{4.15}$$

with the added proviso that $x' \varepsilon\ [1,\infty)$ and $\sigma_p > 0$. The left side of eq. 4.15 is termed the *curvature* of a graph of $y(x')$. The first term on the right side of eq. 4.15 is the curvature induced by the presence of ions in the diffuse swarm, whereas the second term stems from the geometry of the particle surface. The former contribution has the same sign as y, a fact readily proved by noting that anions contribute terms of the form $|Z_i|\exp(|Z_i|y)$ to the sum (including its prefactor coefficient, -1), while cations contribute terms of the form $-|Z_i|\exp(-|Z_i|y)$. The parameter ε_i^2 is subject to the electroneutrality condition,

$$\sum_i \varepsilon_i^2 Z_i = 0 \tag{4.16}$$

that must obtain at the reference point, where $y \equiv 0$. Thus, as an example, for the case of a 2:1:1 mixed electrolyte solution (e.g., Ca^{2+} and Na^{2+} mixed with Cl^-):

$$\begin{aligned}&-\sum_i \varepsilon_i^2 Z_i \exp(-Z_i y) = -2\varepsilon_{\mathrm{Ca}}^2 \exp(-2y) - \varepsilon_{\mathrm{Na}}^2 \exp(-y)\\&+ \varepsilon_{\mathrm{Cl}}^2 \exp(y) = 2\varepsilon_{\mathrm{Ca}}^2[\exp(y) - \exp(-2y)]\\&+ \varepsilon_{\mathrm{Na}}^2[\exp(y) - \exp(-y)]\end{aligned}$$

which always has the same sign as y.

Suppose that the scaled potential at $x' = 1$ were to be negative [i.e., $y(1) < 0$]. Since $dy/dx' < 0$ at $x' = 1$ (eq. 4.13 with $\sigma_p > 0$), while $y = 0$ at the reference point, a negative initial value of y implies the existence of a *minimum* value for y at some $x_0' > 1$. This is evident from the fact that a negative-valued function which initially is decreasing must "turn around" somewhere in order ultimately to increase to the value zero out at the reference point. At the point x'_0, $(dy/dx')_{x'=x'_0} = 0$ by definition of a minimum. Moreover, the assumed negative value of y at x'_0 makes the first term on the right side of eq. 4.15 negative. But this implies that the curvature of y at the hypothesized minimum is also negative (the second, "geometry" term on the right side of eq. 4.14 can make no contribution at $x' = x'_0$). But, by elementary calculus, if the curvature is negative, the value $y(x'_0)$ is then a *maximum*, in contradiction with the hypothesis that a *minimum* value exists at x'_0! Therefore, the initial hypothesis, that $y(1)$ is negative, must be wrong. This argument yields

Theorem 1: *If* $\sigma > 0,\ y(1) > 0$.

QED

The result above can be made even stronger. Given $y(1) > 0$, is it possible, for some $x' \varepsilon\ (1,\infty)$, that there is still somewhere a value $y(x') < 0$? Suppose there is. Then, at a point yet further out from the particle surface, y must have a minimum value at which it "turns around" finally to increase toward the value zero at the reference point. Following the same line of reasoning used to prove PBE Theorem 1, one concludes, once again, that negative curvature is implied by such a minimum value with $y < 0$, which is a contradiction. Therefore, quite generally, we have

Theorem 2: *If* $\sigma > 0,\ y(x') \geq 0$ *for all* $x' \geq 1$.

QED

With the positive-definite character of $y(x')$ now established, the next item of attention is its first derivative, which has been mandated to be negative at the particle surface, $x' = 1$, by eq. 4.13 and the imposed condition, $\sigma_p > 0$. But suppose that dy/dx' were to turn positive for some $x' \varepsilon\ (1,\infty)$. Because $y(x') > 0$ by PBE Theorem 2, there must then also be a *maximum* value of $y(x')$ somewhere in the open interval (x', ∞), in order that $y(x')$ can "turn around" to decrease to the value zero out at the reference point. Let $x'_2 > 1$ be this supposed point of maximum y-value, where $(dy/dx')_{x'=x'_2} = 0$ necessarily. Equation 4.14 and the positive- valuedness of $y(x')$ for all $x' \geq 1$ then imply that the curvature at x'_2 is positive, which means that x'_2 is a point of *minimum* value for $y(x')$. This contradiction shows that positive derivatives of $y(x')$ are impossible, yielding

Theorem 3: *If* $\sigma > 0,\ dy/dx' \leq 0$ *for all* $x' \geq 1$.

QED

The PBE Theorems develop two broad characteristics of the solution of eq. 4.5 subject to a zero value at some reference point and the boundary condition in eq. 4.10: (1) *the sign of* $\psi(x)$ *is the same as that of* σ_p, and (2) $\psi(x)$ *tends monotonically to its zero reference value with increasing* $x > a$. There can be no oscillations of $\psi(x)$ along this pathway and, therefore, $\psi(x) < \psi(a)$ if $\sigma_p > 0$ [or $\psi(x) > \psi(a)$ if $\sigma_p < 0$] for all $x > a$. Note also that, by PBE Theorems 2 and 3, the curvature of $\psi(x)$ always must have the same sign as σ_p. Equation 4.14 with its two terms on the right side indicates that both the ion concentrations in the diffuse swarm and the geometry of the particle surface contribute to this curvature, and these two contributions always have the same sign. For a planar particle surface, of course, there is only a contribution to the curvature of $\psi(x)$ from the diffuse ion swarm.

Additional fundamental insight into the Poisson-Boltzmann equation can be gained by consideration of the special case in which each exponent on the right side of eq. 4.12 is small enough to warrant approximation of each exponential term by the first two terms of its MacLaurin expansion i.e., $[|Z_i|y \lesssim 0.5]$. Under this condition, eq. 4.12 becomes the linear differential equation:

$$\frac{d^2y}{dx'^2} + \frac{(1-2\nu)}{x'}\frac{dy}{dx'} = \beta I a^2 y(x') \tag{4.17}$$

where

$$I = \frac{1}{2}\sum_i c_{i0} Z_i^2 \tag{4.18}$$

is the *ionic strength* in the ion swarm at the reference point and β is defined in eq. 4.6. Equation 4.17, the *Debye-Hückel approximation* to eq. 4.12, has an intrinsic length scale defined by the inverse square-root of

$$\kappa^2 \equiv \beta I = \frac{1}{2}\sum_i \kappa_i^2 Z_i^2 \tag{4.19}$$

where κ_i is defined in eq. 4.7. The general solution of eq. 4.17 which tends to zero as x' tends to a reference point at infinity is:

$$y_\nu^{DH}(z) = A\, z^\nu K_\nu(z) \qquad \left(\nu = \pm\frac{1}{2}, 0\right) \tag{4.20}$$

where $z = \varepsilon x' = \kappa x$ $(\varepsilon = \kappa a)$, $K_\nu(z)$ is termed a *modified Bessel function* of the third kind of order ν,[12] and A is a constant parameter to be determined by eq. 4.13. For the three values of ν under consideration,

$$K_{1/2}(z) = K_{-1/2}(z) = (\pi/2z)^{1/2}\exp(-z) \tag{4.21a}$$

for planar or spherical particle surfaces, and, asymptotically,

$$K_0(z) \underset{z\downarrow 0}{\sim} -\ln(z/2) - 0.577216, \quad K_0(z) \underset{z\uparrow\infty}{\sim} (\pi/2z)^{1/2}\exp(-z) \tag{4.21b}$$

for a cylindrical particle surface.[13]

Equation (4.21) indicates that, in the Debye-Hückel approximation, the electrostatic potential exhibits an *exponential decay* with increasing distance from the particle surface, with κ^{-1} being *the length scale that characterizes this exponential decay*. Note that the decline of $y_\nu^{DH}(z)$ with increasing z is more pronounced, the greater is the curvature of the particle surface: planar < cylindrical < spherical. Thus, at a given $z > \varepsilon$, the electrostatic potential is strongest for a flat surface [exp $(-z)$] and weakest for a spherical surface [exp $(-z)/z$]. These differences arise as a result of the differing values of ν in z^ν, the coefficient of $K_\nu(z)$ in eq. 4.20, since all three modified Bessel functions have the same mathematical form at large z.

The explicit forms of eq. 4.20 that satisfy the boundary condition in eq. 4.10 are ($\sigma_p > 0$, $x \geq a$):

Planar ($\nu = \frac{1}{2}$)

$$y_{1/2}^{\mathrm{DH}}(x) = (\kappa\ell)^{-1}\exp[-\kappa(x-a)] \tag{4.22a}$$

Cylindrical ($\nu = 0$)

$$y_0^{\mathrm{DH}}(x) = (\kappa\ell)^{-1}\frac{K_0(\kappa x)}{K_1(\kappa a)} \tag{4.22b}$$

Spherical ($\nu = -1/2$)

$$y_{-1/2}^{\mathrm{DH}}(x) = (\kappa\ell)^{-1}\frac{\kappa a}{1+\kappa a}\frac{\exp[-\kappa(x-a)]}{(x/a)} \tag{4.22c}$$

where[12]

$$K_1(z) \equiv -\frac{dK_0}{dz} = \begin{cases} z^{-1} & z \downarrow 0 \\ \left(1+\dfrac{1}{2z}\right)K_0(z) & z \uparrow \infty \end{cases} \tag{4.23}$$

corresponding to two the limiting values of $K_0(z)$ in eq. 4.21b. Solutions of eq. 4.17 for $\sigma_p < 0$ are obtained from eq. 4.22 by the transformation in eq. 4.14. The upper bound on the values of $y_\nu^{\mathrm{DH}}(x)$ in eq. 4.22 is $(\kappa\ell)^{-1}$, the ratio of the intrinsic length scale for the spatial variation of $y(x)$ in the diffuse ion swarm to that for its spatial variation at the particle surface. If κ^{-1} takes on values typical for natural waters and ℓ takes on its minimal values for charged natural particles, $0.0015 < \kappa\ell < 0.25$, thus *invalidating* the Debye-Hückel approximation, which requires the inequality,

$$\kappa\ell > 2 \tag{4.24}$$

if the reasonable condition $y_\nu^{\mathrm{DH}}(x) < 0.5$ is to be met.[14] An alternative is to consider values of ℓ that are well above the minimal range, $0.05 < \ell < 0.25$ nm, which implies that the values of $\tilde{\sigma}$ are well below the maximal range. In particular, according to eq. 4.8, $\tilde{\sigma}$ must be less than 0.05 protonic charges per square nanometer if κ lies in the typical range, $0.03 < \kappa < 1\ \mathrm{nm}^{-1}$ and eq. 4.24 is imposed. The root cause of this difficulty with validation of the Debye-Hückel approximation is that the low diffuse-swarm potentials to which eq. 4.22 applies in turn require very high ionic strength ($\kappa > 8\ \mathrm{nm}^{-1}$) in order for there to be enough counterions near a particle surface to screen the surface charge effectively (i.e., large ℓ/κ^{-1}). If there is a natural limit as to how high the ionic strength can be, there may be no range of $y(x)$ that can be described by eq. 4.22. In general, over the normal variation of ionic strength, eq. 4.22 requires $|\sigma_p|$ values below 0.008 C m^{-2} in order to describe $y(x)$ accurately.

Although the Poisson-Boltzmann equation can be solved analytically for the cases $\nu = \frac{1}{2}$, 0, and $-\frac{1}{2}$ in the Debye-Hückel approximation, the same does not hold for eq. 4.12, with the notable exception of the case $\nu = \frac{1}{2}$ (flat particle surface), for which the "geometry" contribution to the curvature of $y(x')$ vanishes.[15,16] The scaled electrostatic potential for this case has been found exactly and analytically for symmetric electrolytes (e.g., NaCl, $CaSO_4$), 2:1 electrolytes (e.g., $CaCl_2$), and 2:1:1 mixed electrolytes (e.g., $CaCl_2$/NaCl mixtures). The scaled potential for the mixed 2:1:1 electrolyte solution, in fact, includes the other two as "end members" of a continuous sequence of mixture compositions. This potential function can be expressed:[17]

$$y_f(x) = \ln\left\{\frac{1+3f}{1+2f}\left(\tanh\left[\frac{1}{2}(1+3f)^{\frac{1}{2}}\left[\kappa(x-a)+A_f\right]^{-2\mathrm{sgn}(\sigma)}\right]\right)-\frac{f}{1+2f}\right\} \tag{4.25}$$

where $f \equiv c_{2+}/c_+$ is the mole fraction of bivalent cations in the mixture at the reference point (equal to the ratio of bivalent to monovalent cation concentrations there), $\kappa_2 \equiv \beta c_{2+}$ (with β given in eq. 4.6),

$$A_f \equiv 2(1+3f)^{-\frac{1}{2}}\tanh^{-1}\left\{\left[\frac{f+(1+2f)\exp\left[y_f(a)\right]}{1+3f}\right]^{-\frac{1}{2}\mathrm{sgn}(\sigma)}\right\} \tag{4.26}$$

with the potential $y_f(a)$ to be determined by the boundary condition at $x = a$ (eq. 4.10),

$$\ell^{-1} = \kappa\left\{\exp\left[\mathrm{sgn}(\sigma_p)y_f(a)\right]-1\right\}\left\{f+(1+2f)\exp\left[-\mathrm{sgn}(\sigma_p)y_f(a)\right]\right\}^{\frac{1}{2}} \tag{4.27}$$

and $\mathrm{sgn}(\sigma)$ is defined in eq. 4.11. The two hyperbolic functions in eqs. 4.25 and 4.26 are defined conventionally by the equations:[18]

$$\tanh(u) \equiv \frac{\exp(u)-\exp(-u)}{\exp(u)+\exp(-u)} \tag{4.28a}$$

$$\tanh^{-1}(v) \equiv \frac{1}{2}\ln\left(\frac{1+v}{1-v}\right) \tag{4.28b}$$

from which follows the inversion relation:

$$\tanh^{-1}[\tanh(u)] = u \tag{4.28c}$$

For a 1:1 electrolyte, $f = 0$ and eq. 4.25 reduces to the expression:[17]

$$\begin{aligned} y_0(x) &= -2\,\mathrm{sgn}(\sigma_p)\ln\left[\tanh\left\{\frac{1}{2}\kappa(x-a)+\tanh^{-1}\left(\exp\left[-\frac{1}{2}\mathrm{sgn}(\sigma_p)y_0(a)\right]\right)\right\}\right] \\ &= 4\tanh^{-1}\left\{\tanh\left[\frac{1}{4}y_0(a)\right]\exp[-\kappa(x-a)]\right\} \end{aligned} \tag{4.29}$$

where the second step is derived after repeated use of eq. 4.28. The potential for a symmetric electrolyte $(Z:Z)$ is then obtained by the transformation: $y_0(x) \to Zy_0(x)$, $\kappa \to Z\kappa$, where Z is the valence of the cation in the electrolyte. The potential for a 2:1 electrolyte is found by letting $f \uparrow \infty$ in eq. 4.25, noting that $f^{\frac{1}{2}}\kappa = \beta c_{2+}$. Because the solutions of the Poisson-Boltzmann equation are invariant under the transformation in eq. 4.14, the potential for a 1:2 electrolyte can be obtained from that for a 2:1 electrolyte by changing the signs of y_f and σ in eq. 4.25 while changing + to − in the subscripts that appear in the definitions of f and κ.[17]

A connection can be made between $y_0(x)$ in eq. 4.29 and $y_{\nu_2}^{DH}(z)$ in eq. 4.22a by considering the MacLaurin series for $\tanh^{-1}(v)$[18] at the lowest order in v [i.e., $\tanh^{-1}(v) \approx v$]:

$$\begin{aligned} y_0(x) &= 4\tanh^{-1}\left\{\tanh\left[\frac{1}{4}y_0(a)\right]\exp[-\kappa(x-a)]\right\} \\ &\approx 4\tanh\left[\frac{1}{4}y_0(a)\right]\exp[-\kappa(x-a)] \end{aligned} \tag{4.30}$$

which shows that, for sufficiently small values of the exponential factor [e.g., if $\exp[-\kappa(x-a)] < 0.4$], the spatial dependence of $y_0(x)$ is identical to that of $y_{\frac{1}{2}}^{DH}(z)$. Therefore, as $x \uparrow \infty$ and $y_0(x) \downarrow 0$, a rescaled Debye-Hückel approximation (with $A \equiv 4\tanh[\frac{1}{4}y_0(a)]$) becomes accurate.[14] Moreover, this result shows that the interpretation of κ^{-1} as the intrinsic length scale for the spatial variation of the diffuse-swarm potential is applicable to $y_0(x)$ at any value of x. Equation 4.30 is a good approximation at spatial points located a distance from the particle surface greater than this intrinsic length scale.

Equation 4.27 can be applied to calculate $y_f(a)$ if κ and ℓ have been specified. If $0.1 < c_+ < 100$ mol m^{-3}, then $0.03 < \kappa < 1$ nm^{-1} and, if $\tilde{\sigma} = 0.78$ nm^{-2}, typical of σ_0 for the clay mineral montmorillonite,[9] then $\ell = 0.14$ nm (eqs. 4.6 to 4.8). The corresponding values of $y_0(a)$ (1:1 electrolyte) then follow:

κ^{-1} (nm)	$y_0(a)$
30	10.73
6	7.48
5	7.12
4	6.67
3	6.10
2	5.29
1	3.93

These values, which show an increase as the intrinsic length scale grows (and c_+ decreases), are much larger than the electrostatic potentials inferred from electrokinetic measurements on smectites, which suggest $y_0(a) \approx 2$ for $\kappa^{-1} < 30$ nm.[19]

This discrepancy brings to mind the question of the inherent accuracy of eq. 4.25 as a description of a diffuse ion swarm imbedded in a dielectric continuum, which can be assessed by comparing the resulting predictions of ion concentrations (eq. 4.4) with the results of exact Monte Carlo calculations for a system of mobile hard-sphere ions immersed in a continuum dielectric fluid that contacts a planar charged surface.[20] Some of these results are reproduced in fig. 4.1, which shows the concentration of cations (c_+) or anions (c_-) in a 1:1 electrolyte plotted against a renormalized, scaled distance from the charged surface (x/d_0, with $a \equiv d_0/2$). The value of the "bulk" solution concentration (c_0) is 100 mol m^{-3}, and σ_p ($\equiv \sigma$) takes on two values (equivalent to $\tilde{\sigma} = 0.54$ and 1.3 nm^{-2}) that bracket the normal range of permanent surface charge density for smectites.[9] (Note that these values are much larger than that for which the Debye-Hückel approximation is valid.) The graphs demonstrate that eq. 4.29 is indeed a self-

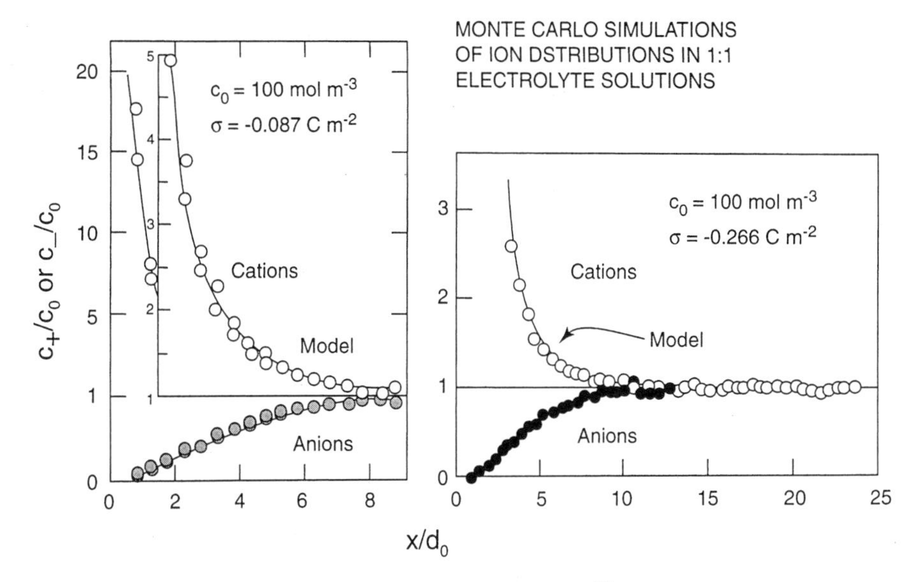

Fig. 4.1 Comparisons between Monte Carlo simulations[20] (circles) of diffuse swarm ion distributions near a planar charged surface with the predictions of eqs. 4.4 and 4.29 (curves). The hard-sphere diameter of the electrolyte ions (d_0) is 0.425 nm.

consistent model of a 1:1 electrolyte solution comprising mobile hard-sphere ions and a continuum dielectric solvent near a uniformly charged plane, at least for c_0 up to 100 mol m^{-3}. Above this concentration, however, the model fails to describe the effect of ion-ion correlations on local ion concentrations. Because of this neglect of ion-ion correlations, eq. 4.25 is completely inaccurate for electrolyte solutions containing bivalent ions at concentrations as low as 5 mol m^{-3}. More comprehensive diffuse ion swarm models that include these correlations provide an excellent description of the swarm on planar surfaces bearing adsorbed bivalent cations.[1,2]

These considerations encourage further exploration of the implications of eq. 4.29 for $c_+ \lesssim 100$ mol m^{-3}. Modified Gouy-Chapman theory in fact provides an accurate description, for 1:1 electrolytes, of two characteristic properties of a diffuse ion swarm, *counterion condensation* and *coion exclusion*, both of which are evident in fig. 4.1. The first property is signaled by the very high concentration of cations (about 20 times c_+) at the distance of closest approach for the surface with $\sigma = -0.087$ C m^{-2} in fig. 4.1. Even at this relatively low surface charge density, the counterion swarm is compressed. In more concrete terms, let $\varphi(x)$ be the fraction of the surface charge density σ that is screened by diffuse swarm counterions occupying a region between the distance of closest approach to the particle surface and a point located at $x > a$. This fraction will of course depend on the surface charge density and the "bulk" electrolyte concentration. But counterion condensation is said to exist if $\varphi(x)$ does not vanish even as the bulk electrolyte concentration goes to zero in a limiting sense.[21] This means that a fraction of the counterions remains close to the charged surface even if the aqueous solution which contacts it becomes infinitely dilute and the intrinsic length scale κ^{-1} becomes infinitely large. Evidently the electrostatic attraction to the charged surface is strong enough to prevent the counterions from moving out to the bulk solution as the latter decreases in concentration. This "condensation" effect pertains strictly to the electrostatic forces that govern the diffuse swarm. Therefore, the "condensed" counterions remain mobile and do *not* form surface complexes.

The quantitative side of counterion condensation is exposed by the definition:[21]

$$\begin{aligned} \varphi(x) &\equiv (F/|\sigma|)\int_a^x [c(s) - c_0]ds \\ &= (Fc_0/|\sigma|)\int_a^x \{\exp|y(s)| - 1\}ds \\ &= \frac{1}{2}(\kappa^2/\ell)\int_{|y_0(a)|}^{|y_0(x)|} \frac{[\exp|y| - 1]}{dy/dx}dy \\ &= \frac{1}{2}(\kappa/\ell)\int_{|y_0(x)|}^{|y_0(a)|} \exp(y/2)dy \\ &= (\kappa/\ell)[\exp|y_0(a)/2| - \exp|y_0(x)/2|] \end{aligned} \tag{4.31}$$

for monovalent counterions (eq. 4.30). Equation 4.4 has been used to obtain the second step, eqs. 4.6 to 4.8 were needed for the third step, and eq. 4.27 (with $f = 0$) was used to get the fourth step. The limiting value of the right side of eq. 4.31 as $\kappa \downarrow 0$ can be deduced with the help of eqs. 4.27 and 4.28:[21]

$$\lim_{\kappa\downarrow 0} \varphi(x) = 1 - \{1 + [(1/2\ell)(x - a)]\}^{-1} \tag{4.32}$$

Values of this limiting fraction of neutralized surface charge appear in the table below for two values of $\tilde{\sigma}$ (corresponding to those in fig. 4.1) and several values of (x-a):

$\lim_{\kappa\downarrow 0} \varphi(x)$		
$\tilde{\sigma} = 0.54\ \text{nm}^{-2}$	$\tilde{\sigma} = 1.3\ \text{nm}^{-2}$	$[x - a]$ (nm)
0.20	0.37	0.1
0.33	0.54	0.2
0.55	0.75	0.5
0.71	0.85	1.0
0.79	0.90	1.5

This table shows that, even at "infinite dilution," the counterions cluster strongly near the particle surface, screening more than 75% of the surface charge within about 1 nm of the distance of closest approach. This result can be appreciated fully after noting that, in respect to "indifferent electrolytes" (section 1.4), the diameter of Cl^- (assumed unsolvated) is 0.362 nm, whereas the diameters of (solvated) Li^+, Na^+, and K^+ are 0.346, 0.374, and 0.410 nm, respectively.[7]

Therefore, between half and three-fourths of the surface charge is screened by counterions within an ionic diameter of the distance of closest approach to the particle surface! This very pronounced clustering of the counterions to effect screening certainly would be as effective as neutralization by surface complexation at balancing the surface charge. Note that, quite generally, half the surface charge is already balanced by diffuse swarm counterions within the rather small distance 2ℓ ($0.17 < 2\ell < 0.41$ nm) from the particle surface.

Coion exclusion is represented in fig. 4.1 by the anion concentrations within 10 ionic diameters of the particle surface, which are seen to fall below the reference-point value, c_0. This depletion of coions by repulsion from a region near a charged particle surface is quantified by measurements of the *exclusion volume*,

$$V_{ex} = -n_{co}^{(w)}/c_0 \tag{4.33}$$

where $n_{co}^{(w)}$ is the (negative) surface excess of coions (eq. 1.4) and c_0 is their "bulk" concentration. Modified Gouy-Chapman theory provides a mathematical model of V_{ex} through the parameter d_{ex}, the *exclusion distance*. For a 1:1 electrolyte:[22]

$$\begin{aligned} d_{ex} &\equiv \int_a^{\infty} \left[1 - \frac{c(x)}{c_0}\right] dx + \frac{1}{2} d_0 \\ &= \int_{y(a)}^{0} \frac{\{1 - \exp[-|y|]\}dy}{dy/dx} + \frac{1}{2} d_0 \\ &= \frac{2}{\kappa}\{1 - \exp[-|y_0(a)|/2]\} + \frac{1}{2} d_0 \end{aligned} \tag{4.34}$$

following a procedure very akin to that used in developing eq. 4.31. By definition, d_{ex} is a weighted sum of distance intervals in the diffuse swarm, with the weighting factor being the fraction of coions *not* to be found in the open interval $(x, x + dx)$, with x itself in the semiclosed interval $[a - d_0/2, \infty)$. In this sense, d_{ex} may be interpreted as an average distance from the geometric boundary of a charged particle [i.e., $a - (d_0/2)$] over which coions are excluded. This average distance is approximately twice the intrinsic length scale, κ^{-1}. It is related to V_{ex} by the definition,

$$V_{ex} \equiv a_s^{ex} d_{ex} \tag{4.35}$$

where a_s^{ex} is the area of the portion of an adsorbent surface which excludes coions.[23]

Values of d_{ex} based on the table of κ^{-1} and $y_0(a)$ given above are readily calculated with eq. 4.34:

κ^{-1}	d_{ex} (nm)
30	60
6	12
5	10
4	8
3	6
2	4
1	2

For these relatively large potentials at the distance of closest approach [$y(a) > 4$], the exclusion distance d_{ex} is equal to $2/\kappa$ ($d_0/2 \approx 0.2$ nm is negligible). Measured values of $d_{ex} = V_{ex}/a_s^{ex}$ based on independent measurements of V_{ex} and a_s^{ex} are in good agreement with eq. 4.34 for 2:1 clay minerals.[24]

These applications further illustrate the significance of the intrinsic scales of electrostatic potential and length that arise from the Poisson-Boltzmann equation. The scale factor RT/F is a criterion for *small potentials*, in that the condition $|\psi(x)| < RT/F$ leads to the Debye-Hückel approximation (eqs. 4.17 and 4.20). The scale factor κ^{-1} (eqs. 4.7 and 4.19) gives a measure of *coion exclusion*, in that $2\kappa^{-1}$ is the average distance from a charged particle surface over which coions are effectively prevented from entry (eq. 4.34). The scale factor ℓ, on the other hand, is a measure of *counterion accumulation*, in that 2ℓ is the distance from a charged particle surface over which half the surface charge is screened by counterions (eq. 4.32). This last scale factor, intimately related to the magnitude of the surface charge (eq. 4.8), is perhaps the best measure of the spatial extent (or "thickness") of the diffusion ion swarm, whereas the length scale κ^{-1} is related only to the bulk population of electrolyte ions.

4.2 Modeling Surface Complexes

As described in chapter 2, abundant and convincing spectroscopic evidence exists for adsorbed ions immobilized into outer-sphere or inner-sphere surface complexes. These surface species, which contribute to the Stern layer charge component of net particle surface charge (eq. 1.42), form on particles analogously to the way they react with solute ligands and, therefore, they are modeled using chemical reactions, as is illustrated throughout chapter 3 in the context of a mathematical description of adsorption kinetics. The modeling of surface complexation equilibria, however, is less complicated than kinetics modeling, because the number of species to be considered is necessarily less than for time-dependent processes, and this simplification perforce leads to a lesser number of parameters required for a chemical description. This advantage notwithstanding, the modeling of surface complexation reactions remains daunted by the onerous task of developing reliable expressions for single-species surface activity coefficients, which are essential both to the measurement and to the application of surface complexation equilibrium constants. The parallel issue of aqueous species activity coefficients that arises in the modeling of soluble complex formation is largely obviated by the existence of excellent semi-empirical equations for the activity coefficients of all species except metal complexes with natural organic ligands.[25]

Current descriptions of surface complexation equilibria thus share a common foundation in the theory of ion association, and they suffer equally under the burden of finding an accurate model of surface species activity coefficients. For the purpose of explicating generic model structure, then, one representative model will suffice for consideration, leaving the details of the others and the criteria used in applying them to the surface chemistry of natural particles to specialized reviews.[26] The pathway to be followed in constructing any of these models is universal: (1) development of *chemical equations* to describe adsorption reactions;

(2) application of *constraints* imposed by mass balance, surface charge balance, and chemical equilibrium; and (3) introduction of *molecular hypotheses* to adduce a model expression for surface species activity coefficients. Available models differ substantially only in respect to steps (1) and (3).[26]

A widely known and frequently utilized surface complexation approach is exemplified by the *Triple Layer Model*,[27] whose initial inspiration may be found in a celebrated paper by O. Stern[27] that introduced what eventually became the Stern layer charge contribution to the electrical double layer. Typical representations of adsorption reactions in this model can be expressed in the same notation as is used in chapter 3:[28]

$$\equiv \mathrm{SOH} + \mathrm{H}^+ \equiv \mathrm{SOH}_2^+ \rightleftarrows \equiv \mathrm{SOH}_2^+ \tag{4.36a}$$

$$\equiv \mathrm{SOH} \rightleftarrows \equiv \mathrm{SO}^- + \mathrm{H}^+ \tag{4.36b}$$

$$\equiv \mathrm{SOH} + \mathrm{M}^{m+} \rightleftarrows \equiv \mathrm{SOM}^{(m-1)} + \mathrm{H}^+ \tag{4.37a}$$

$$2 \equiv \mathrm{SOH} + \mathrm{M}^{m+} \rightleftarrows \equiv (\mathrm{SO})_2\mathrm{M}^{(m-2)} + 2\mathrm{H}^+ \tag{4.37b}$$

$$\equiv \mathrm{SOH} + \mathrm{H}_\ell\mathrm{L} = \equiv \mathrm{SH}_{(\ell-i)}\mathrm{L}^{(1-i)} + \mathrm{H}_2\mathrm{O} + (i-1)\mathrm{H}^+ \quad (i = 1, \cdots, \ell) \tag{4.38a}$$

$$2 \equiv \mathrm{SOH} + \mathrm{H}_\ell\mathrm{L} = \equiv \mathrm{S}_2\mathrm{H}_{(\ell-i)}\mathrm{L}^{(2-i)} + 2\mathrm{H}_2\mathrm{O} + (i-2)\mathrm{H}^+ \quad (i = 1, \cdots, \ell) \tag{4.38b}$$

$$\equiv \mathrm{SOH} + \mathrm{M}^{m+} \rightleftarrows \equiv \mathrm{SO}^- \cdots \mathrm{M}^{m+} + \mathrm{H}^+ \tag{4.39a}$$

$$\equiv \mathrm{SOH} + \mathrm{M}^{m+} + \mathrm{H}_2\mathrm{O} \rightleftarrows \equiv \mathrm{SO}^- \cdots \ \mathrm{MOH}^{(m-1)} + 2\mathrm{H}^+ \tag{4.39b}$$

$$\equiv \mathrm{SOH} + \mathrm{H}^+ + \mathrm{L}^{\ell-} \rightleftarrows \equiv \mathrm{SOH}_2^+ \cdots \mathrm{L}^{\ell-} \tag{4.40a}$$

$$\equiv \mathrm{SOH} + 2\mathrm{H}^+ + \mathrm{L}^{\ell-} \rightleftarrows \equiv \mathrm{SOH}_2^+ \cdots \mathrm{LH}^{(\ell-1)-} \tag{4.40b}$$

$$\equiv \mathrm{SOH} + \mathrm{C}^{c+} \rightleftarrows \equiv \mathrm{SO}^- \cdots \mathrm{C}^{c+} + \mathrm{H}^+ \tag{4.41a}$$

$$\equiv \mathrm{SOH} + \mathrm{H}^+ + \mathrm{A}^{a-} \rightleftarrows \equiv \mathrm{SOH}_2^+ \cdots \mathrm{A}^{a-} \tag{4.41b}$$

where, as in chapters 2 and 3, the three dots separating an adsorbate from a surface site to which it is bound represent outer-sphere complexation. Equations 4.36 are protonation/proton dissociation reactions; eqs. 4.37 and 4.39 are metal adsorption reactions; eqs. 4.38 and 4.40 are ligand adsorption reactions; and eqs. 4.41 are adsorption reactions for the ions in a background electrolyte, $\mathrm{C}_m\mathrm{A}_n$, with $m/n = a/c$. The concentrations of these latter ions, almost universally taken simply to be monovalent species of an indifferent electrolyte (see section 1.4), are assumed much larger than those of the other adsorbing ions, M^{m+} and $\mathrm{L}^{\ell-}$. Each component in the adsorption reactions is then subject to a mass balance constraint:

$$\mathrm{SOH_T} = [\mathrm{SOH}] + \left[\mathrm{SOH_2^+}\right] + [\mathrm{SO^-}] + \sum_{n=1}^{2} n[(\mathrm{SO})_n\mathrm{M}]$$

$$+ \sum_{n=1}^{2}[\mathrm{SO^-} \cdots \mathrm{M(OH)}_n] + \sum_{n=1}^{2}\sum_{i=1}^{\ell} n[S_n\mathrm{H}_{\ell-i}\mathrm{L}] \tag{4.42}$$

$$+ \sum_{n=1}^{2}[\mathrm{SOH_2^+} \cdots \mathrm{H}_n\mathrm{L}] + \left[\mathrm{SO^-} \cdots \mathrm{C^+}\right] + \left[\mathrm{SOH_2^+} \cdots \mathrm{A^-}\right]$$

$$\mathrm{M_T} = \left[\mathrm{M}^{m+}\right] + \sum_{n=1}^{2}\left\{[(\mathrm{SO})_n\mathrm{M}] + \left[\mathrm{SO^-} \cdots \mathrm{M(OH)}_n\right]\right\} \tag{4.43}$$

$$\mathrm{L_T} = \left[\mathrm{L}^{\ell-}\right] + \sum_{n=1}^{2}\left\{\sum_{i=1}^{2}[S_n\mathrm{H}_{\ell-i}\mathrm{L}] + \left[\mathrm{SOH_2^+} \cdots \mathrm{H}_n\mathrm{L}\right]\right\} \tag{4.44}$$

$$\mathrm{C_T} = \left[\mathrm{C^+}\right] + \left[\mathrm{SO^-} \cdots \mathrm{C^+}\right] \tag{4.45}$$

$$\mathrm{A_T} = [\mathrm{A^-}] + \left[\mathrm{SOH_2^+} \cdots \mathrm{A^-}\right] \tag{4.46}$$

where [ ] for surface species is defined as the product of surface excess ($n_i^{(w)}$) with adsorbent solids concentration (c_s). (The subscript "eq" used in chapter 3 is dropped in this chapter to simplify the notation.) Additional concentration terms will appear on the right sides of eqs. 4.43 to 4.46 if the metals and ligands form soluble complexes or solid precipitates.[29]

Equilibrium constants for the reactions in eqs. 4.36 to 4.41 are formulated as is shown in eq. 2.2 for the special case, S = Fe, $\mathrm{M}^{m+} = \mathrm{Cd}^{2+}$. These equilibrium constants can be measured by conventional (if arduous) methods in chemical thermodynamics,[30] but this lengthy process is circumvented in the Triple Layer Model by *defining* the surface species activity coefficients to have a specific parametric model form. While such an approach obviates the need for large data sets to establish the composition dependence of conditional equilibrium constants, it also makes the corresponding equilibrium constants model-dependent and, therefore, not necessarily of strict thermodynamic significance. For this reason, the notation,[28]

$$K_M^1(\mathrm{int}) = \frac{(\equiv \mathrm{SO^-} \cdots \mathrm{M})\left(\mathrm{H^+}\right)}{(\equiv \mathrm{SOH})\left(\mathrm{M}^{m+}\right)} \tag{4.47}$$

will be used to denote the "intrinsic" equilibrium constant for the reaction in eq. 4.39a instead of, say, the more generic form on the left side of eq. 2.2a. (As usual, parentheses around a chemical species signify its thermodynamic activity.)

The Triple Layer Model leaves untouched the single-ion activities of aqueous species that appear in equilibrium constants [e.g., (M^{m+})], with the implicit understanding that the Davies equation or another suitable ion-association model equation will be applied to convert molar concentration into activities.[25,31] For surface species, the Triple Layer Model *ansatz*,[27,32]

$$f_{i\lambda} = f_{i\lambda}^{0} \exp(Z_i F \psi_\lambda / RT) \tag{4.48}$$

is developed for a surface species i, with valence Z_i, residing in a plane λ of the electrical double layer (see fig. 4.2). The first factor on the right side of eq. 4.48 is the value of the activity coefficient $f_{i\lambda}$ when the "surface potential" ψ_λ equals zero. This activity coefficient is then conventionally assumed to be independent of the two indices i and λ (i.e., it is assumed to be the same for all surface species) in the Triple Layer Model and is given the arbitrary value 1.0 simply to reduce the number of adjustable parameters. The "surface potential" ψ_λ in eq. 4.48 is an electrostatic potential assigned to one of three planes near a charged particle surface (fig. 4.2). These planes contain inner-sphere surface complexes ($\lambda \equiv 0$), outer-sphere surface complexes ($\lambda \equiv \beta$), or the boundary of the diffuse ion swarm ($\lambda \equiv d$). The underlying concept here is not unlike mean field theory, which provides the guiding principle for the Gouy-Chapman model.[6,32] In the Triple Layer Model, each adsorbed species is acted upon by an average coulomb field arising from the presence of charged surface sites and the charges of other adsorbed species. The corresponding electrostatic potential of this field takes on discrete values in three hypothesized "planes of adsorption" for surface complexes and diffuse swarm species. Thus the Boltzmann factor on the right side of eq. 4.48 provides an activity-coefficient correction[25,32] for the effects of the mean electrostatic field near a charged particle. Evidently, the first factor $f_{i\lambda}$ contains all other corrections, including those for short-range interactions, and these corrections are considered to be one and the same for all surface species described by the Triple Layer Model, for the sake of simplicity.

Electrostatic potentials in condensed phases (e.g., aqueous solutions) are not measurable.[33] Therefore, the "surface potential" ψ_λ in eq. 4.48 must be itself modeled in terms of measurable quantities. This requirement is met in the Triple Layer Model by invoking two *charge-potential relationships*:[26,27]

$$\sigma_0^{\mathrm{TLM}} = C_1(\psi_0 - \psi_\beta) \qquad \sigma_\mathrm{d} = C_2(\psi_d - \psi_\beta) \tag{4.49}$$

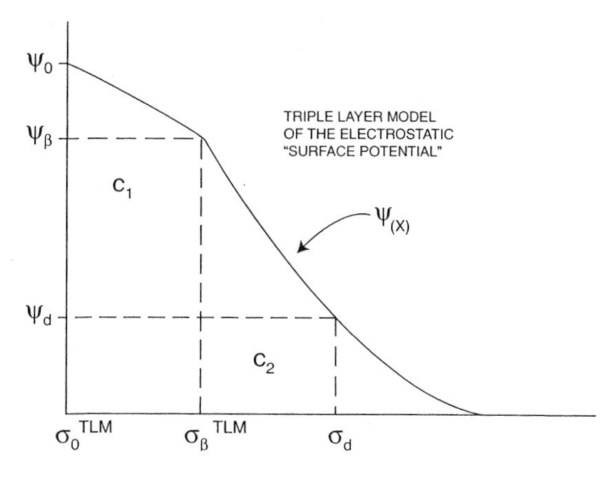

Fig. 4.2 Schematic illustration of electrical double layer structure in the Triple Layer Model. The surface charge density $\sigma_0^{\mathrm{TLM}} = \sigma_{\mathrm{in}} + \sigma_{\mathrm{IS}}$ (eqs. 1.39 and 1.40), while $\sigma_\beta^{\mathrm{TLM}} = \sigma_{\mathrm{OS}}$ (eq. 1.40). The capacitance densities C_1 and C_2 relate σ_0^{TLM} and σ_d (eq. 4.50) to the "surface potentials," ψ_0, ψ_β, and ψ_d (eq. 4.49).

where C_1 and C_2 are adjustable parameters that play the role of interfacial capacitance densities (units of farads per square meter). Equations 4.49 are analogous to the charge-potential relationships for a parallel-plate capacitor, in keeping with the assignment of ψ_λ to discrete planes (fig. 4.2). They are augmented in the Triple Layer Model by imposing the charge-potential relationship in modified Gouy-Chapman theory that can be derived from combining eqs. 4.9 and 4.29 for the case $\sigma_p = -\sigma_d$, $\psi = \psi_d$:

$$\sigma_\mathrm{d} = -(4\varepsilon_0 \mathrm{DRT} c_0)^{\frac{1}{2}} \sinh(F\psi_d/2RT) \tag{4.50}$$

where c_0 is the concentration of a 1:1 background electrolyte at the reference point for the electrostatic potential and all other symbols are defined in section 4.1. (Evidently, eq. 4.50 can be generalized by utilizing eq. 4.25 instead of eq. 4.29.) Equations 4.49 and 4.50, along with the special case of eq. 1.43 that represents charge balance in the Triple Layer Model,[26,27]

$$\sigma_0^{\mathrm{TLM}} + \sigma_\beta^{\mathrm{TLM}} + \sigma_\mathrm{d} = 0 \tag{4.51}$$

provide sufficient conditions for the determination of the model potentials ψ_0, ψ_β, and ψ_d.[29] [Note that $\sigma_\beta^{\mathrm{TLM}}$ is to be identified with σ_{os} (eq. 1.40), while σ_0^{TLM} in eq. 4.51 is to be identified with $(\sigma_{\mathrm{in}} + \sigma_{\mathrm{IS}})$ in eqs. 1.39 and 1.40, and not solely with the structural charge density σ_0 (eq. 1.29)—a small matter of conflicting notation!]

The conditional equilibrium constant associated with eq. 4.47 now can be expressed in the form:[28]

$$K^1_{\mathrm{Mc}} = \frac{[\equiv \mathrm{SO}^- \cdots \mathrm{M}](\mathrm{H}^+)}{[\equiv \mathrm{SOH}](\mathrm{M}^{m+})} = K^1_{\mathrm{M}}(\mathrm{int}) \exp\left[-F\left(m\psi_\beta - \psi_0\right)/RT\right] \tag{4.52}$$

on reference to eq. 4.48:

$$(\equiv \mathrm{SO}^- \cdots \mathrm{M}) = [\equiv \mathrm{SO}^- \cdots \mathrm{M}] \exp(-F\psi_0/RT) \exp(\psi_\beta/RT) \tag{4.53a}$$

$$(\equiv \mathrm{SOH}) = [\equiv \mathrm{SOH}] \tag{4.53b}$$

The Reference State for the activity coefficients in eqs. 4.52 and 4.53 is defined by $\psi_\beta = \psi_0 = 0$ and infinite dilution of the aqueous solution phase.[26,27] Therefore, activities are equal numerically to concentrations (in mol dm^{-3}) under the hypothetical conditions of zero particle charge and ionic strength.[34] By rearrangement of eq. 4.52 and judicious use of eq. 4.48, one finds model equations for the intrinsic equilibrium constants that correspond to the adsorption reactions in eqs. 4.36 to 4.41:[28]

$$K_+(\text{int}) = \frac{[\text{SOH}_2^+]}{[\text{SOH}](\text{H}^+)} \exp[F\psi_0/RT] \tag{4.54a}$$

$$K_-(\text{int}) = \frac{[\text{SO}^-](\text{H}^+)}{[\text{SOH}]} \exp[-F\psi_0/RT] \tag{4.54b}$$

$$K_\text{M}^1(\text{int}) = \frac{\left[\text{SOM}^{(m-1)}\right](\text{H}^+)}{[\text{SOH}](\text{M}^{m+})} \exp[(m-1)F\psi_0/RT] \tag{4.55a}$$

$$K_\text{M}^2(\text{int}) = \frac{\left[(\text{SO})_2\text{M}^{(m-2)}\right](\text{H}^+)^2}{[\text{SOH}]^2(\text{M}^{m+})} \exp[(m-2)F\psi_0/RT] \tag{4.55b}$$

$$K_\text{L}^i(\text{int}) = \frac{\left[\text{SH}_{(\ell-i)}\text{L}^{(1-i)}\right](\text{H}^+)^{(i-1)}}{[\text{SOH}](\text{H}_\ell\text{L})} \exp[(1-i)F\psi_0/RT] \tag{4.56a}$$

$$K_\text{L}^{2i}(\text{int}) = \frac{\left[\text{S}_2\text{H}_{(\ell-i)}\text{L}^{(2-i)}\right](\text{H}^+)^{(i-2)}}{[\text{SOH}]^2(\text{H}_\ell\text{L})} \exp[(2-i)F\psi_0/RT] \tag{4.56b}$$

$$K_\text{M}^1(\text{int}) = \frac{[\text{SO}^- \cdots \text{M}^{m+}](\text{H}^+)}{[\text{SOH}](\text{M}^{m+})} \exp\left[F(m\psi_\beta - \psi_0)/RT\right] \tag{4.57a}$$

$$K_\text{M}^2(\text{int}) = \frac{\left[\text{SO}^- \cdots \text{MOH}^{(m-1)}\right](\text{H}^+)^2}{[\text{SOH}](\text{M}^{m+})} \exp\left[F((m-1)\psi_\beta - \psi_0)/RT\right] \tag{4.57b}$$

$$K_L^1(\text{int}) = \frac{[\text{SOH}_2^+ \cdots \text{L}^{\ell-}]}{[\text{SOH}](\text{H}^+)(\text{L}^{\ell-})} \exp\left[F(\psi_0 - \ell\psi_\beta)/RT\right] \tag{4.58a}$$

$$K_\text{L}^2(\text{int}) = \frac{\left[\text{SOH}_2^+ \cdots \text{LH}^{(\ell-1)^-}\right]}{[\text{SOH}](\text{H}^+)^2(\text{L}^{\ell-})} \exp\left[F(\psi_0 - (\ell-1)\psi_\beta)/RT\right] \tag{4.58b}$$

$$K_\text{C}(\text{int}) = \frac{[\text{SO}^- \cdots \text{C}^+](\text{H}^+)}{[\text{SOH}](\text{C}^+)} \exp\left[F(\psi_\beta - \psi_0)/RT\right] \tag{4.59a}$$

$$K_\text{A}(\text{int}) = \frac{[\text{SOH}_2^+ \cdots \text{A}^-]}{[\text{SOH}](\text{H}^+)(\text{A}^-)} \exp\left[F(\psi_0 - \psi_\beta)/RT\right] \tag{4.59b}$$

Each chemical reaction (eqs. 4.36 to 4.41) yields a surface complex whose formation from ≡SOH and aqueous free ionic species (taken as components in the chemical system under consideration) is constrained by the value of an intrinsic equilibrium constant (eqs. 4.54 to 4.59). The concentrations of the components, in turn, are constrained by mass-balance equations (eqs. 4.42 to 4.46). The "surface potentials" that appear on the right sides of eqs. 4.4 to 4.59 are constrained by the model charge-potential relationships (eqs. 4.49 and 4.50) along with the charge-balance equation (eq. 4.51). Computational closure is ensured by the fact that the surface charge components can be expressed in terms of surface species concentrations.[26,29,35]

Aside from the demonstrated ability of the Triple Layer Model to fit adsorption edge, envelope, or isotherm data,[26,27] there are inherent chemical properties of the model that transcend its application to any particular natural particle system. For example, ≡SOH must represent an "average surface hydroxyl group" in some sense, with the meaning of "average" being strictly operational (i.e., *one* intrinsic equilibrium constant per surface complex is assumed sufficient, even if, on the molecular level, several types of surface hydroxyl may exist on an adsorbent, as discussed in chapter 1). A more subtle property of the model is its supposition that *the background electrolyte always behaves indifferently*. According to PZC Theorem 2, this supposition means p.z.n.c. and p.z.c. are always the same (section 1.4). To see this unambiguously, we note that $\sigma_d = 0$ at the p.z.c. (eq. 1.43 and the definition of p.z.c.). By eqs. 4.50 and 4.9 (with $\sigma_p \equiv -\sigma_d$), both ψ_d and its spatial derivative then vanish at the "d-plane" (fig. 4.2). But this means that $\psi_\beta = 0$ by the second of eqs. 4.49. If $\psi_\beta = 0$, then $\sigma_\beta^{TLM} = 0$, [36] which is the same as $\sigma_S = 0$ because of the Triple Layer Model hypothesis that the background electrolyte ions form only outer-sphere surface complexes (eqs. 4.41). The vanishing of σ_S is the necessary and sufficient condition for equality between p.z.n.c. and p.z.c., according to PZC Theorem 2. Thus, the charge-potential relationships in the Triple Layer Model are sufficient to render any background electrolyte it describes indifferent. Note also that these relationships tacitly assume that diffuse-swarm ions never approach a particle surface more closely than the "d-plane;" i.e., all ions approaching nearer to the surface than the "d-plane" are assumed to be instantly immobilized into surface complexes.

For an adsorbent without structural charge, equality between p.z.c. and p.z.n.c. implies equality between p.z.c. and p.z.n.p.c. (PZC Theorem 1 in section 1.4). In this important special case, which applies to aluminum and iron oxides as well as to humus, a relationship exists between p.z.n.p.c. and the intrinsic equilibrium constants in Eqs. 4.54, given that only a 1:1 background electrolyte is present to provide adsorptive ions other than H^+ or OH^-. Under this condition, since the 1:1 background electrolyte is *always* indifferent, p.z.n.p.c. is termed a *pristine point of zero charge*, p.p.z.c.[37] Equations 4.54 can be combined so as to cancel [≡SOH]:

$$\frac{K_+(\text{int})}{K_-(\text{int})} = \frac{[\equiv \text{SOH}_2^+]}{[\equiv \text{SO}^-]}(\text{H}^+)^{-2}\exp(2F\psi_0/RT) \tag{4.60}$$

At pH = p.p.z.c., $\psi_0 = 0$ because $\sigma_0^{TLM}(=\sigma_H)$ is zero,[36] $(\text{H}^+) = 10^{\text{-p.p.z.c.}}$, and $[\equiv \text{SOH}_2^+] = [\equiv \text{SO}^-]$ because—and *only* because—$\sigma_\beta^{TLM}(=\sigma_{OS})$ is also zero.[38] Therefore, eq. 4.60 can be rearranged to yield the relationship:

$$\text{p.p.z.c.} = \frac{1}{2}[\log K_+(\text{int}) - \log K_-(\text{int})] \tag{4.61}$$

Equation 4.61 is the Triple Layer Model prediction of the pristine point of zero charge in terms of the model protonation/proton dissociation equilibrium constants. Since these latter parameters are independent of ionic strength, so is p.p.z.c. according to the model.[39] Note also that eq. 4.61 implies equality between

K_C(int)/K_(int) and K_A(int)/K_+(int) as an additional constraint on the Triple Layer Model (eqs. 4.54 and 4.59).

In section 1.3 (see also problem 3 in Chapter 1) the relation between Pauling Rule 2 and the Brønsted acidity of surface hydroxyl groups on metal oxides is described, with the inference drawn that protonation of a surface oxygen ion that is singly bonded to a metal cation in the oxide structure is very likely because this configuration leaves the oxygen ion valence highly unsaturated. The degree of unsaturation depends on the bond valence of the metal cation in the oxide. Therefore, one may expect that log K_+(int) would show an inverse relationship with this bond valence. More particularly, the inverse relationship should involve the bond valence of the metal cation divided by its distance to the adsorbed proton, this ratio being a measure of the repulsive coulomb potential acting on the proton.[40] Completing this simple conceptual picture of the affinity of a surface oxygen ion for a proton, one may conjecture that log K_+(int) would show a positive correlation with the inverse of the dielectric constant of the metal oxide on the grounds that small values of this parameter would favor binding of a proton to a surface O as opposed to a water O (i.e., adsorption versus solvation).

Statistical correlations of this character have indeed been discovered for K_+(int) and K_-(int):[40]

$$\log K_+(\text{int}) = 21.1158(1/D_S) - 4.92608(s/R_{SOH}) + 12.9181 \quad (4.62a)$$

$$\log K_-(\text{int}) = -21.1158(1/D_S) + 3.65688(s/R_{SOH}) - 16.4551 \quad (4.62b)$$

where D_S is the dielectric constant of the metal oxide, s is the bond valence of its metal cation (eq. 1.22), and $R_{SOH} = R_{SO} + 0.101$ is the distance between the metal cation and the proton in a ≡SOH group (in nm), where R_{SO} is the metal cation-oxygen ion bond length. As an example of the predictive quality of eqs. 4.62, consider corundum (α-Al_2O_3), whose Al–O bond valence is given by eq. 1.24. There are two Al–O bond lengths in the oxide, leading to an average value of 1.729 nm^{-1} for s/R_{AlOH}. The dielectric constant of corundum is 10.43.[40] Therefore,

$$\log K_+(\text{int}) = 6.4 \qquad \log K_-(\text{int}) = -12.2$$

according to eqs. 4.62. Similarly, for goethite (α-FeOOH), with $D_S = 11.7$ and $S/R_{FeOH} = 1.645$,[40]

$$\log K_+(\text{int}) = 6.6 \qquad \log K_-(\text{int}) = -12.2$$

These latter values may be compared with log K_+(int) = 4.8 ± 0.6 and log K_-(int) = −10.9 ± 0.6, the averages of values available in compilations of Triple Layer Model results.[26] The "order-of-magnitude" quality of eqs. 4.62 is apparent.

Similar correlation results are available for the Triple Layer Model equilibrium constants, K_C(int)/K_-(int) and K_A(int)/K_+(int) (eqs. 4.54 and 4.59). In this case, the only correlating independent variable is D_S, with the two regression parameters then being dependent on the electrolyte ion C^+ or A^-. Some predictions of the linear regression of log[K_C(int)/K_-(int)] or log [K_A(int)/K_+(int)] on $1/D_S$ are listed in the table below (values of D_S in parentheses):[41]

	$\log[K_C(\text{int})/K_-(\text{int})]$			$\log[K_A(\text{int})/K_{+(\text{int})}]$	
Adsorbent oxide	Li^+	Na^+	K^+	Cl^-	NO_3^-
α-FeOOH (11.7)	2.70	2.49	2.30	2.23	1.97
α-Fe_2O_3 (25.0)	3.16	2.87	2.62	2.81	2.25
$Fe(OH)_3$(am) (11.7)	2.70	2.49	2.31	2.24	1.97
α-Al_2O_3 (10.4)	2.59	2.40	2.24	2.10	1.91
SiO_2(am) (3.8)	0.91	1.00	1.11	0.56	1.17
β-MnO_2 (10^4)	3.56	3.20	2.88	3.31	2.49

The ratios of intrinsic equilibrium constants in this table are equal to the Triple Layer Model equilibrium constants for adsorption reactions developed by combining eqs. 4.36 and 4.41:

$$\equiv SO^- + C_+ = \equiv SO^- \cdots C^+ \qquad {}^*K_C(\text{int}) = K_C(\text{int})/K_-(\text{int}) \tag{4.63a}$$

$$\equiv SO_2^+ + A^- = \equiv SOH_2^+ \cdots A^- \qquad {}^*K_A(\text{int}) = K_A(\text{int})/K_+(\text{int}) \tag{4.63b}$$

By way of comparison with the table above, some paired measured values of the two equilibrium constants in eqs. 4.63 can be cited for goethite (α-FeOOH):[26,39]

$\log {}^*K_{Na}(\text{int})$	$\log {}^*K_{Cl}\text{int}$	$\log {}^*K_K(\text{int})$	$\log {}^*K_{Cl}(\text{int})$
0.8	0.6	0.8	0.6
$\log {}^*K_{Na}(\text{int})$	$\log {}^*K_{NO_3}(\text{int})$	$\log {}^*K_K(\text{int})$	$\log {}^*K_{NO_3}(\text{int})$
1.9	2.5	1.5	2.0

The approximate agreement between the correlation model and measured parameters is again evident. Note that the constraint,

$$\log {}^*K_C(\text{int}) = \log {}^*K_A(\text{int}) \tag{4.64}$$

required for eq. 4.61 to be valid is also only approximately satisfied by available estimates or measurements of $K_+(\text{int})$, $K_-(\text{int})$, $K_C(\text{int})$, and $K_A(\text{int})$.[39]

4.3 Modeling Temperature Effects

The effect of temperature on ion adsorption controlled by surface complexation mechanisms is manifest in the temperature dependence of rate coefficients (eq. 3.13) and of intrinsic equilibrium constants, such as those defined in eqs. 4.54 to 4.59. For the latter, chemical thermodynamics provides the van't Hoff equation:[42]

$$\left(\frac{\partial \ln K}{\partial T^0}\right)_{P^0} = \frac{\Delta_r H^0(T^0)}{R(T^0)^2} \tag{4.65}$$

to describe quantitatively how an equilibrium constant K changes with the standard-state temperature T^0 at constant standard-state pressure P^0. The numerator on the right side of eq. 4.65 is the standard enthalpy change of the reaction to which K refers, and the denominator contains the gas constant R (see section 3.1, below eq. 3.13). Integration of eq. 4.65 is required in order to calculate K at a temperature other than the conventional standard state temperature, $T_R^0 = 298.15$ K. This operation, in turn, requires information about the temperature dependence of the standard enthalpy change. In most applications, $\Delta_r H^0(T^0)$ is represented by a two-term Taylor expansion about $T_R^0 = 298.15K$:[43]

$$\Delta_r H^0(T^0) \approx \Delta_r H^0(T_R^0) + \Delta C_P^0(T_R^0)\,(T^0 - T_R^0) \tag{4.66}$$

where

$$\Delta C_P^0 \equiv \left(\frac{\partial \Delta_r H^0}{\partial T^0}\right)_{P^0} \tag{4.67}$$

is the Kirchoff equation for the standard change in heat capacity at constant pressure for the reaction to which K refers.[42,43] Under this approximation, eq. 4.65 becomes

$$\left(\frac{\partial \ln K}{\partial T^0}\right)_{P^0} \approx \frac{\Delta_r H^0(T_R^0)}{R(T^0)^2} + \frac{\Delta_r C_P(T_R^0)}{RT^0}\left(1 - \frac{T_R^0}{T^0}\right) \tag{4.68a}$$

and integration of eq. 4.65 would yield

$$\ln K(T^0) \approx \ln K(T_R^0) + \frac{\Delta_r H^0(T_R^0)}{RT_R^0}\left(1 - \frac{T_R^0}{T^0}\right) + \frac{\Delta_r C_P^0(T_R^0)}{R}\left[\ln\frac{T^0}{T_R^0} - \left(1 - \frac{T_R^0}{T^0}\right)\right] \tag{4.68b}$$

upon integration of both sides of eq. 4.68a between the limits T_R^0 and T^0. The right side of eq. 4.68b contains three parameters, $K(T_R^0)$, $\Delta_r H^0(T_R^0)$, and $\Delta_r C_P(T_R^0)$, whose values at 298.15 K must be known in order to estimate ln K at some other temperature. If $\Delta_r C_P^0(T_R^0)$ is negligibly small, eq. 4.68b reduces to a well known approximate van't Hoff equation featuring a linear dependence of ln $K(T^0)$ on $1/T^0$.[42,43]

As an example of the application of eq. 4.68 to surface complexation reactions, consider the protonation reaction,

$$\equiv SO^- + 2H^+ \rightleftarrows \equiv SOH_2^+ \tag{4.69}$$

which can be obtained from eq. 4.36. On the Triple Layer Model, the equilibrium constant for this reaction is the quotient of K_+(int) and K^-(int), as in eq. 4.60. The temperature dependence of this quotient—under the critical assumption that the two intrinsic equilibrium constants involved are indeed independent of composition (i.e., that they are not *conditional* equilibrium constants because of inadequacies in the Triple Layer Model)—is expressed by eq. 4.68b. In principle, the enthalpy and heat capacity parameters should be obtained from direct calorimetric measurements of $\Delta_r H^0(T^0)$ that validate eq. 4.66.[44] In practice, these two parameters are determined by optimization of eq. 4.68b to fit data on the temperature dependence of the pristine point of zero charge (p.p.z.c.), which is related to K_+(int) and K_-(int) in eq. 4.61:

$$\text{p.p.z.c.}(T^0) \approx \text{p.p.z.c.}(T_R^0) + (2\ln 10)^{-1}\left\{\frac{[\Delta_r H^0(T_R^0) - T_R^0\Delta_r C_P^0(T_R^0)]}{RT_R^0}\right.$$

$$\left.\times\left(1 - \frac{T_R^0}{T^0}\right)\right\} + \frac{\Delta_r C_P^0(T_R^0)}{2R}\log\frac{T^0}{T_R^0} \tag{4.70}$$

Figure 4.3 shows an application of eq. 4.70 to rutile (T_iO_2).[45] The optimized values of the parameters involved are:

$$\text{p.p.z.c.}(298.15\text{ K}) = 5.4 \qquad \Delta_r H^0(2.98.15\text{ K}) = -52.8\text{ kJ mol}^{-1}$$

$$\Delta_r C_P^0(298.15\text{ K}) = 251\text{ J mol}^{-1}\text{K}^{-1}$$

for 298.15 K $< T^0 <$ 568.15 K. Values of $\Delta_r H^0$ (298.15 K) that have been reported for rutile[44–46] range from −40 to −53 kJ mol^{-1}, while those reported for $\Delta_r C_P^0$ (298.15 K) range from 0 to 377 J mol^{-1} K^{-1}. A calorimetric measurement of $\Delta_r H^0$ (298.15 K) for rutile yields −44 kJ mol^{-1}, in agreement with the range of values inferred from statistical optimization.[44–46]

Figure 4.3 indicates a drop in p.p.z.c. by more than 0.5 over the temperature range 25 to 100°C. For other metal oxides, the decreases in p.p.z.c. are comparable,[44–46] with $\Delta_r H^0$ (298.15 K) ranging from −32 to −100 kJ mol^{-1}. The impact of $\Delta_r C_P^0$ (298.15 K) on this trend in p.p.z.c. is rather small, because the two terms in eq. 4.68b containing this parameter as a factor will cancel so long as the approximation, $\ln(T^0/T_R{}^0) \approx 1 - (T_R{}^0/T^0)$ is accurate (i.e., if $|T^0 - T_R^0| \lesssim 0.1\, T^0$). Thus $\Delta_r C_P^0$ (298.15 K) is not a sensitive parameter in eq. 4.70.

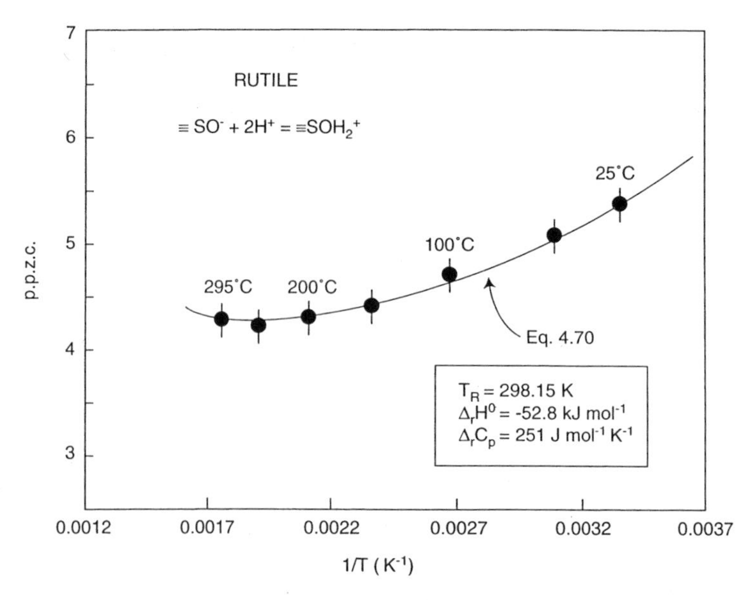

Fig. 4.3 Application of eq. 4.70 to measurements[46] of the p.p.z.c. of rutile (TiO_2) over the temperature range 25 to 295°C.[45]

Although eq. 4.70 depends for its validity on the Taylor expansion of $\Delta_r H^0(T^0)$ in eq. 4.66, it does not depend particularly on the form of the protonation reaction in eq. 4.69. For example, if a more detailed picture of a protonatable surface site than $\equiv SO^-$ is used [say, one based on applying bond valence concepts to surface OH groups singly-coordinated to a trivalent metal cation (see problem 3 in chapter 1)], then eq. 4.69 could be replaced by the protonation reaction:[47]

$$\equiv SOH^{\frac{1}{2}} + H^+ \rightleftarrows \equiv SOH_2^{+\frac{1}{2}} \tag{4.69a}$$

and eq. 4.61 would become:

$$\text{p.p.z.c.} = \log K_H(\text{int}) \tag{4.61a}$$

$K_H(\text{int})$ being the intrinsic equilibrium constant for the new reaction in eq. 4.69a. It is evident that $K_H(\text{int})$ must formally be equal to the square-root of K_+/K_-, if Eqs. 4.70 and 4.69a describe the same protonation process as observed in a titration experiment, and, therefore, that $\Delta_r H^0$ for the latter reaction will equal one-half $\Delta_r H^0$ for the former reaction. The key point here is that p.p.z.c. will always be proportional to the logarithm of a *model* equilibrium constant whose temperature dependence is expressed by eq. 4.65, or by the three-parameter approximation in eq. 4.70, irrespective of any underlying molecular hypotheses. Indeed, eq. 4.70 is *independent* of the surface complexation model selected to describe the p.p.z.c. so long as the associated intrinsic equilibrium constant is not composition dependent.

This conclusion can be epitomized by generalizing eq. 4.70 to relate the p.p.z.c. and $\Delta_r H^0$ at any temperature:

$$\Delta_r H^0(T^0) - T^0 \Delta_r S^0(T^0) = -n \ln 10 R T^0 \text{ p.p.z.c.} \tag{4.71}$$

upon introducing $\Delta_r G^0 \equiv \Delta_r H^0 - T^0 \Delta_r S^0$ and noting that the intrinsic equilibrium constant associated with both $\Delta_r G^0$ and p.p.z.c. may be derived from an "*n-pK* model" ($n = 1, 2$).[47] Evidently $\Delta_r S^0(T^0)$ can be estimated if $\Delta_r H^0(T^0)$ and a corresponding p.p.z.c. value are available. Compilations of these estimates[44–46] indicate that $\Delta_r S^0(298.15) > 0$ and that it ranges from 20 to 80 J mol^{-1} K^{-1}, thus prompting the suggestion[48] that $\Delta_r S^0(298.15) \approx 50$ J mol^{-1} K^{-1} might be a useful average value to introduce into eq. 4.71. This approximation would make $\Delta_r H^0$ depend on the p.p.z.c., with surface protonation becoming an increasingly more exothermic process as p.p.z.c. increases. Perusal of $\Delta_r H^0$ (298.15) data for a broad variety of minerals[45] suggests that this correlation is roughly correct. Thus, strong bond formation between a proton and a mineral surface is signaled by a high p.p.z.c. value—a not unreasonable physical concordance.

Besides the conceptual measurability issues that attend p.p.z.c., the details of which are discussed in sections 1.4 and 4.2, specific experimental difficulties arise when it is measured at temperatures in excess of 90°C.[49] Therefore, it is pertinent to explore whether eq. 4.68 is justifiable on fundamental grounds for the reactions in eqs. 4.69 and 4.69a, so as to have confidence in the accuracy of enthalpy and heat capacity parameters determined almost always by goodness-of-fit criteria. In general terms, $\Delta_r C_P^0$ is expected to be minimal—and the Taylor expansion in eq.

4.66 more accurate—if the reactants comprise the same number and types of chemical species as do the products of a chemical reaction.[43] The reason for this expectation is that changes in the degrees of motional freedom (the molecular basis for changes in heat capacity) should be minimal when the products of a reaction are similar in nature to the reactants.

This condition is not particularly well met in eq. 4.69 (or in eq. 4.69a), but the situation can be improved by adding to it the ionization reaction of liquid water to yield the composite protonation reaction[43,49]

$$\equiv SO^- + H^+ + H_2O \Longleftrightarrow \equiv SOH_2^+ + OH^- \quad (4.72)$$

This "isocoulombic" reaction contains like numbers of each type of charged species on both sides. Its equilibrium constant can be expressed in the logarithmic form:

$$\log K = \log[K_+(\text{int})/K_-(\text{int})] + \log K_W \quad (4.73)$$

where K_W is the ion product for liquid water, with temperature dependence:

$$\log K_W(T^0) = -13.995 + 14.996\left(1 - \frac{298.15}{T^0}\right) - 0.01706(T^0 - 298.15) \quad (4.74)$$

over the range 273.15 K $< T^0 <$ 373.15 K. The right side of eq. 4.73 should be a better candidate for modeling with eq. 4.68b than is $\log[K_+(\text{int})/K_-(\text{int})]$ alone.[43,49] Indeed, if $|\Delta_r H^0(298.15)|$ and $|\Delta_r C_p^0(298.15)|$ were close in value to $|\Delta_r H_w^0(298.15)| = 55.8$ kJ mol^{-1} and $|\Delta_r C_{pw}^0(298.15\text{ K})| = 223.8$ J mol^{-1} K^{-1} for the ionization reaction of liquid water, then the temperature variation of log K_{iso} would be essentially negligible, and the approximation:

$$\log[K_+(\text{int})/K_-(\text{int})] + \log K_W(T^0) = \text{constant} = 2\text{ p.p.z.c.}(298.15\text{ K}) - 13.995 \quad (4.75)$$

would be accurate at *any* temperature T^0. Equation 4.75 can be tested readily, since data on the temperature dependence of p.p.z.c. are available:[46,50]

Temperature (°C)	p.p.z.c.	2 p.p.z.c. + log K_W
Rutile		
25	5.4	−3.2
50	5.1	−3.1
100	4.7	−2.9
Goethite		
25	8.3	2.6
40	8.1	2.7
55	7.9	2.7
70	7.7	2.6

Constancy of [p.p.z.c. + $(1/n)$log K_W] for either $n = 1$ or 2 has been observed for a variety of oxide minerals,[44,46,50] but not all.[49] When it does obtain, as in the table above, it signals the fact that the temperature dependence of p.p.z.c. is congruent with the temperature shift of "neutral pH" in liquid water.

Given the exothermic nature of surface protonation reactions, increases in temperature must decrease p.p.z.c. values and thereby broaden the range of pH in a Schindler diagram (section 1.5) over which an adsorbent functions as a cation exchanger. This conclusion alone is sufficient to make the prediction that *cation adsorption should be enhanced, and anion adsorption should be diminished, as temperature increases.* Figures 4.4 and 4.5 illustrate the essential correctness of this broad prediction. Figure 4.4 shows the effect of increasing temperature on the adsorption edge for Cd^{2+} reacted with goethite.[51] The value of the parameter pH_{50} (see eq. 1.12 and fig. 1.1) decreases from 6.4 at 5°C to 5.9 at 35°C, indicating the typical shift of an adsorption edge to smaller pH values as temperature is increased.[44,50–52] Figure 4.5, which can be compared with fig. 1.2, displays adsorption envelopes for borate reacted with an Alfisol soil at three temperatures.[53] A decrease in the adsorption maximum as the temperature increases from 10°C to 40°C is apparent. Similar trends obtain for other oxyanions on diverse adsorbents.[44,54] Thus, quite generally, increasing temperature leads to greater cation adsorption and lesser anion adsorption under given conditions.

Adsorption edges and envelopes are influenced, of course, by other variables than temperature. For example, at fixed temperature and ionic strength, adsorption edges shift to higher pH values (i.e., pH_{50} increases) as the initial concentration of adsorptive cation is increased.[51] At fixed temperatures, ionic strength, and initial adsorptive concentration, adsorption edges shift to lower pH values (i.e., pH_{50} decreases) as the time of reaction between the adsorptive and adsorbent increases.[51] This effect of reaction time offsets that of initial concentration, but complements that of temperature. For example, $pH_{50} = 5.85$ for the adsorption edge of Cd^{2+} on goethite at 35°C (fig. 4.4) determined after 7 days' reaction with 10 mmol Cd m^{-3} initial concentration. But essentially the same value of pH_{50} (5.83) is observed for the Cd^{2+} adsorption edge at 5°C after 21 days reaction with 1 mmol Cd m^{-3} initial concentration.[51] Similarly, ionic strength effects on an adsorption edge can offset the shift caused by temperature changes, as illustrated in fig. 4.6 for Ca^{2+} adsorption by rutile.[55] Adsorption edges determined in 0.03 mol NaCl dm^{-3} background electrolyte solution (filled symbols) are shifted

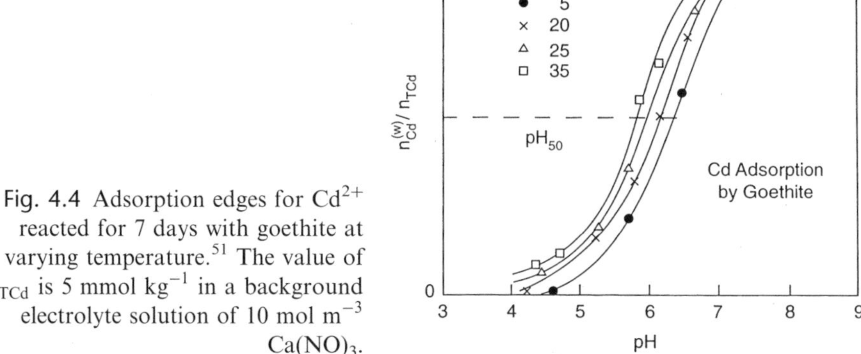

Fig. 4.4 Adsorption edges for Cd^{2+} reacted for 7 days with goethite at varying temperature.[51] The value of n_{TCd} is 5 mmol kg^{-1} in a background electrolyte solution of 10 mol m^{-3} $Ca(NO)_3$.

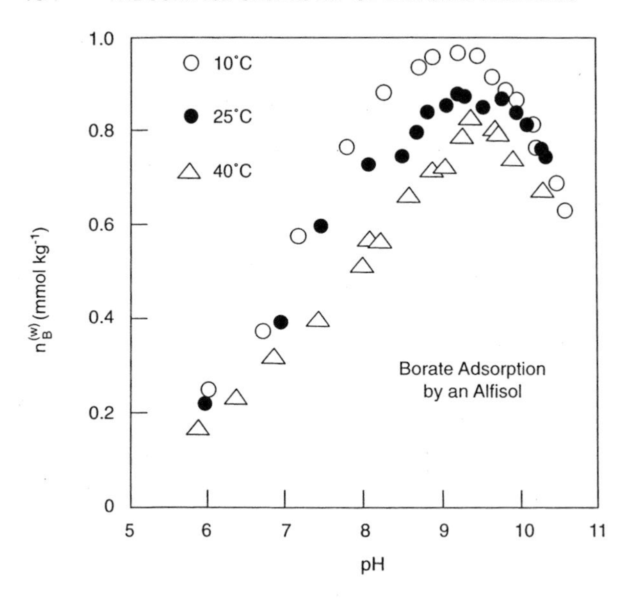

Fig. 4.5 Adsorption envelopes for borate reacted for 2 h at varying temperature with an Alfisol suspended in NaCl solution.[53]

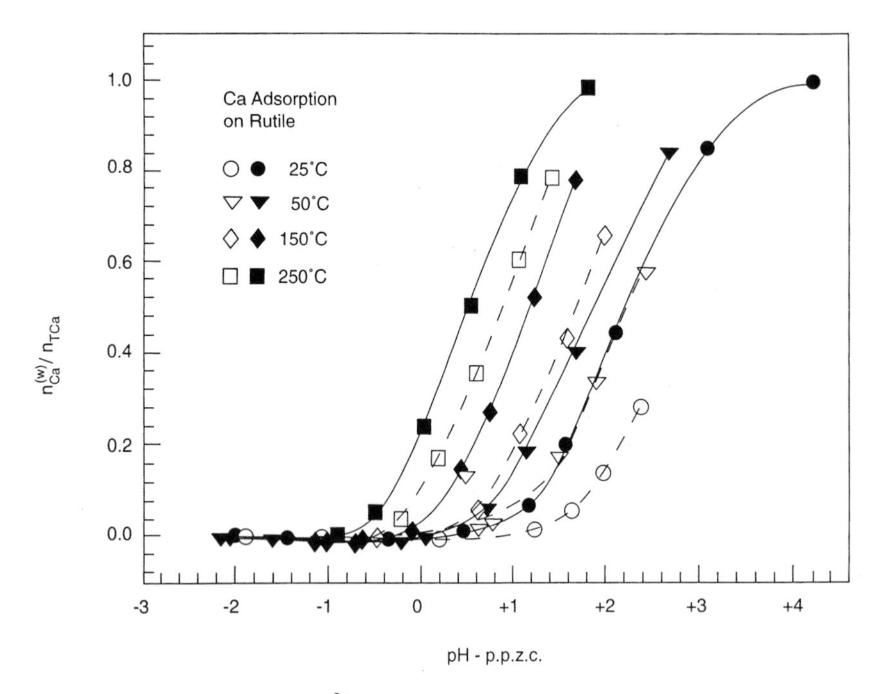

Fig. 4.6 Adsorption edges for Ca^{2+} on rutile (TiO_2) at varying temperature and two ionic strengths (NaCl), 30 mol m^{-3} (filled symbols) and 300 mol m^{-3} (open symbols).[55] The values of p.p.z.c. are: 5.3 (25°C), 5.4 (50°C), 4.3 (150°C), and 4.1 (250°C).

to higher pH values (open symbols) in 0.3 mol NaCl dm^{-3} background solution. Thus, the adsorption edge at 25°C in the lower ionic strength medium is congruent with the adsorption edge at 50°C in the higher ionic strength medium. Note also that plotting the adsorption edges as a function of the renormalized variable, pH − p.p.z.c., does not make them superpose. Evidently the effect of temperature on Ca^{2+} adsorption is more complex than merely a result of decreasing p.p.z.c. values.

Temperature dependence in ion adsorption also is manifest in adsorption isotherms, as illustrated in fig. 4.7 for Pb^{2+} adsorption by goethite at pH < p.z.n.c.[56] The remarkable increase in the surface excess of Pb between 55°C and 70°C is a signature of the logarithmic relationship between an equilibrium constant and temperature realized in the van't Hoff equation (eq. 4.65). This trend is apparent in the temperature dependence of the four empirical parameters that appear in the generic two-term isotherm equation in eq. 1.13:[56]

t(°C)	b_1 (mmol kg^{-1})	K_1 (m^3 mol^{-1})	b_2 (mmol kg^{-1})	K_2 (m^3 mol^{-1})
10	9.63	227	55.0	3.3
25	7.76	185	55.0	4.4
40	8.31	269	60.5	4.8
55	10.78	290	55.0	5.4
70	13.86	332	66.0	6.5

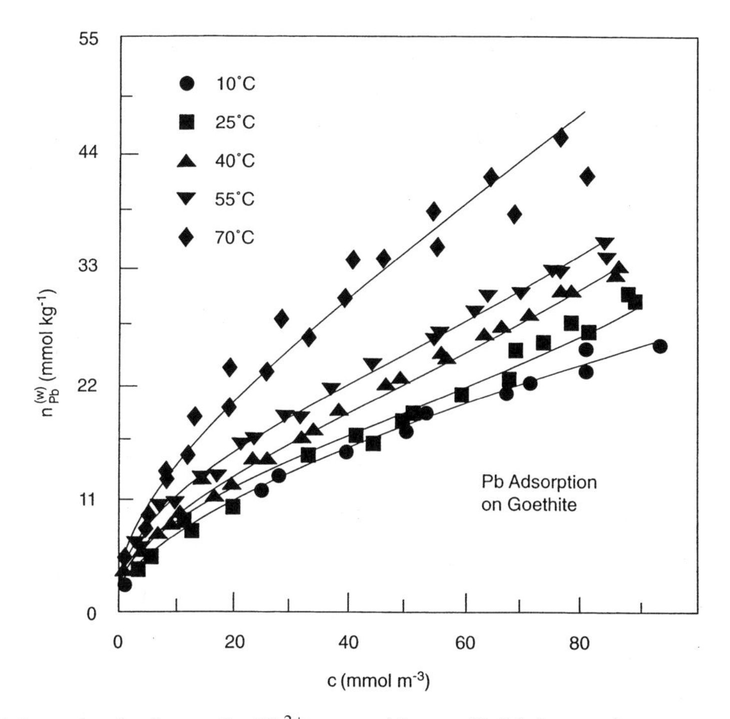

Fig. 4.7 Adsorption isotherms for Pb^{2+} on goethite at pH 5.5 for varying temperatures.[56] The solids concentration is 6.7 g L^{-1} in 10 mol m^{-3} KNO_3 solution. The curves through the data points are fits of eq. 1.13.

The temperature dependence of the two "affinity" parameters can be fit to eq. 4.68b for the case $\Delta C_P^0 \equiv 0$:

$$\ln K_1 = 14.99 - (789.8/T^0) \quad r^2 = 0.69$$
$$\ln K_2 = 11.724 - (1015/T^0) \quad r^2 = 0.97$$

which imply $\Delta_r H_1^0 \approx 6.6$ kJ mol^{-1} and $\Delta_r H_2^0 \approx 8.4$ kJ mol^{-1}, respectively, at 298.15 K.[56]

Unlike the empirical models in eqs. 1.12 and 1.13, surface complexation models can describe adsorption edges or envelopes and isotherms simultaneously.[26] The temperature dependence of the resulting intrinsic equilibrium constants then can be fit to eq. 4.68b similarly to the overall protonation intrinsic equilibrium constant for the reaction in eq. 4.69. This has been done, for example, in the case of Cd^{2+} adsorption by kaolinite over the temperature range 10°C to 70°C.[57] At pH > p.z.n.p.c. Cd^{2+} adsorption was modeled by eq. 4.37b,

$$2 \equiv \mathrm{SOH} + \mathrm{Cd}^2 + \rightleftarrows (\mathrm{SO})_2\mathrm{Cd}^0 + 2\mathrm{H}^+ \tag{4.76}$$

with an intrinsic equilibrium constant $K_{\mathrm{Cd}}^2(\mathrm{int})$:

$$\ln K_{\mathrm{Cd}}^2(\mathrm{int}) = -18.108 + 29.988\left(1 - \frac{298.15}{T^0}\right) \tag{4.77}$$

for 283.15 K < T^0 < 343.15 K. It follows from a comparison between eqs. 4.68b and 4.77 that $\Delta_r H^0$ (298.15) = 74 kJ mol^{-1} for the surface complexation reaction in eq. 4.76.[57] In parallel with the treatment of the surface protonation reaction in eq. 4.69, the surface complexation reaction in eq. 4.76 is transformed to an "isocoulombic" reaction by subtracting from it the hydrolysis reaction:

$$\mathrm{Cd}^{2+} + 2\mathrm{H_2O} \rightleftarrows \mathrm{Cd(OH)}_2^0 + 2\mathrm{H}^+ \tag{4.78}$$

to achieve the composite reaction:

$$2 \equiv \mathrm{SOH} + \mathrm{Cd(OH)}_2^0 \rightleftarrows \equiv (\mathrm{SO})_2\mathrm{Cd}^0 + 2\mathrm{H_2O} \tag{4.79}$$

with intrinsic equilibrium constant $K_{\mathrm{Cd}}^2(\mathrm{int})/{}^*\beta_2$, where ${}^*\beta_2$ is the equilibrium constant for the hydrolysis reaction. Following the same line of reasoning used to arrive at eq. 4.75, one infers that $\log K_{\mathrm{Cd}}^2(\mathrm{int}) - \log {}^*\beta_2$ may show little or no temperature dependence. A direct calculation with eq. 4.77 and data on the temperature variation of $\log {}^*\beta_2$ shows that this conjecture is approximately correct:[57]

t (°C)	$\log K_{\mathrm{Cd}}^2(\mathrm{int}) - \log {}^*\beta_2$
10	12.9
25	12.5
40	12.1
55	12.1
70	11.8

Equation 4.79 exposes the chemical symmetry between aqueous and adsorbed species [$\equiv$SOH $\leftrightarrow$ H_2O, $\equiv(SO)_2Cd^0$ $\leftrightarrow$ $Cd(OH)_2^0$] implied by the positive correlation observed commonly between intrinsic equilibrium constants for metal cation

surface complexation by hydroxylated adsorbents and thermodynamic equilibrium constants for metal cation hydrolysis (section 1.5). The same chemical symmetry appears in eq. 4.72 [$\equiv SO^- \leftrightarrow OH^-$, $\equiv SOH_2^+ \leftrightarrow H^+ + H_2O = H_3O^+$] implying a positive correlation between p.p.z.c. and "neutral pH" as temperature varies.

4.4 Modeling Affinity

Given the polyfunctionality of natural particle surfaces, adsorption reactions involving even a single type of aqueous species usually will result in a variety of adsorbate species partitioned concurrently among the diffuse ion swarm, outer-sphere surface complexes, and inner-sphere surface complexes. For example, Cd^{2+} in solution may adsorb as Cd^{2+} and $CdOH^+$ in the diffuse swarm; as $\equiv SO^- \cdots Cd^{2+}$ and $\equiv SO^- \cdots CdOH^+$ in outer-sphere surface complexes; and as $\equiv (SO)_2Cd^0$ and $\equiv SOCdOH^0$ in inner-sphere surface complexes, to enumerate but a few possibilities. Moreover, the bonding characteristics of the surface complexes may vary spatially on the adsorbent surface if it is highly heterogeneous in either chemical composition or molecular-scale topography. This multifarious surface speciation would imply the need for a corresponding multiplicity of intrinsic equilibrium constants—and perforce a daunting parametric model complexity—in order to apply the theoretical approaches described in sections 4.1 and 4.2.

An alternative methodology, which has its origins in models of ion adsorption by macromolecules,[58] replaces the problem of devising chemical models for legions of adsorbate species by the problem of determining a sufficient number of "affinity classes" with which to describe adsorption processes quantitatively over desired ranges of pH and aqueous solution composition. Each "affinity class" is characterized by a capacity parameter b and a conditional affinity parameter K in the spirit of an isotherm equation, such as that in eq. 1.13. However, an "affinity class" may have no simple relationship to the surface functional groups on an adsorbent and indeed may represent a variety of adsorbate structures, since all three kinds of adsorbed species may be contributing to any one capacity or affinity parameter. Thus, the adsorption behavior of an ion is modeled as if it derived from interactions between the ion and a suite of *hypothetical* noninteracting surface functional groups, each class of which is distinguished by the two parameters b, K.[59] No a priori molecular-scale interpretation can be associated with "affinity classes", although ancillary surface functional group analysis by chemical methods may permit mechanistic inferences to be made. In adopting this approach, therefore, one concedes to an exchange of molecular verisimilitude for model simplicity.

The fundamental hypothesis made in the modeling of "affinity classes" is expressed mathematically by the integral representation:[59]

$$n(c) = \int_0^\infty \frac{f(K)Kc}{1 + Kc} dK \tag{4.80}$$

where n is the surface excess of an adsorptive whose concentration is c (eq. 1.4) and $f(K)dK$ is the maximum surface excess of the adsorptive when bound to sites whose affinity parameters lie in the range K to $K + dK$, such that

$$n_{\max} = \int_0^\infty f(K)dK \tag{4.81}$$

is the maximum surface excess attainable by adsorption onto all sites, under the conditions that obtain when $n(c)$ is measured. Equation 4.80 assigns to each affinity class a Langmuir adsorption isotherm (cf. eq. 1.13) with affinity parameter K and a capacity parameter determined by the value of $f(K)$. Equation 1.13 is then the special case of eq. 4.80 obtained after introducing the specific mathematical representation:

$$f(K) = b_1\delta(K - K_1) + b_2\delta(K - K_2) \tag{4.82}$$

where[60]

$$\int_{-\infty}^{\infty} g(x)\delta(x - a)dx \equiv g(a) \tag{4.83}$$

defines the delta-"function" $\delta(x)$ and its effect on an integrable function $g(x)$. The *affinity spectrum* $f(K)$ in eq. 4.82 may be pictured as two vertical spikes, positioned at $K = K_1$ and $K = K_2$, respectively, representing the contributions of two "affinity classes," such that, by eq. 4.83,

$$\begin{aligned} n_{\max} &= \int_0^\infty [b_1\delta(K - K_1) + b_2\delta(K - K_2)]dK \\ &= b_1 + b_2 = b \end{aligned} \tag{4.84}$$

gives the contribution of each class to the maximum surface excess possible according to eq. 1.13. [Note that eq. 4.83 has been used in eq. 4.84 as the special case of constant $g(x)$ for $x > 0$ and $g(x) \equiv 0$ for $x < 0$.]

Equation 4.80 is a mathematical representation of the surface excess as a function of adsorptive concentration in terms of an integrated Langmuir isotherm equation, whose derivation in statistical thermodynamics[61] shows it to describe adsorption onto independent sites of uniform affinity, with no interactions among adsorbate molecules. This representation of $n(c)$ is known[62] to be unique [i.e., a one-to-one relationship exists between $n(c)$ and $f(K)$] if $n(c)$ satisfies two asymptotic conditions:[63]

$$\lim_{c\downarrow 0}\left[n(c)/c^\beta\right] = \text{constant} \qquad (0 < \beta \leq 1) \tag{4.85a}$$

$$\lim_{c\uparrow\infty}[n(c)/c] = 0 \tag{4.85b}$$

The first condition applies at very low surface excess, with $n(c)$ permitted to vary with concentration as no higher power than the first, whereas the second condition stipulates that this variation with concentration must be less than as the first power at very high surface excess. Equation 4.85a is consistent with a finite value of the distribution coefficient (eq. 1.8) as $c\downarrow 0$, although it does not require this property, while eq. 4.85b is consistent with a plateau in the adsorption isotherm as

$c\uparrow\infty$, but also does not require it. Taken together, these two conditions imply that the graph of $n(c)$ must be *concave* to the c-axis, that is,

$$\frac{dn}{dc} \le \frac{n(c)}{c} \qquad (0 < c < \infty) \tag{4.85c}$$

as is evident also from a comparison between the first derivative of the right side of eq. 4.80 with respect to c and the ratio of the right side to c. Thus, eq. 4.80 cannot describe an S-curve adsorption isotherm (see fig. 1.3), which is convex to the c-axis.

In the special case that $\beta = 1$ in eq. 4.85a and $n(c)$ achieves a plateau value as $c\uparrow\infty$, the limiting mathematical forms of $n(c)$ at very low and very high concentrations can be calculated *exactly*.[63] At very low concentrations,

$$n(c) \approx n_{\max}\left[\langle K\rangle c - \langle K^2\rangle c^2\right] \tag{4.86a}$$

whereas at very high concentrations,

$$n(c) \approx n_{\max}\left[1 - \frac{\langle K^{-1}\rangle}{c}\right] \tag{4.86b}$$

where $n_{\max}$ is qiven by eq. 4.81 and

$$\langle K^n\rangle \equiv \frac{\int_0^\infty K^n f(K) dK}{\int_0^\infty f(K)\, dK} \qquad (n = \pm 1, \cdots) \tag{4.87}$$

defines the n normalized moments of the affinity spectrum $f(K)$. [Note that $n_{\max} = \lim_{c\uparrow\infty} n(c)$ is assumed here to be finite and equal to the plateau value of $n(c)$ at high concentrations. Equation 4.81, the zeroth moment of $f(K)$, then follows by evaluating the limit of the right side of eq. 4.80 as $c\uparrow\infty$.] Therefore, adsorption isotherm data obtained at very low and very high concentrations of the adsorptive can be used to estimate four of the moments of $f(K)$. Moreover, because the model adsorption isotherm in eq. 1.13 contains four parameters and has the limiting behavior assumed for $n(c)$, it can be fit *uniquely* to any adsorption isotherm data having this limiting behavior. The fit will be *exact* at very low and very high concentrations, and will interpolate the isotherm data at all concentrations between. In the context of this model, eq. 4.87 becomes:

$$\langle K^n\rangle = \frac{1}{b}(b_1 K_1^n + b_2 K_2^n) \quad (n = 0, \pm 1, 2) \tag{4.88}$$

leading, along with eq. 4.84, to four algebraic equations equivalent to eqs. 1.17, 1.18, 1.20, and 1.21, which can be solved for the four adjustable parameters, b_1, b_2, K_1, and K_2.[63]

In current practice, however, nonlinear least-squares statistical methods are applied to adsorption isotherm data to determine optimized values of model parameters based on an entire data set, instead of on data pertaining only to the extremes of very low and very high adsorptive concentration, which are difficult to obtain precisely and accurately. The numerical problem to be solved thus is shifted from solving algebraic equations based on eqs. 4.84 and 4.88 to minimizing the *chi-square statistic*,[64]

$$\chi^2 = \sum_{i=1}^{N} \left[\frac{n_i - n(c_i)}{\sigma_i} \right]^2 \tag{4.89}$$

where n_i, one of N measured values of a surface excess, corresponds to the adsorptive concentration c_i, σ_i is the composite standard deviation associated with imprecision in the measurements of both n_i and c_i, and $n(c_i)$ is a model expression for the adsorption isotherm evaluated at the concentration c_i. The chi-square statistic can be shown to approach the value of N within a standard deviation $(2N)^{\frac{1}{2}}$ if $N \uparrow \infty$ and the deviations of the n_i from the $n(c_i)$ follow a gaussian probability distribution.[64]

A variety of computational methods is available to perform the minimization of χ^2 using eq. 4.80 or a specific parametric model of $n(c)$, such as eq. 1.13.[64–66] It is observed typically, however, that this approach is not sufficient to yield a unique result; that is, a number of χ^2 values can be found that meet the criterion,

$$N - (2N)^{\frac{1}{2}} \leq \chi^2 \leq N + (2N)^{\frac{1}{2}} \tag{4.93}$$

but correspond to rather different values of model parameters, or even to different models themselves.[65] This fact becomes especially pertinent in view of the typial lack of features in adsorption isotherm data sets that might be used to select models on the basis of qualitative criteria. Given this situation, recourse is made to additional constraints on the minimization process, a subjective practice termed *regularization*,[64] that will yield a narrower choice of model parameters or characteristics while meeting the criterion in eq. 4.93.

Uner regularization, the numerical problem to be solved in minimizationof the sum,[64,67]

$$\chi^2[\{s_\ell\}] + \sum_{j=1}^{M} \lambda_j R_j[s_\ell\}]$$

where $\{s_\ell\}$ is a set of M parameters that are to describe the adsorption data quantitatively in the context of the model, $n(c)$, and $\{\lambda_j\}$ is a set of M optimizing "multipliers" whose values are to be adjusted so as to ensure that eq. 4.93 is satisfied while providing an optimal influence of the *regularizing function*, $R_j[\{s_\ell\}]$, on the minimization process. The choice of mathematical form for $R_j[\{s_\ell\}]$ is subjective, being based on a priori ideas about what the model, $n(c)$, should be like. Three typical examples are:[64,67]

$R_j[\{s_\ell\}]$	Constraint Imposed
$s_j \sum_{\ell-1}^{j-1} s_\ell$	Small number of parameters s_j
$(s_{j+2} - 2s_{j+1} + s_j)^2$	Smooth variation among the s_j
$s_j \ln(s_j/s_j^0)$	Close approximation of s_j to s_j^0

The first example in the table is a function of $\{s_\ell\}$ that increases in magnitude as the number of parameters M increases. Therefore, it penalizes any attempt to minimize χ^2 plus the sum of $R_j[\{s_\ell\}]$ that utilizes a large number of parameters.

This kind of regularization would be useful for an adsorption isotherm model based on an M-term generalization of eq. 1.13:[66,67]

$$n(c) = \sum_{j=1}^{M} s_j \frac{K_j c}{1 + K_j c} \tag{4.94}$$

in which the parameters to be optimized are the M Langmuir capacity parameters, each of which is associated with an affinity parameter K_j whose values are prearranged on a fixed grid over a domain that is appropriate to the adsorption data set being modeled.[67] Regularization of this kind would assure that M in eq. 4.94 is no larger than necessary to fit the model under the criterion in eq. 4.93.

The second example is a finite-difference approximation to the second derivative of a function whose values are the model parameters.[64] Its use in regularization penalizes any model dependence on the parameters that results in strong oscillations of their values as the index j is changed. Therefore, it selects for smooth variation among the M parameters. This kind of regularization is useful when $n(c)$ is given by eq. 4.80, with the right side then approximated by a suitable quadrature algorithm involving discretization of the affinity spectrum, $f(K)$. The third example is an "entropy function" that penalizes any value of s_j that deviates significantly from an a priori value, s_j^0. The a priori set $\{s_\ell^0\}$ may be selected from a preliminary fitting of a model to the adsorption data set or it may be simply a set of M equal numbers serving as initial guesses for $\{s_\ell^0\}$.[64,67,68] This kind of regularization can be useful if $\{s_\ell^0\}$ is indeed close to the optional $\{s_\ell\}$, since it ensures that any deviations of the s_j from the s_j^0 will occur only if they are required to fit the model to the data set. The subjective nature of regularization is especially apparent in this example.

Figure 4.8 illustrates the regularization approach to determining the affinity spectrum for the adsorption isotherm of a cationic surfactant compound on soil particles suspended in 10 mol m^{-3} NaCl solution.[67] The log-log plot of the adsorption isotherm data at the top of the figure exemplifies the rather featureless concave character of data that meet the condition in eq. 4.85c, which is equivalent to a slope ≤ 1.0 in the log-log plot. Below the plot is the affinity spectrum obtained by minimization under regularization for a small number of parameters using eq. 4.94 as the model isotherm equation. Without regularization, the number of Langmuir terms required to fit the data was 8 instead of 4, with both minimizations yielding acceptable χ^2 values. The bottom graph in the figure is an affinity spectrum based on eq. 4.80 as the model isotherm equation under regularization for smoothness. In this case, the two delta-"functions" near pK $\approx$ 3.5 in the discrete affinity spectrum are merged into a broad resonance that accounts for most of the integral of $f(K)$ (eq. 4.81). This integral was equal to 122 mmol kg^{-1} for the discrete affinity spectrum (i.e., the sum of the four s_j-values), whereas for the continuous spectrum it was equal to 146 mmol kg^{-1}. Both fits of the data gave the curve drawn through the data points in fig. 4.8.

Equation 4.80 is readily generalized to model the concurrent adsorption of two species:[69]

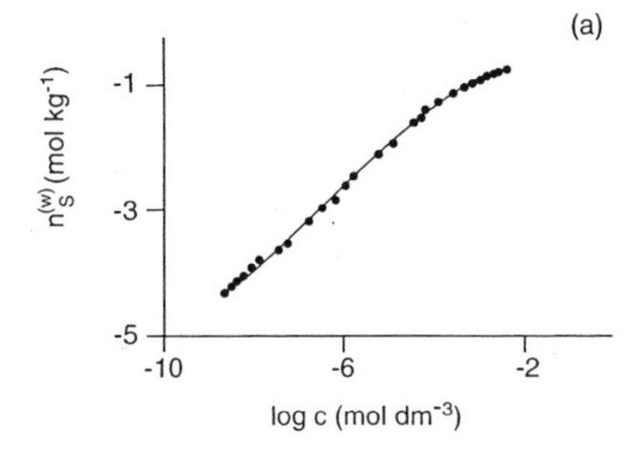

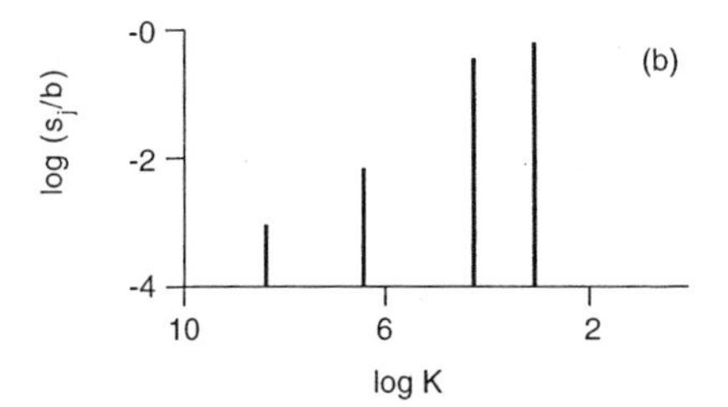

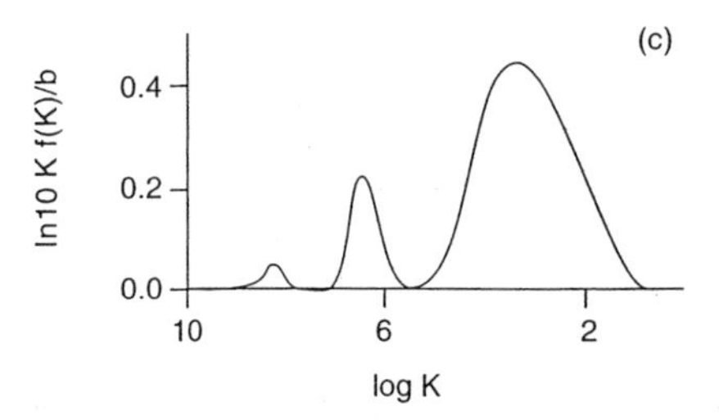

Fig. 4.8 (a) Log-log plot of an adsorption isotherm for dodecylpyridinium on a soil suspended in 10 mol m^{-3} NaCl solution. Data from B. J. Brownawell, H. Chen, J. M. Collier, and J. C. Westall, *Environ. Sci. Technol.* **24**:1234 (1990). (b) Affinity spectrum obtained by fitting the data in (a) to eq. 4.94 with regularization favoring the smallest possible value of M.[67] (c) Affinity spectrum obtained by fitting the data in (a) to eq. 4.80 with regularization for smooth variation of $f(K)$.[67] The y-axis variable is the affinity spectrum for log K, which is a more convenient variate than K when the range of the latter is large. The two peaks on the left of the principal peak have been magnified 30× to enhance their visibility.

$$n_i(c_1, c_2) = \int_0^\infty \int_0^\infty \frac{f(K_1, K_2)\, K_i c_i}{1 + K_1 c_1 + K_2 c_2} dK_1 dK_2 \tag{4.95}$$

where n_i is the surface excess of an adsorptive ion whose concentration is c_i $(i = 1, 2)$ and $f(K_1,K_2)dK_1dK_2$ is the maximum surface excess of either adsorptive 1 or 2 when bound to sites whose affinity parameters lie in the range K_1 to $K_1 + dK_1$, for adsorptive "1" or K_2 to $K_2 + dK_2$ for adsorptive "2", such that

$$n_{\max} = \int_0^\infty \int_0^\infty f(K_1, K_2) dK_1 dK_2 \tag{4.96}$$

is the maximum surface excess attainable by adsorption onto all sites under the conditions that apply to measurements of $n_1(c_1, c_2)$ and $n_2(c_1, c_2)$. The denominator in the integrand on the right side of eq. 4.95 is a direct generalization of that on the right side of eq. 4.80, to permit an isolated adsorption site to bind "no species," species "1," or species "2."[70] Figure 4.9 shows an example of $f(K_1,K_2)$ for the concurrent adsorption of protons and Ca^{2+} ions by a peat humic acid under the conditions $6 < pH < 10$ and $[Ca^{2+}]$ in the range 10^{-6} to 10^{-2} mol dm^{-3}.[69]

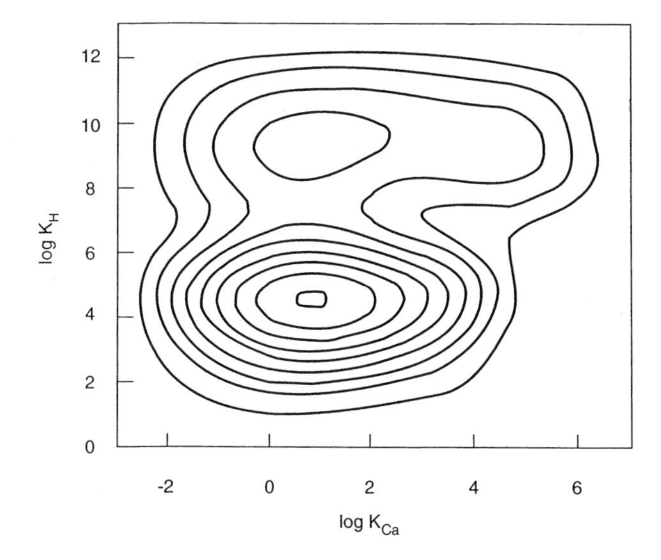

Fig. 4.9 Contour plot of the affinity spectrum $[(\ln 10)^2\, K_H K_{Ca} f(K_H,\ K_{Ca})]$ obtained by fitting eq. 4.95 to data on the concurrent adsorption of H^+ and Ca^{2+} by peat humic acid.[69] Data from M. F. Benedetti, C. J. Milne, D. G. Kinniburgh, W. H. van Riemsdijk, and L.K. Koopal, *Environ. Sci. Technol.* **29**:446 (1995). The contours, which indicate a bimodal distribution of log K_H values, both a peak in log K_{Ca} near 1.0, can be interpreted as evidence for carboxyl and phenolic OH groups in humic acid reacting with equal affinity for Ca^{2+}.

The bimodal shape of the affinity spectrum reflects the existence of two broad classes of adsorption site, both with a log K_{Ca} mode near 1, but with two log K_H modes, near 5 and 9, respectively characteristic of carboxyl and phenolic OH functional groups. Evidently there are considerably fewer of the latter than the former in the humic acid sample. As is true for eq. 4.80, it is possible to calculate $f(K_1, K_2)$ analytically if a specific model adsorption isotherm equation is adopted to describe $n_i(c_1, c_2)$.[69]

4.5 Modeling Natural Particle Adsorption Reactions

Studies of the morphology and chemical composition of natural particles in the clay and silt size fractions indicate clearly their convoluted, heterogeneous nature. Mineral particles having a variety of irregular shapes are clustered together in a morass of organic tendrils and spheroids whose conformations are labile, depending on conditions of pH, ionic strength, and the identity of attendant counterions.[71] Faced with this daunting complexity, the modeling of ion adsorption becomes an exercise, constrained by parsimony, whose foremost challenge is to strike a balance between the multiplicity of chemical parameters necessary to describe adsorbate speciation accurately and the varieties of surface reactivity that must be considered—even for one type of surface species—simply to capture the vagaries of natural particle topography.[72]

One approach to this conundrum has been to postulate a *pair* of overall reactions having the form of eq. 4.37a, with the two corresponding intrinsic equilibrium constants then being expressed as in eq. 4.55a. Noting that, for natural particles, the acid-base reactions in eqs. 4.36a and 4.36b typically make very small contributions to surface speciation, relative to the uncharged reactant surface species in eq. 4.37a, and that, *for strongly adsorbing metals*, ionic strength effects are comparably small, one can neglect the surface species activity coefficient that appears in eq. 4.55a.[73] Under these simplifying conditions, the number of adjustable model parameters is just three: two equilibrium constants and the fraction of adsorption sites that is attributed to one of the two complexation reactions invoked. (If the total quantity of adsorption sites is not known, then the number of adjustable parameters increases to four.) This concept of a "two-site, non-electrostatic" model of metal adsorption is mathematically equivalent to the use of eq. 1.13 for describing the surface excess.[74] The "strong site" and "weak site" in the model[73] correspond to the affinity classes associated with the larger and smaller values of the two affinity parameters, K_1 and K_2, respectively. The success of this model at both laboratory and field scales[73] then can be understood in terms of the uniqueness of the four parameters that can be deduced from fitting eq. 1.13 to experimental adsorption data, and the ubiquity of such data exhibiting a finite distribution coefficient as the surface excess approaches zero. Evidently parameter optimization of this model can be enhanced by discrete-site regularization, as discussed in section 4.4 for eq. 4.94.[67]

The "two-site, non-electrostatic" model thus is associated with the affinity spectrum in eq. 4.82, a condition for the success of which now can be seen in the key assumption that surface charge and ionic strength variations have little impact on ion adsorption. Although this assumption may well be appropriate for natural particles low in organic carbon, it is apparent from a large body of scientific work that the opposite is true for natural particles comprising significant amounts of humus.[75] For these latter particles, the development of surface charge as pH changes and the evolution of particle morphology with variations in both pH and ionic strength are major factors influencing their behavior as adsorbents. This additional complexity can be incorporated into tractable modeling by endowing the delta-"functions" in eq. 4.82 with finite breadth, thus passing from a discrete to a continuous affinity spectrum in the spirit of the two lower graphs in fig. 4.8. Broadening the two delta-"functions" may be interpreted as one way to capture the "smearing-out" effect of electrostatic interactions on conditional equilibrium constants, which in surface complexation models is handled parametrically by the exponential factors containing "surface potentials."

Inspection of the bottom graph in fig. 4.8 reveals the rather symmetric shape of the resonances centered on the three modal values of pK for adsorption of a surfactant cation by soil particles in suspension with NaCl solution. This distinguishing feature of a resonance can be mimicked well by the *Sips affinity spectrum*,[76]

$$p(y) = \frac{b}{2\pi} \frac{\sin(\pi\beta)}{\cos(\pi\beta) + \cosh[\beta(y - y_m)]} \tag{4.97}$$

where $y \equiv \ln K$,

$$b \equiv \int_{-\infty}^{\infty} p(y)dy \tag{4.98}$$

is the maximum surface excess, analogous to $n_{\max}$ in eq. 4.84, and β is an adjustable parameter that must lie the open interval (0, 1) in order that the integral in eq. 4.98 converge. The Sips affinity spectrum is a log transform of $f(K)$, such that $p(y) = Kf(K)$ while $p(y)dy = f(K)dK$. The mode of the spectrum occurs at $y = y_m$, where $p(y_m) = (b/2\pi)\tan(\pi\beta/2)$. The shape of $p(y)$ is very similar to that of a normal density function (i.e., a gaussian distribution of the variate ln K), differing significantly only in the "wings" for $|\beta(y - y_m)| > 5$.[76] The parameter β determines the breadth of the Sips affinity spectrum (fig. 4.10), the sharpness of the resonance at the mode increasing as $\beta \uparrow 1$. In the limit $\beta \uparrow 1$, $p(y) = 0$ if $y \neq y_m$, but is proportional to $(1 - \beta)^{-1}$ if $y = y_m$. Thus $p(y)$ has the properties of $\delta(y - y_m)$ in this limit, given eqs. 4.83 and 4.98.

The introduction of eq. 4.97 into eq. 4.80 yields the model adsorption isotherm equation:[76]

$$n(c) = \frac{b(\tilde{K}c)^{\beta}}{1 + (\tilde{K}c)^{\beta}} \quad (0 < \beta < 1) \tag{4.99}$$

where $\ln \tilde{K} \equiv y_m$, the mode of the Sips distribution. Equation 4.99 is an evident generalization of the Langmuir model adsorption isotherm, termed the *Langmuir-*

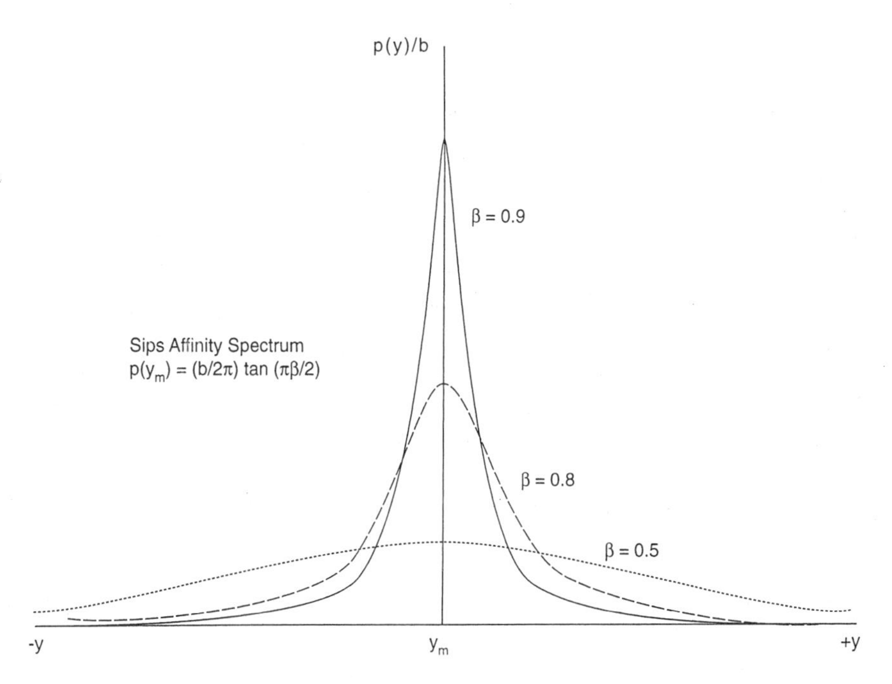

Fig. 4.10 Graph of the Sips affinity spectrum for varying "heterogeneity parameter," β. The mode y_m corresponds to ln $\tilde{K}$ in eq. 4.99.

Freundlich model. At very low adsorptive concentrations, $n(c) \underset{c\downarrow 0}{\sim} \tilde{K}^{\beta} c^{\beta}$ $(0 < \beta < 1)$, which implies that the distribution coefficient corresponding to the Langmuir-Freundlich model is not finite in this limit. This behavior is a direct reflection of breadth in the resonance in the affinity spectrum as manifested through the parameter β.

The generalization of eq. 1.13 that corresponds to the Sips affinity spectrum is:

$$n_i^{(w)} = \frac{b_1 \left(\tilde{K}_1 c_i\right)^{\beta_1}}{1 + \left(\tilde{K}_1 c_i\right)^{\beta_1}} + \frac{b_2 \left(\tilde{K}_2 c_i\right)^{\beta_2}}{1 + \left(\tilde{K}_2 c_i\right)^{\beta_2}} \tag{4.100}$$

where $b = b_1 + b_2$ as in eq. 4.84. This model adsorption isotherm equation describes the surface complexation of an adsorptive ion in the same spirit as eq. 1.13 when it is applied as the "two-site, nonelectrostatic" model of strong ion adsorption. However, in the present context ionic strength effects on adsorption are deemed important and, therefore, a diffuse swarm of adsorbed ions should exist along with surface complexes.

According to eq. 1.43 and the concepts used to derive eqs. 4.31 and 4.34,[77]

$$\sigma_{\mathrm{p}} + V_D \sum\nolimits_i Z_i(\bar{c}_i - c_{0i}) = 0 \tag{4.101}$$

expresses the contribution of the diffuse ion swarm to surface charge balance, where σ_p is in mol$_c$ kg^{-1}, Z_i is the valence of a diffuse-swarm ion of type i whose bulk concentration is c_{0i}, and

$$\bar{c}_i \equiv (1/V_D) \int_{V_D} c_i(x) d^3x \tag{4.102}$$

is the average concentration (mol dm^{-3}) of a diffuse-swarm ion in an aqueous solution volume V_D (L kg^{-1}). This latter parameter can be estimated in modified Gouy-Chapman theory following the method used to calculate the exclusion distance (eq. 4.34). Thus V_D can be equated to the average volume of aqueous solution from which coions are excluded in the vicinity of a charged particle surface. For monovalent coions near a negatively charged planar surface,

$$V_D \equiv a_s \int_a^{\infty} \left[1 - \frac{c(x)}{c_0}\right] dx = (2a_s/\kappa)\left\{1 - \exp\left[-\left|y_0(a)\right|/2\right]\right\} \tag{4.103}$$

where a_s is specific surface area, κ is defined by eq. 4.7, and $y_0(a)$ is given by eq. 4.27 with $f = 0$ and $\sigma < 0$. Since κ^2 is proportional to ionic strength (eq. 4.19), eq. 4.103 can be expressed as the logarithmic relation:[77]

$$\log V_D = \alpha - \frac{1}{2} \log I \qquad (I \text{ in mol dm}^{-3}) \tag{4.104}$$

where the constant parameter α encapsulates the logarithm of the factors other than ionic strength. Equation 4.104 indicates an increasing V_D with decreasing ionic strength, as is observed experimentally for humic substances in aqueous solutions with ionic strength varying from 0.001 to 2 mol dm^{-3}.[78] The parameter α in eq. 4.104 can be determined conveniently from extrapolation of data on V_D(I) to its value at I = 10 mol dm^{-3}: $\alpha = \log\ V_D(10) + \frac{1}{2}$. An extrapolation of this kind for a number of humic and fulvic acids indicates $\alpha \approx -0.53 \pm 0.01$, which is essentially the same as the coefficient of log I.[78]

Figure 4.11 shows the affinity spectra for fulvic and humic acid as determined by applying eqs. 4.100 and 4.101 to 49 data sets on proton adsorption by humic substances.[79] In this application, the concentration variable in eq. 4.100 was equated with $\bar{c}_i$ in eq. 4.101, which then is computed during the course of parameter optimization. (This procedure is quite analogous to the iterative computation of a "surface potential" during the course of surface speciation calculations based on, say, the Triple Layer Model.[3]) Thus a diffuse swarm concentration (typically one to two orders of magnitude greater than the bulk proton concentration, c_0) is assumed to determine strong proton adsorption, modeled by the Langmuir-Freundlich isotherms in eq. 4.100. The optimized parameters for the humic substances are:[79]

	b_1 ($mol_c\ kg^{-1}$)	$\log \tilde{K}_1$	β_1	b_2 ($mol_c\ kg^{-1}$)	$\log \tilde{K}_2$	β_2
Fulvic	5.88	2.34	0.38	1.86	8.60	0.53
Humic	3.15	2.93	0.50	2.55	8.00	0.26

However, the coefficient of log I in eq. 4.104 also was optimized while the parameter $\alpha \equiv [\log V_D(10) - \text{"coefficient"}]$ was set equal to $[-1 - \text{"coefficient"}]$, thus leaving eq. 4.104 with a single adjustable parameter. The resulting absolute values of "coefficient" were 0.57 (fulvic acid) and 0.49 (humic acid), which do not differ

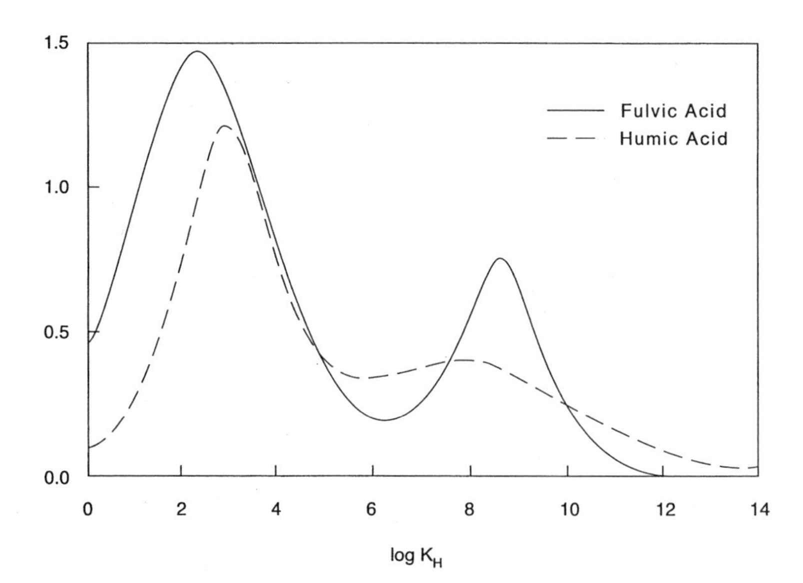

Fig. 4.11 Affinity spectra for proton adsorption by humic and fulvic acid based on fitting eqs. 4.100 to 4.104 to titration data on a broad variety of humic substances.[79] The y-axis is the affinity spectrum for log K_H, modeled as the sum of two Sips affinity spectra (eq. 4.97).

statistically from the theoretical value of 0.5. Notable in fig. 4.11 and in the parameter set above are the similarity in log $\tilde{K}_1$ and log $\tilde{K}_2$ for both humic substances, whereas the corresponding b_1 values are quite different, indicating a much larger population of carboxyl groups in fulvic acid.

Equation 4.100 can be generalized to describe the concurrent adsorption of protons and metal cations following the line of reasoning that led to eq. 4.95.[80] Given the success of eq. 4.100 as a model adsorption isotherm equation for $n_{\mathrm{H}}^{(w)}$, two terms are assumed to be necessary for modeling $n_M^{(w)}$ as well, where M refers to a metal cation, such as Cd^{2+}. For each term, the competition between protons and metal cations for the class of adsorption sites to which the term refers is taken into account by generalizing eq. 4.99 to have the form:

$$n_i(c_H, c_M) = b\frac{\left(\tilde{K}_i c_i\right)^{\beta_i}}{\left(\tilde{K}_H c_H\right)^{\beta_H} + \left(\tilde{K}_M c_M\right)^{\beta_M}} \frac{\left[\left(\tilde{K}_H c_H\right)^{\beta_H} + \left(\tilde{K}_M c_M\right)^{\beta_M}\right]^p}{1 + \left[\left(\tilde{K}_H c_H\right)^{\beta_H} + \left(\tilde{K}_M c_M\right)^{\beta_M}\right]^p} \quad (4.105)$$

where $i = H$ or M. The first factor on the right side of eq. 4.105 is the maximum surface excess of protons and metal cations for the class of adsorption sites whose model affinity parameters for these two adsorptives are $\tilde{K}_H$ and $\tilde{K}_M$, respectively. The second factor gives the fraction of these adsorption sites occupied by species i ($i = H$ or M). The exponents β_i ($i = H, M$) reflect the "smearing-out" of the affinities of these adsorption sits for the two adsorptives, evidently by electrostatic effects. Thus the first two factors on the right side of eq. 4.105 combine to describe a capacity factor for the adsorption of species i ($i = H$ or M). The third factor is the result of introducing the competitive-adsorption model isotherm,

$$\frac{K\left[(k_H c_H)^{\beta_H} + (k_M c_M)^{\beta_M}\right]}{1 + K\left[(k_H c_H)^{\beta_H} + (k_M c_M)^{\beta_M}\right]}$$

into eq. 4.80 in place of the Langmuir adsorption isotherm equation, then integrating over all K with $f(K)dK$ given by a Sips affinity spectrum (eq. 4.97) whose breadth parameter is p. The modal affinity parameters in eq. 4.105 then are defined as $\tilde{K}_i \equiv \tilde{K}^{\frac{1}{\beta_i}} k_i$ ($i = H, M$), where ln $\tilde{K}$ is the mode of the Sips distribution.[80] This *ansatz* is tantamount to assuming that the affinity parameter for a species i ($i = H$ or M) factorizes into a part that is species-specific and subject to "smearing-out" $\left(k_i^{\beta_i}\right)$ times a part that is adsorption site-specific $\left(\tilde{K}\right)$ and subject to site heterogeneity as reflected in the breadth parameter p.

The generalization of eq. 4.100 that corresponds to eq. 4.105 is:[81]

$$n_i^{(w)} = b_1 \frac{\left(\tilde{K}_{1i} c_i\right)^{\beta_{1i}}}{\left(\tilde{K}_{1\mathrm{H}} c_{\mathrm{H}}\right)^{\beta_{1H}} + \left(\tilde{K}_{1\mathrm{M}} c_{\mathrm{M}}\right)^{\beta_{1M}}} \times \frac{\left[\left(\tilde{K}_{1\mathrm{H}} c_{\mathrm{H}}\right)^{\beta_{1H}} + \left(\tilde{K}_{1\mathrm{M}} c_{\mathrm{M}}\right)^{\beta_{1M}}\right]^{p_1}}{1 + \left[\left(\tilde{K}_{1\mathrm{H}} c_{\mathrm{H}}\right)^{\beta_{1H}} + \left(\tilde{K}_{1\mathrm{M}} c_{\mathrm{M}}\right)^{\beta_{1M}}\right]^{p_1}}$$

$$+ b_2 \frac{\left(\tilde{K}_{2i}c_i\right)^{\beta_{2i}}}{\left(\tilde{K}_{2\mathrm{H}}c_{\mathrm{H}}\right)^{\beta_{2H}} + \left(\tilde{K}_{2\mathrm{M}}c_{\mathrm{M}}\right)^{\beta_{2M}}} \times \frac{\left[\left(\tilde{K}_{2\mathrm{H}}c_{\mathrm{H}}\right)^{\beta_{2H}} + \left(\tilde{K}_{2\mathrm{M}}c_{\mathrm{M}}\right)^{\beta_{2M}}\right]^{p_2}}{1 + \left[\left(\tilde{K}_{2\mathrm{H}}c_{\mathrm{H}}\right)^{\beta_{2H}} + \left(\tilde{K}_{2\mathrm{M}}c_{\mathrm{M}}\right)^{\beta_{2M}}\right]^{p_2}} \tag{4.106}$$

In the special case of proton adsorption only, eq. 4.106 reduces to eq. 4.100 with the definition $\beta_\ell \equiv \beta_{\ell\mathrm{H}} p_\ell \quad (\ell = 1,\ 2)$. Otherwise, eq. 4.106 is a twelve-parameter (!) model adsorption isotherm equation whose optimization is performed along with that of eq. 4.104, typically involving the coefficient of log I instead of α as the adjustable parameter. Figure 4.12 shows an application of eq. 4.106 to the concurrent adsorption of protons and Cd^{2+} by a peat humic acid.[81] The curves through the data points for two values of pH and ionic strength were calculated with the coefficient of log I in eq. 4.104 equal to $-$ 0.43 ($\alpha = -$ 0.57) and the parameter values:

log $\tilde{K}_{1i}$		β_{1i}	log $\tilde{K}_{2i}$	β_{2i}
H:	2.98	0.86	8.73	0.57
Cd:	0.10	0.81	2.03	0.48
$b_l = 2.74$ mol kg^{-1}		$p_1 = 0.54$	$b_2 = 3.54$ mol kg^{-1}	$p_2 = 0.54$

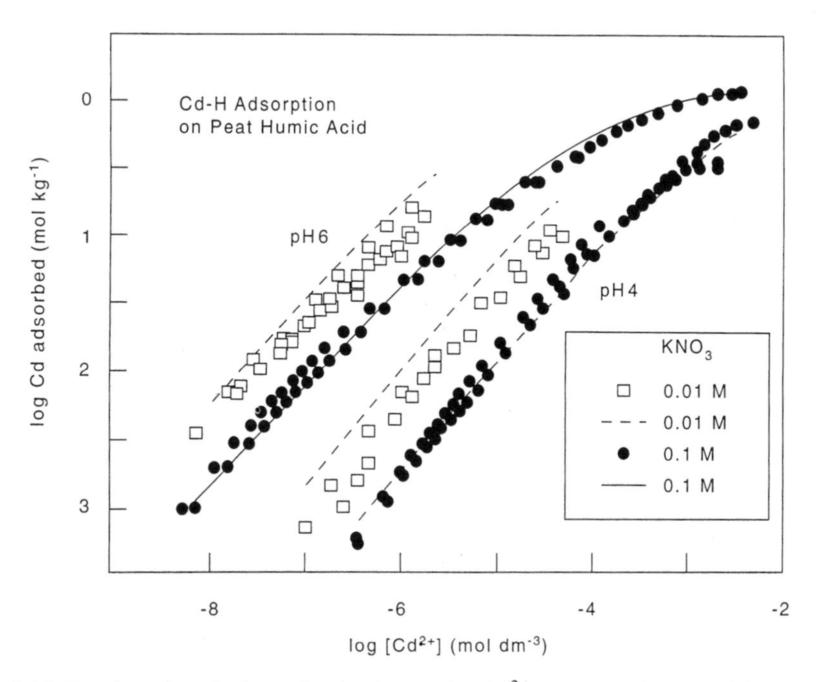

Fig. 4.12 Log-log plot of adsorption isotherms for Cd^{2+} on a peat humic acid at two pH values and ionic strengths.[81] The curves through the data points represent fits of eq. 4.106 constrained by eqs. 4.101 and 4.104.

The results for log $\tilde{K}_{1i}$ and log $\tilde{K}_{2i}$ support the identification of corresponding affinity classes as "carboxylic" and "phenolic OH" groups, respectively. This attribution suggests that b_1, log $\tilde{K}_{1H}$ and b_2, log $\tilde{K}_{2H}$ can be determined with proton titration data alone, but attempts to use this approach in an effort to reduce the number of adjustable parameters in eq. 4.106 when applied to concurrent proton and metal cation binding have not proved successful.[82] On the other hand, the quality of the data fitting in fig. 4.12 and the concomitant partitioning of adsorbed metal cations into concurrent surface-complex and diffuse-swarm species implied by eqs. 4.104 and 4.106 indicate the need for multifaceted adsorption models with at least several parameters in order to model the reactions of natural particles with aqueous solutions having wide variations of pH and ionic strength.

Notes

1. Recent effort toward developing a molecular model of the diffuse ion swarm is described by L. Blum and D. Henderson, pp. 239–276 in *Fundamentals of Inhomogeneous Fluids*, D. Henderson (Ed.), Marcel Dekker, New York, 1992. Experimental studies designed to probe the structure of the diffuse swarm are detailed in P. Fenter, L. Cheng, S. Rihs, M. Machesky, M.J. Bedzyk, and N.C. Sturchio, *J. Colloid Interface Sci.* **225**:154 (2000) and P. Fenter, H. Teng, P. Geissbühler, J.M. Hanchar, K.L. Nagy, and N.C. Sturchio, *Geochim. Cosmochim. Acta* **64**:3663 (2000). See also the other references cited in note 27 of chapter 2.
2. The classic papers on the Gouy-Chapman model are: M. Gouy, *J. Phys.* **9**:457 (1910); D.L. Chapman, *Phil. Mag.* **25**:475 (1915); D.C. Grahame, *Chem. Rev.* **41**:441 (1947); and E.C. Childs, *Trans. Faraday Soc.* **50**:1356 (1954). See also E.J.W. Verwey and J.Th.G. Overbeek, *Theory of the Stability of Lyophobic Colloids*, Dover, New York, 1999. More complete models of the diffuse ion swarm are reviewed comprehensively by L. Blum and D. Henderson, pp. 239–276 in *Fundamentals of Inhomogeneous Fluids*, Di Henderson (Ed.), Marcel Dekker, New York, 1992; S.L. Carnie and G.M. Torrie, *Advan. Chem. Phys.* **56**:141 (1984); and P. Attard, *Advan. Chem. Phys.* **92**:1 (1996).
3. See, e.g., W.D. Schecher and D.C. McAvoy, *MINEQL+, A Chemical Equilibrium Modeling System*, Environmental Research Software, Hallowell, ME, 2001, and D.A. Dzombak and F.M.M. Morel, *Surface Complexation Modeling*, John Wiley, New York, 1990.
4. These concepts are discussed in detail by S.L. Carnie and G.L. Torrie, *Advan. Chem. Phys.* **56**:141(1984). The Poisson equation is a partial differential equation that describes the spatial variability of the mean electrostatic potential resulting from a swarm of ions interacting through the coulomb force. The potential of mean force is the average potential energy associated with the average force acting on a selected ion. This force includes not only coulomb interactions, but also interactions with the solvent and all varieties of short-range (van der Waals) interaction.
5. See S.J. Miklavcic, pp. 81–134 in *Interfacial Forces and Fields*, J.-P. Hsu (Ed.), Marcel Dekker, New York, 1999, for a comprehensive review of these issues.
6. This premise underlies a broad variety of models used in the molecular theory of condensed phases and known collectively as "mean field theory." Examples include the Bragg-Williams model of solid solutions, the Curie-Weiss model of a ferromagnet, and the van der Waals model of a liquid. For a review of mean field theory, see, e.g., T.L. Hill, *An Introduction to Statistical Thermodynamics*,

Dover, New York, 1986. The fundamental notion in mean field theory is that any ion or molecule in a condensed phase can be portrayed as an individual species acted upon by an average long-range force created by all other molecular species in the phase. In the present example, the distribution of ions in the diffuse swarm is described solely in terms of a mean electrostatic potential created by the coulomb interactions between all of the ions in the swarm.

7. The parameter $d_0/2$ is equal to the radius of the ion (either counterion or coion) plus twice the radius of a water molecule (0.14 nm). Values of the ion-water molecule distance of closest approach in aqueous solution have been catalogued for many inorganic ions [see, e.g., H. Ohtaki and T. Radnai, *Chem. Rev.* **93**:1157 (1993)] and shown to be simply the sum of the radius of a water molecule given above, plus the crystallographic radius of the ion, to a high degree of statistical inference [see G. Sposito, *Metal Ions Biol. Sys.* **20**:1 (1986)]. The case of unequal radii for the counterion and coion is discussed by J.P. Valleau and G.M. Torrie, *J. Chem. Phys.* **76**:4623 (1982), **81**:6291 (1984).
8. An introductory discussion of eq. 4.5 and its principal implications for diffuse swarm behavior is given in chapters 3 and 5 of D.F. Evans and H. Wennerström, *The Colloidal Domain,* VCH Publishers, New York, 1994. The left side of eq. 4.5 is the Laplacian operator acting on $\psi(x)$, the various forms of which under differing symmetry are described in detail by P. Moon and D.E. Spencer, *Field Theory Handbook*, Springer-Verlag, New York, 1988.
9. See, e.g., table 2.3 in G. Sposito, *The Chemistry of Soils*, Oxford University Press, New York, 1989 (clay minerals) and S. Pivovarov, *J. Colloid Interface Sci.* **196**:321 (1997) (oxide minerals) for estimates of σ^*.
10. These basic concepts and many salient features of eq. 4.5 are discussed authoritatively in chapter 7 of R.J. Hunter, *Foundations of Colloid Science*, Oxford University Press, New York, 2001. Note that the use of eq. 4.9 assumes that σ is a constant, predetermined by the structure of the adsorbent and the composition of the aqueous solution with which the adsorbent has equilibrated. Thus ψ(a) is determined by eq. 4.9 and the imposed value of σ, as follows from the uniqueness of solutions of eq. 4.5. See, e.g., A.D. MacGillivray, *J. Chem. Phys.* **57**:407 (1972) for a brief discussion of this point.
11. The PBE Theorems were first demonstrated for a special case of eq. 4.12, when $\nu = 0$ (cylindrical surface), $|Z_i| = 1$, and only two ion types are present, by A.D. MacGillivray, *J. Chem. Phys.* **56**:80 (1972).
12. The properties of modified Bessel functions are described in chapters 9 and 10 of M. Abramowitz and I. Stegun, *Handbook of Mathematical Functions*, Dover, New York, 1972. The standard mathematical reference is G.N. Watson, *A Treatise on the Theory of Bessel Functions*, Cambridge University Press, New York, 1995.
13. The modified Bessel function of order zero has a logarithmic singularity at the origin, but decays super-exponentially as its argument increases. It has the integral representation (see note 12):[12]

$$K_0(\kappa x) = \int_0^\infty \frac{\exp\left[-\kappa\left(x^2+s^2\right)^{\frac{1}{2}}\right]}{(x^2+s)^{\frac{1}{2}}}\,ds$$

which is a superposition of potentials for a spherical particle surface in the Debye-Hückel approximation (eq. 4.22c), the cylinder thus being conceptualized as a stacking of circles of radius x along an axis whose coordinate is the variable s.

14. Equation 4.24 is obviated if eq. 4.20 is interpreted strictly as an *asymptotic* (large z) solution of eq. 4.12, with the constant parameter A to be determined by matching $y_\nu(z)$ to an exact solution of eq. 4.12 at sufficiently large z. In this situation, A is only a matching parameter, with no constraint imposed on its

value by the lower boundary condition in eq. 4.13. Instead, a constraint is imposed on the smallest value of z for which eq. 4.20 is an accurate description of $y(z)$.

15. The case $\nu = 0$ (cylinder) can be solved analytically for a symmetric electrolyte if the right side of eq. 4.5 is approximated by deleting the exponential term for the coion. This approximation amounts to complete coion exclusion from the diffuse swarm, a situation that may obtain for sufficiently large values of $-Zy$ (e.g., $-Zy > 1.2$), where Z is the valence of the counterion. The solution for $y(x)$ in this case is given by R. Fuoss, A. Katchalsky, and S. Lifson, *Proc. Natl. Acad. Sci. USA* **37**:579 (1951) and T. Alfrey, P.W. Berg, and H. Morawetz, *J. Polymer Sci.* **7**:543 (1951). Comparisons between this approximate analytical solution and the results of numerical integration of eq. 4.12 for 1:1 electrolytes (e.g., NaCl) are made by M. Le Bret and B.H. Zimm, *Biopolymers* **23**:271, 287 (1984). Comparisons between eq. 4.22b (Debye-Hückel approximation) and the results of numerical integration of eq. 4.5 for a 1:1 electrolyte are made by D. Stigter, *J. Colloid Interface Sci.* **53**:296 (1975), *J. Phys. Chem.* **82**:1603 (1978), who provides values of a scale factor that relates $y_0(x)$ to the potential obtained by numerical integration. This factor is within 10% of the value 1.0 for $\kappa a < 0.03$ and $y_0(a) < 4$. A.D. MacGillivray and J.J. Winkleman [*J. Chem. Phys.* **45**:2184 (1966)] also show that eq. 4.22b is a good approximation to the exact solution of eq. 4.12 for the 1:1 electrolyte under the same conditions on κa and $y_0(a)$ as given above. A.D. MacGillivray [*J. Chem. Phys.* **57**:4075 (1972)] supports these observations with a formal proof that the exact solution tends asymptotically to $y_0(x)$ in the limit $\kappa a \downarrow 0$ if $y_0(a) < 4$. H. Ohshima [*J. Colloid Interface Sci.* **200**:291 (1998)] derives an approximate solution of eq. 4.12 for a 1:1 electrolyte with $\kappa a > 0.1$ which is accurate to within 10% at $x = a$. Approximate solutions also are presented for 2:1 and 3:1 electrolytes.
16. The case $\nu = -1/2$ (sphere) has not been solved analytically except in the Debye-Hückel approximation. Extensive tables of the results of numerical integration of eq. 4.5 have been published by A.L. Loeb, J.T.G. Overbeek, and P.H. Wiersma, *The Electrical Double Layer Around a Spherical Colloid Particle*, M.I.T. Press, Cambridge, MA, 1961, and D. Stigter [*Electroanal. Chem. Interfacial Electrochem.* **37**:61 (1972)] has tabulated values of a scale factor that relates $y_{-\frac{1}{2}}(\varepsilon)$ to the potential at $z = \varepsilon$ obtained by numerical integration and a scale factor relating $(dy_{-\frac{1}{2}}/dz)_{z=\varepsilon}$ to $(\kappa\ell)^{-1}$ (These scale factors are required if $y_{-\frac{1}{2}}(z)$ is made to match the numerical results for large values of z.) Approximate analytical solutions of eq. 4.5 for $\nu = -1/2$ based on replacing dy/dx' with the derivative for the case $\nu = \frac{1}{2}$ (i.e., a flat surface) are given by L.R. White [*J. Chem. Soc. Faraday. Trans. II* **73**:577 (1977)] and H. Ohshima, T.W. Healy, and L.R. White [*J. Colloid Interface Sci.* **90**:17 (1982)]. These approximate solutions, which have been shown by A.N. Stokes [*J. Chem. Phys.* **65**:261 (1976)] to be the first terms of an asymptotic expansion for large x', are accurate to within 5% for $y(a) < 10$ and $\kappa a > 0.1$.
17. B. Abraham-Shrauner, *J. Math. Biol.* **2**:333 (1975), **4**:201 (1977). See also D.C. Grahame, *J. Chem. Phys.* **21**:1054 (1953) and section 5.1 in G. Sposito, *The Surface Chemistry of Soils*, Oxford University Press, New York, 1984, for explicit equations describing the solutions of eq. 4.12 for 1:1, 2:1, and 1:2 electrolytes. Z. Chen and R.K. Singh [*J. Colloid Interface Sci.* **245**:301 (2002)] have rediscovered the results of Abraham-Shrauner by induction for the cases of 1:1 and 2:1 electrolytes.
18. See chapter 4 in M. Abramowitz and I. Stegun, *Handbook of Mathematiacl Functions*, Dover, New York, 1972, for a comprehensive summary of the properties of hyperbolic functions.
19. J. B. Harsh and S. Xu, *Advan. Agron.* **14**:131 (1990), R.F. Giese, W. Wu, and C.J. van Oss, *J. Dispersion Sci. Technol.* **17**:527 (1996). There are several reasons for this discrepancy, all of which relate to the model through which an electrostatic

potential is calculated from an electrophoretic mobility measurement. For a comprehensive discussion of electrokinetics measurements, see R.J. Hunger, *Zeta Potential in Colloid Science*. Academic Press, New York, 1981.

20. G.M. Torrie and J.P. Valleau, *J. Phys. Chem.* **86**:3251 (1982). At bulk concentrations greater than 100 mol m^{-3}, modified Gouy-Chapman theory cannot describe the effects of ion-ion correlations on ion concentrations in the diffuse swarm. Even at lower bulk concentrations, eq. 4.29 *overestimates* the scaled potential for $\tilde{\sigma} > 0.6$ nm^{-2}. Thus, eq. 4.29 provides a reasonably—but not perfectly—accurate description of the physical system it is supposed to model, if the electrolyte is 1:1 and its bulk concentration is below 100 mol m^{-3}. See also S.L. Carnie and G.M. Torrie, *Advan Chem. Phys.* **56**:141 (1984).
21. This definition of $\varphi(x)$ is given by B.H. Zimm and M. Le Bret [*J. Biomolecular Structure Dynamics* **1**:461 (1983)], who also conjecture that the limit of $\varphi(x)$ as $\kappa \downarrow 0$ is sensitively dependent on $|\sigma|$ for cylindrical particles and is equal to zero for spherical particles, a trend in keeping with the weakening of the scaled potential at any $x > a$ as the particle surface becomes more spherical, noted in connection with the Debye-Hückel approximation (eq. 4.22). For the flat particle surface, the very compressed counterion swarm implied by eq. 4.32 stands in sharp contradiction to the common misnomer in referring loosely to κ^{-1} as the "thickness of the diffuse double layer," since κ^{-1} is infinitely large in the limit of infinite dilution.
22. See, e.g., R.M. Pashley and J.P. Quirk, *Soil Sci. Soc. Am. J.* **61**:58 (1997) for a derivation of eq. 4.34 and for similar results applicable to 2:1 electrolytes.
23. The coion exclusion specific surface area $a_s^{\mathbf{ex}}$ will differ from the specific surface area as measured by, say, gas adsorption or microscope techniques if the adsorbent surface that repels coions is not the same as that accessible to a vapor adsorptive or revealed to an electron microscope. For example, although the clay mineral, illite, has the same specific surface area as determined by N_2 adsorption (eq. 1.5) when bearing either Li^+ or Cs^+ as adsorbed cations on its siloxane surfaces, the value of a_s^{ex} for Li-illite is many times larger than for Cs-illite because strong surface complexation of Cs^+ by the mineral reduces σ essentially to zero, thus vitiating coion repulsion. For a discussion of $a_s^{ex}$and other specific surface areas of natural particles, see section 1.4 in G. Sposito, *The Surface Chemistry of Soils*, Oxford University Press, New York, 1984.
24. See, e.g., G. Sposito, pp. 127–155 in *Clay-Water Interface and its Rheological Implications*, N. Güven and R.M. Pollastro (Eds.), The Clay Minerals Society, Boulder, CO, 1992. Extension of eq. 4.34 to the case of a disk-shaped particle with edges whose dimensions and surface charge differ from those of its flat circular faces is described by F.-R.C. Chang and G. Sposito, *J. Colloid Interface Sci.* **163**:19 (1994), **178**:555 (1996). The values of d_{ex} calculated with a numerical solution of the Poisson-Boltzmann equation for a particle with a large radius (radius $\gg \kappa^{-1}$) are similar to those obtained from eq. 4.34 because coion exclusion from the region of electric field "spillover" outside the coion-attracting edge surface is relatively smaller than that from above the coion-repelling circular face. Very large deviations from eq. 4.34, however, are predicted by the disk model for small particles (radius $\lesssim \kappa^{-1}$) in dilute 1:1 electrolyte solutions, because coion depletion outside the edge surface becomes very important.
25. D.R. Turner, chapter 4 in *Metal Speciation and Bioavailability in Aquatic Systems*, A. Tessier and D.R. Turner (Eds.), John Wiley, New York, 1995. A discussion of single-species activity coefficients is given by G. Sposito, *Soil Sci. Soc. Am. J.* **48**:531 (1984).
26. Seminal reviews of surface complexation modeling are given by S. Goldberg, *Advan. Agron.* **47**:233 (1992) and chapter 10 in *Structure and Surface Reactions of Soil Particles*, P.M. Huang, N. Senesi, and J. Buffle (Eds.), John Wiley, New York, 1998. See also J.A. Davis and D.B. Kent, *Rev. Mineral* **23**:177 (1990). Issues related to the application of surface complexation models to natural par-

ticles are discussed by J.A. Davis, J.A. Coston, D.B. Kent, and C.C. Fuller, *Environ. Sci. Technol.* **32**:2820 (1998) and D.B. Kent, R.H. Abrams, J.A. Davis, J.A. Coston, and D.R. LeBlanc, *Water Resour. Res.* **36**:3411 (2000).

27. The pathway of development of the Triple Layer Model is traced in: O. Stern, *Z. Elektrochem.* **30**:508 (1924); D.E. Yates, S. Levine, and T.W. Healy, *J.C.S. Faraday I* **70**:1807 (1974); J.A. Davis, R.O. James, and J.O. Leckie, *J. Colloid Interface Sci.* **63**:480 (1978); J.A. Davis and J.O Leckie, *J. Colloid Interface Sci.* **67**:90 (1978), *J. Colloid Interface Sci.* **74**:32 (1980), pp. 299–317 in *Chemical Modeling in Aqueous Systems*, E.A. Jenne (Ed.), American Chemical Society, Washington, DC, 1979; R.O. James and G.A. Parks, *Surface Colloid Sci.* **12**:119 (1982); and K.F. Hayes and J.O. Leckie, *J. Colloid Interface Sci.* **115**:564 (1987).
28. The notation in the present section conforms to that in S. Goldberg, *Advan. Agron.* **47**:233 (1992). The appellation, "intrinsic," was invoked by W. Stumm, H. Hohl, and F. Dalang, *Croat. Chem. Acta* **48**:491 (1976).
29. Implementation of the Triple Layer Model for speciation calculations in which soluble complexation, surface complexation, and precipitation may occur is discussed by W.D. Schecher and D.C. McAvoy, *MINEQL+, A Chemical Equilibrium Modeling System*, Environmental Research Software, Hallowell, ME, 2001.
30. See, e.g., Sections 4.1 and 5.3 in G. Sposito, *Chemical Equilibria and Kinetics in Soils*, Oxford University Press, New York, 1994. Surface species activity coefficients, like their counterparts for aqueous solution species, can be calculated by applying the Gibbs-Duhem equation to constrain the composition dependence of a conditional equilibrium constant at fixed T and P. The observed composition dependence of the latter is then introduced into conventional expressions for the species activity coefficients to compute their composition dependence.
31. An introductory discussion of the Davies equation is given in Section 4.5 of G. Sposito, *The Chemistry of Soils*, Oxford University Press, 1989. For a comparison of the ion-association and specific interaction models of aqueous species activity coefficients, see G. Sposito, *Soil Sci. Am. J.* **48**:531 (1984).
32. Equation 4.48 is derived in the context of mean field theory by G. Sposito, *J. Colloid Interface Sci.* **91**:329 (1983). For a summary discussion of the mean field theory approach to surface complexation modeling, see section 5.2 in G. Sposito, *The Surface Chemistry of Soils*, Oxford University Press, 1984. M. Borkovec [*Langmuir* **13**:2608 (1997)] has illustrated the accuracy of the mean field theory approximation for protonation reactions on a planar particle surface.
33. This fact, first stated by J. W. Gibbs and emphasized by W. Stumm [*Croat. Chem. Acta* **48**:491(1976)] is discussed in Special topic 2 of G. Sposito, *Chemical Equilibria and Kinetics in Soils*, Oxford University Press, 1994, and in section 3.3 of G. Sposito, *The Surface Chemistry of Soils*, Oxford University Press, 1984.
34. The experimental difficulties inherent to extrapolation of data obtained at non-zero surface charge and ionic strength to the Infinite Dilution Reference State as a method of calculating intrinsic equilibrium constants is discussed by P.W. Schindler, pp. 1–49 in *Adsorption of Inorganics at Solid-Liquid Interfaces*, M.A. Anderson and A.J. Rubin (Eds.), Ann Arbor Science, Ann Arbor, MI, 1981; R.O. James and G.A. Parks, *Surf. Colloid Sci.* **12**:119 (1982); and K.F. Hayes, G. Redden, W. Ela, and J.O. Leckie, *J. Colloid Interface Sci.* **142**:448 (1991).
35. Parameter estimation for the Triple Layer Model (or any surface complexation model) is a non-trivial optimization problem because of a lack of independence (e.g., SOH_T, a sensitive parameter that is difficult to measure accurately, and the intrinsic equilibrium constants are correlated negatively) and differing sensitivity (e.g., SOH_T versus the two capacitance densities) among its parameters. Detailed analyses of these important issues are given by S. Goldberg, *Advan. Agron.* **47**:233 (1992); K.F. Hayes et al., *J. Colloid Interfac. Sci.* **142**:448 (1991); L.K. Koopal, W.H. van Riemsdijk, and M.G. Roffey, *J. Colloid Interface Sci.* **118**:117 (1987); J.

Lützenkirchen, *Environ. Sci. Technol.* **32**:3149 (1998); and C. J. Tadanier and M. J. Eick, *Soil Sci. Soc. Am. J.* 66:1505 (2002). L.E. Katz and K.F. Hayes [*J. Colloid Interface Sci.* **170**:477 (1995)] note that a broad range of adsorption data obtained under widely-varying conditions of pH, ionic strength, surface loading, and solids concentration can be most valuable in constraining the parameter estimation problem. Infrared spectroscopy also has proved to be useful in reducing the number of surface species that must be considered when modeling the adsorption of oxyanions. See J. Nordin, P. Persson, E. Laiti, and S. Sjöberg, *Langmuir* **13**:4085 (1997) (*phthalate*); P. Persson, J. Nordin, J. Rosenqvist, L. Lövgren, L.-O. Öhman, and S. Sjöberg, *J. Colloid Interface Sci.* **206**:252 (1998) (*phthalate*); J. Nordin, P. Persson, A. Nordin, and S. Sjöberg, *Langmuir* **14**:3655 (1998) (*pyromellitate*); S. Goldberg, *Soil Sci. Soc. Am. J.* **63**:823 (1999) (*borate*); O.S. Pokrovsky, J.A. Mielczarski, O. Barres, and J. Schott, *Langmuir* **16**:2677 (2000) (*carbonate*); J.-F. Boily, N. Nilsson, P. Persson, and S. Sjöberg, *Langmuir* **16**:5719 (2000) (*pyromellitate*); J.-F. Boily, P. Persson, and S. Sjöberg, *Geochim. Cosmochim. Acta* **64**:3453 (2000) (*phthalate, trimellitate, pyromellitate*); S. Goldberg and C.T. Johnston, *J. Colloid Interface Sci.* **234**:204 (2001) (*arsenite, arsenate*); M. Villalobos and J.O. Leckie, *J. Colloid Interface Sci.* **235**:15 (2001) (*carbonate*).

36. Vanishing "surface potential" does not imply vanishing surface charge if the two ions in the electrolyte have different radii, for then the "surface potential" depends on the charge separation induced near the particle surface by differing ion size. See, e.g., J.P. Valleau and G.M. Torrie, *J. Chem. Phys.* **76**:4623 (1982).
37. This term was coined by M.A.F. Pyman, J.W. Bowden, and A.M. Posner, *Aust. J. Soil Res.* **17**:191–195 (1979).
38. On the Triple Layer Model, $[\equiv \mathrm{SOH}_2^+]$ and $[\equiv \mathrm{SO}^-]$ are concentrations of protonated and deprotonated surface OH groups whose charge is balanced *by diffuse-swarm ions only*. Thus σ_{H}, which, in principle, can be measured as described in section 1.3, is expressed by the concentration difference:

$$\sigma_{\mathrm{H}} = \{[\equiv \mathrm{SOH}_2^+] + [\equiv \mathrm{SOH}_2^+ \cdots \mathrm{A}^-] - [\equiv \mathrm{SO}^-] - [\equiv \mathrm{SO}^- \cdots \mathrm{C}^+]\}/c_s = \sigma_0^{\mathrm{TLM}}$$

in the Triple Layer Model when *only* the 1:1 background electrolyte is present. At the p.p.z.c.,

$$\sigma_{\mathrm{OS}} = \{[\equiv \mathrm{SO}^- \cdots \mathrm{C}^+] - [\equiv \mathrm{SOH}_2^+ \cdots \mathrm{A}^-]\}/c_s = \sigma_\beta^{\mathrm{TLM}}$$

is equal to zero and, because σ_{H} is also zero, $[\equiv \mathrm{SOH}_2^+] = [\equiv \mathrm{SO}^-]$, as stated following eq. 4.60. The key insight here is that $[\equiv \mathrm{SOH}_2^+]$ and $[\equiv \mathrm{SO}^-]$ *are not directly accessible to measurement*. This attribute of the Triple Layer Model—a result of its use of the Infinite Dilution Reference State for both surface and aqueous species[26,27]—is also an evident cause of experimental difficulties in obtaining accurate values of its parameters (see note 34).
39. See, e.g., table 2 in J. Lützenkirchen, P. Magnico, and P. Behra, *J. Colloid Interface Sci.* **170**:326 (1995), wherein explicit calculations of p.p.z.c. as a function of ionic strength for several metal oxides suspended in 1:1 electrolyte solutions are presented based on the Triple Layer Model. It is instructive to apply eq. 4.61 to the values of log K_+(int) and –log K_-(int), listed in the first two columns of table 1 in this paper, to compute p.p.z.c., then compare the results with the p.p.z.c. values listed in table 2. The discrepancies among these latter values—both in respect to ionic strength variation and comparison with eq. 4.61—are typical of the precision of data fitting using the Triple Layer Model.
40. D.A. Sverjensky, *Nature* **364**:776 (1993), *Geochim. Cosmochim. Acta* **58**:3123 (1994); D.A. Sverjensky and N. Sahai, *Geochim. Cosmochim. Acta* **60**:3773 (1996), **62**:3703 (1998); N. Sahai and D.A. Sverjensky, *Geochim. Cosmochim. Acta* **61**:2801, 2827 (1997); L.J. Criscenti and D.A. Sverjensky, *Am. J. Sci.* **299**:828 (1999); N. Sahai, *Geochim. Cosmochim. Acta* **64**:3629 (2000). Equations 4.62 are developed in the second and fourth papers cited, whereas the others

develop modeling approaches to log K_C(int), log K_A(int), log K_M(int), and log K_L(int) in Eqs. 4.55–4.59, as well as to enthalpies of adsorption. See also D.A. Kulik, *Geochim. Cosmochim. Acta* **64**:3161 (2000), **65**:2027 (2001) for a complementary approach to adsorption enthalpies in the Triple Layer Model.

41. N. Sahai and D.A. Sverjensky, *Geochim. Cosmochim. Acta* **61**:2801 (1997). The correlation equations for $*K_{Li}$(int), $*K_{Na}$(int), and $*K_K$(int), $*K_{Cl}$(int), and $*K_{NO_3}$(int) are:

$$\log {}^*K_{Li}(\text{int}) = -10.047(1/D_s) + 3.56$$
$$\log {}^*K_{Na}(\text{int}) = -8.383(1/D_s) + 3.20$$
$$\log {}^*K_{K}(\text{int}) = -6.755(1/D_s) + 2.88$$
$$\log {}^*K_{Cl}(\text{int}) = -12.622(1/D_s) + 3.31$$
$$\log {}^*K_{NO_3}(\text{int}) = -6.050(1/D_s) + 2.49$$

D.A. Sverjensky [*Geochim. Cosmochim. Acta* **65**:3643 (2001)] also has developed correlation equations for the capacitance density C_1 (eq. 4.49) in terms of the hydrated radius of the cation in an indifferent electrolyte.

42. See, e.g., section 1.4 in G. Sposito, *Chemical Equilibria and Kinetics in Soils*, Oxford University Press, New York, 1994, for a derivation of eq. 4.65 and a discussion of the standard enthalpy change of a reaction. A detailed exposition of the temperature dependence of equilibrium constants is given by K. Denbigh, *The Principles of Chemical Equilibrium*, Cambridge University Press, New York, 1981.
43. Equation 4.66 is discussed in detail by G.M. Anderson and D.A. Crerar, *Thermodynamics in Geochemistry*, Oxford University Press, New York, 1933, Section 13.3, if an "Arrhenius plot" of log K versus 1/T is linear, only the first term on the right side of eq. 4.66 is needed to describe the temperature dependence of ln K in eq. 4.65. The second term, therefore, describes the curvature in an "Arrhenius plot."
44. M.L. Machesky, pp. 282–292 in *Chemical Modeling of Aqueous Systems II*, D.C. Melchior and R.L. Bassett (Eds.), American Chemical Society, Washington, DC, 1990; A. de Keizer, L.G.J. Fokkink, and J. Lyklema, *Colloids Surf.* **49**:149 (1990); L.G.J. Fokkink, A. de Keizer, and J. Lyklema, *J. Colloid Interface Sci.* **127**:116 (1989); M.A. Blesa, A.J.G. Maroto, and A.E. Regazzoni, *J. Colloid Interface Sci.* **140**:287 (1990).
45. D.A. Sverjensky and N. Sahai, *Geochim. Cosmochim. Acta.* **61**:2801 (1997).
46. M.L. Machesky, D.J. Wesolowski, D.A. Palmer, and K. Ichiro-Hayashi, *J. Colloid Interface Sci.* **200**:298 (1998).
47. Comparisons between the "2-pK" approach to eq. 4.69 and the "1-pK" approach inherent to eq. 4.69a are made by J. Lützenkirchen, *Environ. Sci. Technol.* **32**:3149 (1998). As noted in this latter article, both approaches, of course, can be accommodated by the Triple Layer Model. For a discussion of the "1-pK" approach in terms of its molecular hypotheses, see W.H. van Riemsdijk and T. Hiemstra, pp. 68–87, in *Mineral-Water Interfacial Reactions*, D.L. Sparks and T.J. Grundl (Eds.), American Chemical Society, Washington, DC, 1998.
48. D.A. Kulik, *Geochim. Cosmochim. Acta* **64**:3161 (2000).
49. M.A.A. Schoonen, *Geochim. Cosmochim. Acta* **58**:2845 (1994).
50. D.P. Rodda, B.B. Johnson, and J.D. Wells, *J. Colloid Interface Sci.* **161**:57 (1993).
51. G.W. Bruemmer, J. Gerth, and K.G. Tiller, *J. Soil Sci.* **39**:37 (1988). See also P. Trivedi and L. Axe, *Environ. Sci. Technol.* **34**:2215 (2000); **35**:1779, 1892 (2001).
52. B.B. Johnson, *Environ. Sci. Technol.* **24**:112 (1990); P.V. Brady, *Geochim. Cosmochim. Acta* **56**:2941 (1992), **58**:1213 (1994); M. Kosmulski, *J. Colloid Interface Sci.* **192**:215 (1997); M.J. Angrove, B.B. Johnson, and J.D. Wells, *J. Colloid Interface Sci.* **204**:93 (1998).

53. S. Goldberg, H.S. Forster, and E.L. Heick, *Soil Sci.* **156**:316 (1993).
54. M.L. Machesky, B.L. Bischoff, and M.A. Anderson, *Environ. Sci. Technol.* **23**:580 (1989); P. Benoit, J.G. Hering, and W. Stumm, *Appl. Geochem.* **8**:127 (1993).
55. M.K. Ridley, M.L. Machesky, D.J. Weslowski, and D.A. Palmer, *Geochim. Cosmochim. Acta* **63**:3087 (1999).
56. D.P. Rodda, B.B. Johnson, and J.D. Wells, *J. Colloid Interface Sci.* **184**:365 (1996). The four empirical parameters were obtained from nonlinear least-squares adjustment of eq. 1.13, with the second term simplified to $b_2K_2c_{Pb}$ and b_2 estimated from a plot of the adsorption data like that in fig. 1.4. The apparent increase in $b_1 + b_2$ with temperature may be an artifact of a least-squares adjustment without imposing the constraint that $b_1 + b_2$ remain constant.
57. M.J. Angrove et al., *J. Colloid Interface Sci.* **204**:93 (1998). The adsorption of Cd^{2+} is modeled as a cation exchange process for pH < p.z.n.p.c., whereas for pH > p.z.n.p.c. it is modeled with eq. 4.76. The values of log $^*\beta_2$ reported by Angrove et al. can be fit to eq. 4.68b to yield the regression equation:

$$\log {}^*\beta_2 = -0.9548 - (2727.18/T^\circ) \quad r^2 = 0.997$$

for 283.15 K < T° < 343.15 K.
58. A classic example of the "affinity class" approach to modeling ion adsorption in physical biochemistry is given by G. Scatchard, J.S. Coleman, and A.L. Shen, *J. Am. Chem. Soc.* **79**:12 (1957). Figure 3 in that article can be compared with fig. 1.4 in the present book.
59. A comprehensive survey of models based on "affinity classes" is in chapter 5 of J. Buffle, *Complexation Reactions in Aquatic Systems*, John Wiley, New York, 1988. Shorter reviews may be found in G. Sposito, *Crit. Rev. Environ. Control* **16**:193 (1986) and W.H. van Riemsdijk and L.K. Koopal, pp. 455–495 in *Environmental Particles*, Vol. 1, J. Buffle and H.P. van Leeuwen (Eds.), CRC Press (Lewis), Boca Ratón, FL, 1992. Note that eq. 4.80 can be generalized by replacing Kc/(1 + Kc) in the integrand with any other one-parameter isotherm equation.
60. More generally,

$$\int_{-\infty}^{\infty} g(x)\,\delta[\psi(x)]dx \equiv \sum\nolimits_n \int_{-\infty}^{\infty} \frac{g(x)\delta(x - x_n)dx}{|\psi'(x_n)|} = \sum\nolimits_n \left[g(x_n)\,|\psi'(x_n)|^{-1} \right]$$

where $\psi(x)$ is a differentiable function with the value 0 at a set of points $\{x_n\}$ [i.e., $\psi(x_n) = 0$, $n = 1, 2, \cdots$] while $\psi'(x_n) \equiv (d\psi/dx)_{x=x_n} \neq 0$ $(n = 1, 2, \cdots)$. Equation 4.83 is the special case, $\psi(x) = x - a$, which has the value 0 at only one point $(x_1 = a)$ with $\psi'(a) = 1$. The classic introduction to the delta-"function" is M.J. Lighthill, *Introduction to Fourier Analysis and Generalized Functions*, Cambridge University Press, Cambridge, 1958.
61. See, e.g., chapter 7 in T.L. Hill, *An Introduction to Statistical Thermodynamics*, Dover, New York, 1986, for a thorough discussion of the statistical thermodynamics approach to adsorption isotherm equations.
62. Mathematical properties of eq. 4.80 are discussed by R. Sips, *J. Chem. Phys.* **16**:490 (1948), **18**:1024 (1950); U. Landman and E.W. Montroll, *J. Chem. Phys.* **64**:1762 (1976); J. Jagiełło and J.A. Schwarz, *J. Colloid Interface Sci.* **146**:415 (1991); and G.J.M. Koper and M. Borkovec, *J. Chem. Phys.* **104**:4204 (1996). Comprehensive discussions of the issues involved with integral representations of adsorption isotherms are given in chapter 5 of W. Rudzinski and D.H. Everett, *Adsorption of Gases on Heterogeneous Surfaces*, Academic Press, San Diego, CA, 1992, and in M. Borkovec, B. Jönsson, and G.J.M. Koper, *Surface Colloid Sci.* **16**:99 (2001).
63. G. Sposito, *Soil Sci. Soc. Am. J.* **46**:1147 (1982).
64. See, e.g., W.H. Press, S.A. Teukolsky, W.T. Vetterling, and B.P. Flannery, *Numerical Recipes in FORTRAN*, Cambridge University Press, New York,

1992. Sections 15.1 and 18.5–18.7 in this charming book provide useful summaries of regularization methods for the fitting of experimental data to model equations, with no apologies for the subjective nature of these methods.

65. Examples of different adsorption isotherm equations and, therefore, different affinity spectra exhibiting equally good fits to experimental adsorption data are given by D.G. Kinniburgh, J.A. Barker, and M. Whitfield, *J. Colloid Interface Sci.* **95**:370 (1983) and D.G. Kinniburgh, *Environ. Sci. Technol.* **20**:895 (1986). See also problem 5 in chapter 1.
66. M. Borkovec, U. Rusch, M. Černik, G.J.M. Koper, and J.C. Westall, *Colloids Surfaces* **107**A:285 (1996), **122**A:267 (1997). Figures 1 and 2 in this useful review compare the affinity spectrum obtained by fitting eq. 4.80 to proton binding data with either a continuous or delta-"function" f(K) model to the affinity spectrum as known from independent experiments on the acidic functional groups of a protein and citrate, respectively.
67. The use of these three regularizing functions is discussed in detail by M. Černik, M. Borkovec, and J.C. Westall, *Environ. Sci. Technol.* **29**:413 (1995). Often the special case, $\lambda_j = \lambda$, is invoked to reduce the number of adjustable parameters. Note that λ_j serves as a "weighting function" for determining the degree of influence of $R_j[\{s_\ell\}]$ in the minimization process. Usually it is selected to be large as needed to ensure that the criterion in eq. 4.93 is met. D.S. Smith and F.G. Ferris [*Environ. Sci. Technol.* **35**:4637 (2001)], however, have developed an algorithm that optimizes λ instead of imposing its value.
68. J.L. Garcés, F. Mas, and J. Puy, *Environ. Sci. Technol.* **32**:539 (1998) describe the use of an "entropy function" in regularization for model adsorption isotherm equations. The principal advantage of this approach accrues when one has a solid notion of what the set of adjustable parameters $\{s_\ell^0\}$ should be.
69. U. Rusch, M. Borkovec, J. Daicic, and W.H. van Riemsdijk, *J. Colloid Interface Sci.* **191**:247 (1997). See also M. Cernik, M. Borkovec, and J.C. Westall, *Langmuir* **12**:6127.
70. For a derivation of this denominator, known in molecular theory as an "independent site grand partition function," see section 7–2 in T.L. Hill, *An Introduction to Statistical Thermodynamics*, Dover, New York, 1986. Note that "no species" in the context of this model means either an unoccupied (other than by water molecules) adsorption site or a site occupied by a species other than "1" or "2." For the data shown in fig. 4.9, the latter condition applies, since the measurements were made in the presence of KNO_3 background electrolyte solution.
71. Electron micrographs and chemical composition data showing these characteristics can be found in J. Buffle and G.G. Leppard, *Environ. Sci. Technol.* **29**:2169, 2176 (1995); G.A. Jackson and A.B. Bird, *Environ. Sci. Technol.* **32**:2805 (1998); J. Buffle, K.J. Wilkinson, S. Stoll, M. Filella, and J. Zhang, *Environ. Sci. Technol.* **32**:2887 (1998); D. Perret, J.-F. Gaillard, J. Dominik, and O. Atteia, *Environ. Sci. Technol.* **34**:3540 (2000).
72. Examples of this exercise are provided by H. Radovanovic and A.E. Koelmans, *Environ. Sci. Technol.* **32**:753 (1998) and by T.D. Small, L.A. Warren, E.E. Roden, and F.G. Ferris, *Environ. Sci. Technol.* **33**:4465 (1999). Once a model with a few parameters has been optimized on a large number of data sets, an attempt can be made to correlate equilibrium constants or affinity parameters in the model with composition data pertaining to the natural particle adsorbent. See, e.g., S.-Z. Lee, H.E. Allen, C.P. Huang, D.L. Sparks, P.F. Sanders, and W.J.G.M. Peijnenburg, *Environ. Sci. Technol.* **30**:3418 (1996); F. Wang, J. Chen, and W. Forsling, *Environ. Sci. Technol.* **31**:448 (1997); S. Goldberg, S.M. Lesch, and D.L. Suarez, *Soil Sci. Soc. Am. J.* **64**:1356 (2000).
73. For a review and applications of this "two-site" modeling approach, see J.A. Davis et al., *Environ. Sci. Technol.* **32**:2820 (1998), and D.B. Kent et al., *Water Resour. Res* **36**:3411 (2000). A similar approach has been described by M.H.

Bradbury and B. Bayens, *J. Contam. Hydrol.* **27**:199, 223 (1997), *Geochim. Cosmochim. Acta* **63**:325 (1999). Notable in the extensive discussion of this latter modeling approach that ensued [*J. Contam. Hydrol.* **28**:1 (1997)] is the absence of any remark on its relation to eq. 4.94.

74. The relation between eq. 4.55a and either term in eq. 1.13 becomes transparent after making the associations:

$$K_j \leftrightarrow K_M^1(\text{int}) \quad b_j \leftrightarrow [\text{SOH}] + [\text{SOM}^{m-1}] \quad c_i \leftrightarrow (\text{M}^{m+})/(\text{H}^+)$$

where j indexes one of the terms in eq. 1.13 and i indexes an adsorptive ion. See also D.G. Kinniburgh et al., *J. Colloid Interface Sci.* **95**:370 (1983) for an explicit derivation of a "competitive" Langmuir adsorption isotherm equation.

75. See, e.g., chapter 6 in J. Buffle, *Complexation Reactions in Aquatic Systems*, John Wiley, New York, 1988.

76. R. Sips, *J. Chem. Phys.* **16**:490 (1948). Application of eq. 4.97 to adsorption by soils was introduced in G. Sposito, *Soil Sci. Soc. Am. J.* **44**:652 (1979). See also D.G. Kinniburgh *J. Colloid Interface Sci.* **95**: 370 (1983), for a comparison of the Sips affinity spectrum to other models. Note that the mode of the Sips affinity spectrum, ln $\tilde{K}$ in eq. 4.99, is dependent on the values of the parameters b and β in eq. 4.97: $\ln \tilde{K} = (b/2\pi) \tan (\pi\beta/2)$. Thus, the mode shifts to larger values as the maximum surface excess and the sharpness of the resonance in the affinity spectrum increase. Conversely, a given value of $\tilde{K}$ can be generated either by a broad affinity spectrum and a large number of adsorption sites or by a narrow spectrum and few sites.

77. This concept of including the diffuse swarm in models of proton adsorption by natural particles was introduced by J.A. Marinsky, pp. 49–81 in *Aquatic Surface Chemistry*, W. Stumm (Ed.), Wiley-Interscience, New York, 1987. Equation 4.104 is discussed by D.G. Kinniburgh, W.H. van Riemsdijk, L.K. Koopal, M. Borkovec, M.F. Benedetti, and M.J. Avena, *Colloids Surf.* **151**A:147 (1999). The parameter V_D is an average volume of aqueous solution around a charged particle in which all counterions and coions in the diffuse swarm are included. See also section 5.1 in G. Sposito, *The Surface Chemistry of Soils*, Oxford University Press, New York.

78. M.F. Benedetti, W.H. van Riemsdijk, and L.K. Koopal, *Environ. Sci. Technol.* **30**:1805 (1996).

79. C.J. Milne, D.G. Kinniburgh, and E. Tipping, *Environ. Sci. Technol.* **35**:2049 (2001). For a comparison among alternative models of proton adsorption by humic acid, see C.J. Milne, D.G. Kinniburgh, J.C.M. De Wit, W.H. van Riemsdijk, and L.K. Koopal, *Geochim. Cosmochim. Acta* **59**:1101 (1995). Applications to metal cation absorption are given in C.J. Milne, D.G. Kinniburgh, W.H. van Riemsdijk, and E. Tipping, *Environ. Sci. Technol.* **37**:958 (2003).

80. L.K. Koopal, W.H. van Riemsdijk, J.C.M. de Wit, and M.F. Benedetti, *J. Colloid Interface Sci.* **166**:51 (1994).

81. D.G. Kinniburgh, C.J. Milne, M.F. Benedetti, J.P. Pinheiro, J. Filius, L.K. Koopal, and W.H. van Riemsdijk, *Environ. Sci. Technol.* **30**:1687 (1996). For additional discussion of the vicissitudes of model optimization based on eqs. 4.104 and 4.106, see M.F. Benedetti, W.H. van Riemsdijk, L.K. Koopal, D.G. Kinniburgh, D.C. Gooddy, and C.J. Milne, *Geochim. Cosmochim. Acta* **60**:2503 (1996) and D.G. Kinniburgh et al., *Colloid Surf.*. 151A:147 (1999).

82. I. Christl, C.J. Milne, D.G. Kinniburgh, and R. Kretzschmar, *Environ. Sci. Technol.* **35**:2505 (2001).

For Further Reading

S. Goldberg, *Advan. Agron.* **47**:233 (1992). This classic review of surface complexation modeling as applied to ion adsorption by natural particles is an excellent introduction to the concepts and practice of surface speciation calculations. The implementation of the models described in this review is conveniently portrayed by W.D. Schecher and D.C. McAvoy, *MINEQL+, A Chemical Equilibrium Modeling System*, Environmental Research Software, Hallowell, ME, 2001.

L.E. Katz and E.J. Boyle-Wright, pp. 213–255 in *Physical and Chemical Processes of Water and Solute Transport/Retention in Soil*, H.M. Selim and D.L. Sparks (Eds.), Soil Science Society of America, Madison, WI, 2001. This chapter presents a careful working discussion of how surface spectroscopy can be used to inform surface complexation modeling.

D. Langmuir, *Aqueous Environmental Geochemistry*, Prentice Hall, Upper Saddle River, NJ, 1997. Section 10.4 of this advanced textbook provides detailed examples of surface complexation modeling as used in applied mineral geochemistry. A similar perspective is taken for bacterial surface geochemistry by J.B. Fein, *Chem. Geol.* **169**:265 (2000).

M. Borkovec et al., *Colloids Surf.* **107**A:285 (1996). This review of modeling ion adsorption with affinity spectra emphasizes practical matters, such as regularization and faithful inversion of isotherm data to recover known affinity spectra.

E. M. Murphy and J. M. Zachara, *Geoderma* **67**:103 (1995). This review of the complexity in ion adsorption by organo-mineral particles is a most useful companion to the present chapter.

Research Matters

1. 1. M. F. Benedetti, W. H. van Riemsdijk, and L. K. Koopal [*Environ. Sci. Technol.* **30**:1805 (1996)] have used proton titration data on humic substances at varying ionic strength to determine V_D using eq. 4.101 in the special case of a 1:1 electrolyte (their eq. 5). The results (their fig. 4) can be summarized in a power-law relationship, $V_D = cI^{-d}$ where c and d are constant parameters.
   (a) Calculate the values of the two parameters using the data in table 2 of the article by Benedetti et al. Place a 95% confidence interval on each parameter value and examine these for d to determine whether it differs significantly from 0.5.
   (b) What assumptions are involved in using eq. 4.103 as a model of the power-law relation between V_D and I? Are they consistent with the data in fig. 3 of the article by Benedetti et al.?
   (c) Typical values of a_s for humic substances are in the range 500 to 800 $m^2\ g^{-1}$. Use this information to estimate the value of the parameter c in the relation between V_D and I above. Compare your estimate with the values of c obtained in (a).
2. J. Lützenkirchen, P. Magnico, and P. Behra [*J. Colloid Interface Sci.* **170**:326 (1995)] have examined the internal consistency of the Triple Layer Model as applied to calculate the p.z.n.p.c. for oxide adsorbents. In the notation of the present chapter, they derive the equation (their eq. 9a): p.z.n.p.c. = $\frac{1}{2}\left[\log\ K_+ - \log\ K_-(\text{int})\right] + \frac{1}{2}\log\left\{1 + {}^*K_A(\text{int})[A^-]/1 + {}^*K_C(\text{int})[C^+]\right\}$, where the four intrinsic equilibrium constants are defined in eqs. 4.54, 4.59, and 4.63.
   (a) Apply the surface charge balance condition at the p.z..n.p.c., $\left[\text{SOH}_2^+\right] + \left[\text{SOH}_2^+ \cdots \text{A}^-\right] = [\text{SO}^-] + \left[\text{SO}^- \cdots \text{C}^+\right]$ to derive an equation for p.z.n.p.c. in terms of the four intrinsic equilibrium constants. Use eqs. 4.49 to 4.51 to reduce your result to the expression given above.
   (b) What conditions are required for the equation derived in (a) to be reduced further to eq. 4.61? Use the values of log K_+(int), log K_-(int), log K_C(int), and log K_A(int) compiled in table IV of Goldberg, *Advan. Agron.* **47**:233 (1992) to test the conditions you develop.

3. P. G. Wightman, J. B. Fein, D. J. Weslowski, T. J. Phelps, P. Bénézeth, and D. A. Palmer [*Geochim. Cosmochim. Acta* **65**:3657 (2001)] have measured the intrinsic equilibrium constants for the deprotonation reaction, $\equiv$SOH = $\equiv SO^- + H^+$ on the surface of *Bacillus subtilis*, a bacterium found in groundwater aquifers. The bacterial surface is not charged at pH ≤ 2.2, but takes on increasingly negative charge at higher pH values. Proton titration data for this surface require three deprotonation reactions for accurate chemical modeling, with $\equiv$SOH then identified as carboxyl, phosphate, and hydroxyl in the three reactions. At 30°C the values of log K_{-}(int) for the reactions are -4.4 ± 0.7, -6.1 ± 0.8, and -8.3 ± 0.6, respectively.
   (a) Use the data in table 2 of the article by Wightman et al. to develop eq. 4.68b as a model for the temperature dependence of log K_{-}(int) over the range 30 to 75°C for each of the three deprotonation reactions.
   (b) Develop an "isocoulombic" reaction for surface deprotonation by analogy with the development of the reaction in eq. 4.72. Express log K for this "isocoulombic" deprotonation reaction in terms of log K_{-}(int) and log K_W, then compute its temperature dependence over the range 30 to 75°C using the data in table 2 of Wightman et al. for the three deprotonation reactions along with eq. 4.74.
4. J. S. Cox, D. Scott Smith, L. A. Warren, and F. G. Ferris [*Environ. Sci. Technol.* **33**:4514 (1999)] have determined the affinity spectrum for proton desorption from the surface of *B. subtilis* based on the model adsorption isotherm in eq. 4.94 with $M \leq 7$ (their fig. 3). Ultimately $M = 5$ was selected, with the corresponding log K values: -5.17 ± 0.01, -5.88 ± 0.01, -6.91 ± 0.05, -7.88 ± 0.04, and -9.24 ± 0.07 determined at I = 0.025 mol dm^{-3} (their table 2).
   (a) Examine the optimization procedure used by Cox et al. in light of the concepts for guiding this procedure that were developed by M. Cernik, M. Borkovec, and J. C. Westall, *Environ. Sci. Technol.* **29**:413 (1995). In particular, discuss the relationship between fig. 3 in the article by Cox et al. and fig. 3 in the article by Cernik et al. Include in your discussion the results for log K_{-}(int) given in problem 3 above and the remarks about the chemical nature of the surface functional groups on *B. subtilis* that appear in the articles by Cox et al. and Wightman et al. (problem 3).
   (b) Given an average imprecision of 0.7 log units in the determination of log K, consider representing the affinity spectrum for proton desorption from *B. subtilis* by three Sips spectra (eq. 4.97) with modes at ln $\tilde{K}$ values equivalent to the log K_{-}(int) values reported by Wightman et al. (see, e.g., fig. 4.8). Estimate a value for the Sips parameter β using the formula, $\cosh(\beta\Delta) = 2 + \cos(\pi\beta)$, where Δ equals the imprecision in ln $\tilde{K}$ (i.e., $1.612 = 0.7 \ln 10$). Then prepare a graph similar to fig. 4.10 and note whether the spread of the peak is sufficient to include more than one of the log K values reported by Cox et al.
5. J. A. Davis, J. A. Coston, D. B. Kent, and C. C. Fuller [*Environ. Sci. Technol.* **32**:2820 (1998)] have been able to model both adsorption isotherms and adsorption edges (their figs. 1 and 2) for Zn^{2+} on aquifer sediments by invoking the reaction in eq. 4.37a for two kinds of surface site. However, the intrinsic equilibrium constants for these reactions, nominally expressed in eq. 4.55a, do not contain the factor in the "surface potential" ψ_0. The values of these latter parameters and the associated values of SOH_T (eq. 4.42 with only the first and fourth terms retained) are listed in Table 2 of the article by Davis et al.
   (a) Show that the "two-site—one proton" model described by Davis et al. for Zn^{2+} adsorption is equivalent to eq. 1.13.
   (b) Use the result in (a) to calculate distribution curves for Zn^{2+} adsorbed on the aquifer sediments corresponding to those in fig. 6 in D. B. Kent, R. H. Abrams, J. A. Davis, J. A. Coston, and D. R. LeBlanc, *Water Resour. Res.* **36**:3411 (2000).

5

Colloidal Phenomena

5.1 Probing Particle Structure

The natural particles whose surface reactions figure most importantly in the geochemistry of soils and sediments range in size from approximately 10 nm to 10 μm—from fine clay to fine silt—while comprising a heterogeneous mixture of crystalline minerals, amorphous weathering residues, humus, and microbial biomass. The size range just given, which can be broadened in certain circumstances by as much as an order of magnitude from either endpoint, encompasses typical groundwater colloids and viruses in its lower portion, with riverine suspended solids and phytoplankton included in its upper portion. The processes forming these particles involve the life cycles of organisms, the geochemical weathering of terrestrial materials, and mass transport.[1] Once they are formed, the particles themselves may be consumed by the biota; react with dissolved solutes and, therefore, undergo further transformation; or be moved by advective transport through the atmosphere, surface water, or subsurface water. This last mode of natural particle evolution can be an effective mechanism for the movement of adsorbed nutrients or contaminants, a phenomenon termed *colloid-facilitated transport* of adsorbed species.[2] Reaction with dissolved solutes, on the other hand, which mediates surface charge, can facilitate the settling of natural particles out of the aqueous phase in which they are suspended, thus reducing the mobility of adsorbed species.

The abundant literature on the morphology of natural particles, while clearly revealing their complexity, still permits a few broad structural conclusions to be drawn.[3] In biologically active earth materials and water bodies, particles comprising clay minerals and metal oxides, including amorphous silica and metal-hydroxy polymers, typically bear adsorbed humus. Particle morphology may be roughly spherical or disk-shaped, with tendrils of extracellular products and cell wall compounds emanating from fine-clay-sized inorganic cores. These particles tend to coalesce rather slowly to form larger, relatively compact units when in aqueous suspension (time scales of weeks or more) unless the ionic strength is high. Inorganic particles also can associate with smaller organic colloids by coalescence, with the latter particles possibly changing their conformation as a result of interactions with the charged surface of the inorganic partner. If the organic colloids are instead larger than the inorganic particles, or if they have relatively long tendrils, the organic colloids may bind the smaller inorganic particles into a fibrous network of whorls having a complex overall morphology. These particles, in turn, may settle out of suspension quickly. Even these brief summarizing remarks should make clear the point that natural particle structures are inherently irregular and complex, with particle surface charge and the composition of the suspending aqueous solution playing important roles in mediating the formation of these structures (fig. 5.1).

Complementary to the microscopic examination of natural particle structure,[3] the experimental technique of *photon scattering* has proven to be a most useful approach to gathering information about the organization of very small colloids that have coalesced into aggregates.[4] The physical basis of this methodology can be understood in terms of the scattering diagram in fig. 5.2. Photons with momentum $\hbar \boldsymbol{k}$, where $\hbar$ is the Dirac constant and k is the wavenumber vector ($\boldsymbol{k} = k\,\hat{\boldsymbol{k}}$, $k \equiv 2\pi/\lambda$, the unit vector $\hat{\boldsymbol{k}}$ denoting the direction and λ denoting the wavelength of the radiation propagated by the photons[5]), are incident upon a particle, then are scattered by one of its colloidal constituents through an angle θ, after which the momentum becomes $\hbar k'$. Two scattering events are depicted in fig. 5.2, one located by the vector $\boldsymbol{r}_1$ and the other by $\boldsymbol{r}_2$. The path of a photon scattered at $\boldsymbol{r}_2$ differs from that scattered at $\boldsymbol{r}_1$ by the two line segments denoted AB and BC. This path difference gives rise to the possibility of *interference* between the scattered photons. Evidently $\overline{AB} = \hat{\boldsymbol{k}} \cdot (\boldsymbol{r}_2 - \boldsymbol{r}_1)$ and $\overline{BC} = \hat{\boldsymbol{k}}' \cdot (\boldsymbol{r}_1 - \boldsymbol{r}_2)$, such that the total difference in path length between the two scattering events is $\left(\hat{\boldsymbol{k}} - \hat{\boldsymbol{k}}'\right) \cdot (\boldsymbol{r}_2 - \boldsymbol{r}_1)$. It is this scalar product that determines quantitatively the extent of photon interference.[6]

For the case of photon scattering by particles in aqueous suspension, several conditions must be met in order to interpret measurements of the intensity of the scattered photons as a function of the angle of scattering θ:[4,7]

1. The particles must be *transparent* to the photons (i.e., no photon absorption) and *non-refractive*, so that the intensity and momentum are the same for all incident photons illuminating scattering centers within a suspended particle. The condition of no significant refraction requires the inequality,

$$|m - 1| \ll 1 \tag{5.1}$$

to be satisfied, where $m = n_s/n_w$, n_s being the refractive index of the scattering particle and n_w being that of the aqueous phase. For example, $n_s \approx 1.5$ for clay minerals and $n_w \approx 1.33$, yielding $m \approx 1.13$, which satisfies eq. 5.1 reasonably well.

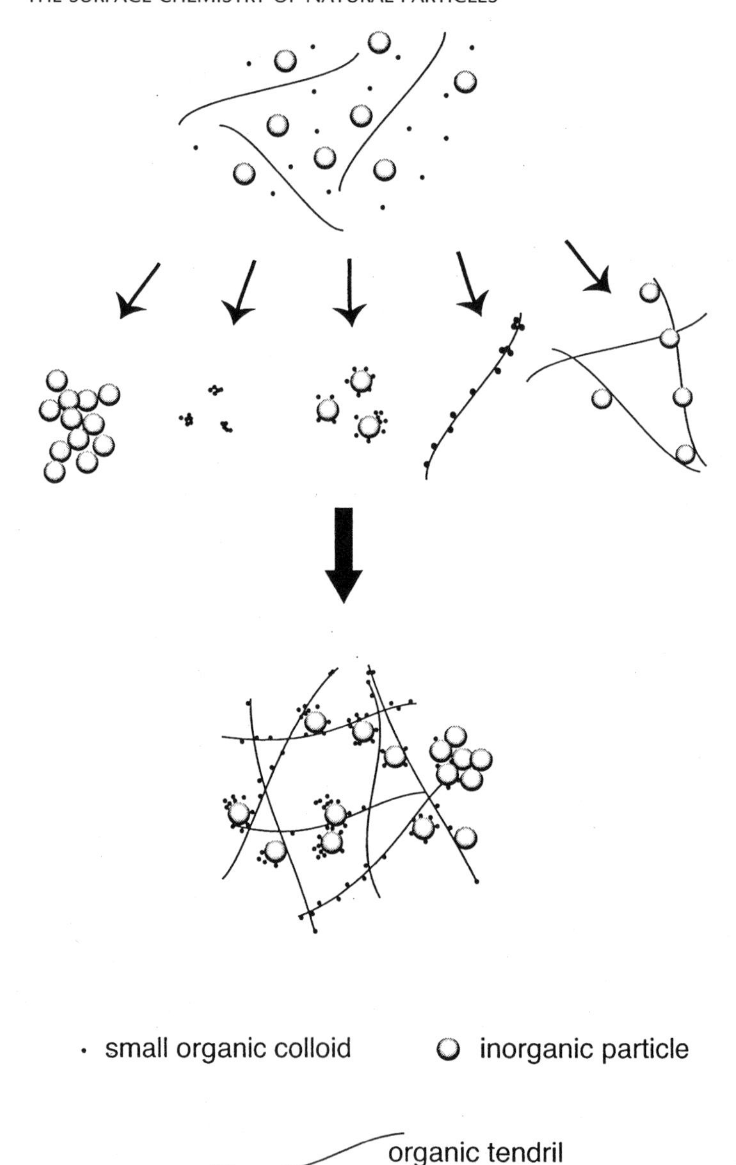

Fig. 5.1 Schematic cartoon of the key processes involved in natural particle formation (J. Buffle et al.[3]). Note the porous irregular structure of the aggregate comprising inorganic colloids, organic colloids, and organic biopolymers (bottom of figure).

2. *Multiple* scattering events do not occur. Moreover, the dominant effect of photon scattering by a single particle in suspension is to change the *direction* of the photon wavenumber vector, not its magnitude; that is, elastic scattering. (The photon energy $\hbar ck$, is not changed by scattering, where c is the photon speed.[5]) In practical terms, elastic scattering occurs if the inequality,[7]

$$|k' - k| \ll (2R)^{-1} \tag{5.2a}$$

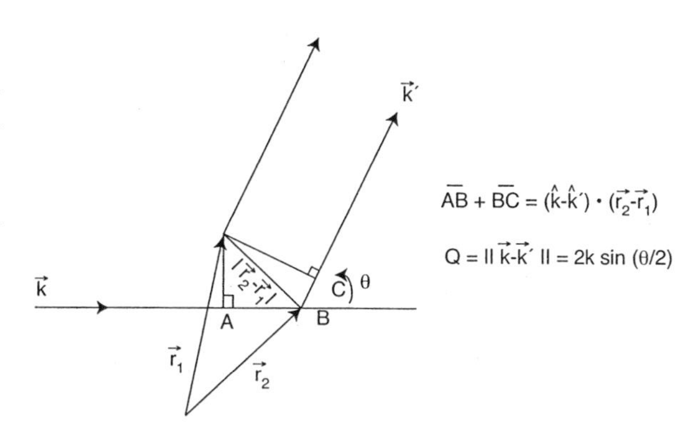

Fig. 5.2 Geometric relationships attendant to coherent photon scattering by two scattering centers located at the points $\boldsymbol{r}_1$ and $\boldsymbol{r}_2$. The additional distance traveled by the photon scattered at $\boldsymbol{r}_2$ relative to that scattered at $\boldsymbol{r}_1$ is $(\overline{AB} + \overline{BC})$, which must equal an integral number of photon wavelengths if photon scattering by the two centers is to be coherent. The magnitude of the change in the photon wavenumber vector is Q, which depends on the photon wavelength and the scattering angle θ.

holds for some characteristic length scale R that typifies the radius of the scattering particle (e.g., its radius of gyration,[8] the cube root of its volume, or the geometric mean of its longitudinal and transverse dimensions). Equation 5.2a assures that interference among scattered photons is not obliterated simply because they are propagating radiation having significantly different wavelengths, even if the scattering events have occurred at points widely separated within the particle. Given the definitions of wavenumber and the ratio m, the constraint in eq. 5.2a can be expressed in an alternate form:

$$2kR|m - 1 \ll 1 \qquad (5.2b)$$

since $k' = mk$. Thus, for example, if $m \approx 1.13$ then $R/\lambda \ll 0.6$ is stipulated by eq. 5.2b. Since 10 nm $<$ R $<$ 10 μm for the scattering particles, *the wavelength of the incident radiation must lie in the nanometer to micrometer range (visible light to infrared radiation)*. Conversely, if, say, He-Ne laser light of wavelength 632.8 nm in vacuo is incident on a particle, then $\lambda = 632.8/1.33 = 475.8$ nm and R $\ll 0.3$ μm (fine clay) in order to satisfy eq. 5.2b.

The magnitude of the vector difference, $\hat{\boldsymbol{k}} - \hat{\boldsymbol{k}}'$, which figures importantly in the interference produced among scattered photons and, therefore, their resultant intensity, depends on the scattering angle θ (fig. 5.2). This relationship follows from an application of a standard trigonometric identity to calculate $||\hat{\boldsymbol{k}} - \hat{\boldsymbol{k}}'||$:

$$\begin{aligned} \left\|\hat{\boldsymbol{k}} - \hat{\boldsymbol{k}}'\right\|^2 &\equiv \left(\hat{\boldsymbol{k}} - \hat{\boldsymbol{k}}'\right) \cdot \left(\hat{\boldsymbol{k}} - \hat{\boldsymbol{k}}'\right) = 1 - 2\cos\theta + 1 \\ &= 2(1 - \cos\theta) = 4\sin^2(\theta/2) \end{aligned} \qquad (5.3)$$

where the second step takes note of the unit length of $\hat{\boldsymbol{k}}$ and $\hat{\boldsymbol{k}}$, as well as the included angle θ between them. It follows from Eqs. 5.2 and 5.3 that the magnitude of $\boldsymbol{k} - \boldsymbol{k}'$ (i.e., the change in wavenumber vector produced by photon scattering) is:

$$Q \equiv \left\| \boldsymbol{k} - \boldsymbol{k}' \right\| \approx k \left\| \hat{\boldsymbol{k}} - \hat{\boldsymbol{k}}' \right\| = 2k \sin(\theta/2) \tag{5.4}$$

where $k = 2\pi n_w/\lambda_0$, λ_0 being the wavelength of the incident radiation in vacuo. Physically speaking, $\hbar Q$ is the magnitude of the change in photon momentum provoked by the scattering events depicted in fig. 5.2 when eqs. 5.1 and 5.2 are applicable. This momentum change, imparted to the particle from which the photon is scattered, leads to constructive photon interference if the length scale L of the separation between scattering centers satisfies the condition:[6,7,9]

$$QL \approx 1 \tag{5.5}$$

Thus Q^{-1} is the length scale probed in photon scattering by a particle. *Measurement of the intensity of the scattered photon beam reveals the length scales that characterize particle structure.*

After suitable calibration for: (1) photon absorption, (2) scattering by the apparatus in which a particle suspension is presented to the incident beam, (3) multiple scattering processes, and (4) photon scattering by the individual colloids making up a scattering particle or by the individual particles in a suspension, the intensity of photons detected at the scattering wave number $\boldsymbol{Q}$ can be modeled by the equation:

$$I(\boldsymbol{Q}, t) = \sum_q n(q, t) q^2 S_q(\boldsymbol{Q}) \tag{5.6}$$

where q is the number of colloids that have coalesced to form a particle (henceforth termed a "q-mer"), $n(q, t)$ is the number of q-mers per unit volume at the time t in a suspension, and $S_q(\boldsymbol{Q})$ is the *structure factor* for a q-mer, that is, the scattering intensity contributed solely by the spatial arrangement of colloids in the q-mer. Equation 5.6 is a model expression that pictures each particle in a suspension as an aggregate formed by q colloids, with each such aggregate then scattering photons independently of the others. The total intensity of photon scattering $I(\boldsymbol{Q}, t)$ is thereby rendered a weighted sum over all possible aggregate sizes (with size represented by q), the weighting factor then being the number density, $n(q, t)$. The q^2 factor in eq. 5.6 arises because the photon scattering intensity for a single aggregate is proportional to the square of its volume,[7] hence to q^2, irrespective of the details of its structure.

The structure factor models the Q-dependence of photon scattering caused by the internal structural features of a q-mer. If the q-mer is isotropic, with the number density $\rho_q(r)$ of its constituents[8] varying spatially along a radial line outward from its center-of-mass, then[10]

$$S_q(Q) = q^{-1} + 4\pi\left(1 - q^{-1}\right) \int_0^\infty \rho_q(r) \frac{\sin(Qr)}{Qr} r^2 dr \tag{5.7}$$

The spatial distribution of the colloids and, therefore, the structure of the aggregate, is modeled by $\rho_q(r)$, which is subject to the normalization condition:[8]

$$4\pi \int_0^\infty \rho_q(r)\, r^2 dr = 1 \tag{5.8}$$

The trigonometric factor in the integral term of eq. 5.7 [sin(Qr)/Qr] models the effects of interference between photons scattered simultaneously from one colloid placed at the center-of-mass of the aggregate and from another placed at a distance r away from the center-of-mass. This interference will *not* be manifest in $S_q(Q)$, however, if the colloids making up the aggregate are uniformly distributed at points a very small distance apart—smaller than Q^{-1} for any experimentally accessible Q-value—or if Q^{-1} is much larger than any possible distance between two colloids in the aggregate. Photon scattering with interference effects, therefore, is the result of $\rho_q(r)$ exhibiting significant spatial variability on a length scale approximately equal to Q^{-1}. It is in this sense that eq. 5.7 is said to represent[10] "Bragg diffraction by spatial variations of density with periodicity Q^{-1}."

For values of Qr small enough to justify the two-term MacLaurin expansion,

$$\frac{\sin(Qr)}{Qr} \approx 1 - \frac{1}{6}(Qr)^2 \tag{5.9}$$

eq. 5.7 takes on a universal form

$$Sq(Q) \approx 1 - \frac{1}{3}(QR_{\mathrm{Gq}})^2 \tag{5.10}$$

known as the *Guinier approximation*. The second term in eq. 5.10 is the result of incorporating the relation between $\rho_q(r)$ and the *radius of gyration* R_{Gq},[8,10]

$$R_{\mathrm{Gq}}^2 = 2\pi \int_0^{\infty} r^4 \rho_q(r) dr \tag{5.11}$$

while neglecting q^{-1} relative to 1, consistent with the large aggregate size implied by the concept of the radius of gyration. The corresponding Guinier approximation for the photon scattering intensity is:

$$I(Q, t) \approx \sum_q n(q, t)q^2 \left\{1 - \frac{1}{3}[QR_{\mathrm{GZ}}(t)]^2\right\} \tag{5.12}$$

where

$$[R_{\mathrm{GZ}}(t)]^2 \equiv \frac{\sum_q n(q, t)q^2 R_{\mathrm{Gq}}^2}{\sum_q n(q, t)q^2} \tag{5.13}$$

is termed the *"z-average" radius of gyration*.[4,10,11] Equation 5.12 provides the basis for measurement of the average size of an aggregate in a suspension, irrespective of particle shape or internal structure.

Figure 5.3 illustrates an application of eq. 5.12 to determine the z-average radius of gyration for carbonaceous soot aerosols.[12] This "Guinier plot" features $I(0)/I(Q)$ as the ordinate and Q^2 as the abscissa; hence, R_{GZ}^2 is equal to 3 times the slope of the linear graph. The experimental arrangement is such that the graphs from the bottom to the top of the figure correspond to increasing numbers of soot colloids having coalesced to form aggregates, thus producing increasing values of R_{GZ} ranging from 15 to 117 nm. The photon scattering experiment involved radiation from an Ar laser ($\lambda_0 = 488$ nm) and observations of scattering at angles between 10 and 110°. Therefore, by eq. 5.4, $2.2 < Q < 21\ \mu\mathrm{m}^{-1}$ and $5 < Q^2 < 450\ \mu\mathrm{m}^{-2}$, as shown in fig. 5.3. Note that, according to

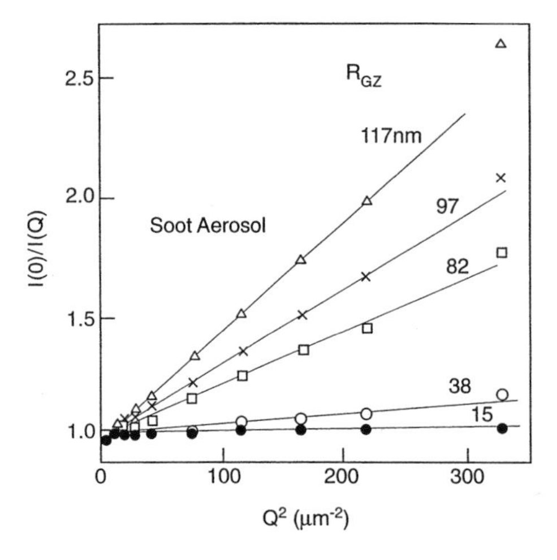

Fig. 5.3 Guinier plot of photon scattering data for a soot aerosol (S. Gangopadhyay et al.[12]). The slope of the plot is equal to $\frac{1}{3}R_{GZ}^2$

the criterion of smallness underlying the derivation of eq. 5.12, $I(0)/I(Q) \ll 4/3$ for the Guinier approximation to be accurate. This condition is met by the data plotted in the two lower graphs in fig. 5.3, but is satisfied only for $Q^2 \lesssim 100\ \mu m^{-2}$ for the other data, despite the fact that the graphs themselves remain linear for Q^2 values up to about 300 μm^{-2}, or for $I(0)/I(Q) \lesssim 2$, an example of a fairly widespread observation.[12]

Given that the refractive index of the aerosol colloids is about 1.6, the condition for using eq. 5.6 to interpret photon scattering data is, according to eq. 5.2b, $R \ll 65$ nm, which again is not met by the data in the upper three graphs in fig. 5.3. However, this interpretation of eq. 5.2b assumes that the scattering particle has no porosity, which is highly unlikely given the particle formation scenario in fig. 5.1. In practice,[12,13] the factor $|m-1|$ in eq. 5.2b is observed to scale with the number density of colloids in the scattering particle, such that the constraint in eq. 5.2b should be replaced by[13]

$$2kR(1-\phi)|m-1| \ll 1 \tag{5.2c}$$

where ϕ is the porosity of the scattering particle (volume fraction of pore space). Given $\phi \approx 0.95$ for soot particles,[13] the constraint in eq. 5.2c becomes $R \ll 1.3\ \mu m$, which is easily met by the data in fig. 5.3. As a general rule, eq. 5.2c may be applied with a default value $\phi \approx 2/3$ to estimate R.[12,14]

Photon scattering at Q-values larger than those consistent with the Guinier approximation is expected to probe the internal structure of aggregates in suspension. In the context provided by eq. 5.6, these larger values of Q should correspond to length scales that are small when compared to the size of an aggregate, but large when compared to the size of a colloid making up an aggregate. If ξ represents the length scale of aggregate size and r_0 represents that of a constituent colloid, then Q must satisfy the condition:

$$r_0 \ll Q^{-1} \ll \xi \tag{5.14}$$

in order to reveal integral aggregate structure through photon scattering. Equation 5.14 is to be added to eqs. 5.1 and 5.2c as a criterion for the interpretation of $I(\boldsymbol{Q}, t)$ in terms of natural particle structure.

Published studies of photon scattering by suspensions of mineral particles typically show the trend illustrated in fig. 5.4, which is a log-log plot of $I(Q, t)$ versus Q measured at successive values of t during the formation of aggregates by illite colloids suspended in NaCl solution at pH 8.[15] As time passes and larger aggregates are formed, the log-log plot exhibits a *linear* portion whose domain in Q continually increases. This trend can be understood as an effect of increasing particle size, which enhances the photon-scattering intensity (q^2 term in eq. 5.6) and moves it out of the Guinier approximation (eq. 5.12), whose slope in a log-log plot is always very small. The photon-scattering intensity in eq. 5.6 is thus a composite of scattering by particles small enough to satisfy eq. 5.12, together with those large enough to yield a linear log-log plot. Eventually, most of the particles grow large enough to scatter photons according to a power-law relation between $I(Q, t)$ and Q, which then leads to a linear log-log plot as seen in fig. 5.4. Figures 5.5 and 5.6 further illustrate linearity in log-log plots of $I(Q, t)$ versus Q, the example being hematite colloids suspended in either KCl or KH_2PO_4 solution at pH 6.[16] These data indicate that the slope of the log-log plot can depend on solute concentration. Note that Cl^- and $H_2PO_4^-$ are counterions for hematite at the low pH value of the experiments (p.z.n.c. $\approx$ 9).

A power-law relationship between $I(Q, t)$ and Q subject to eq. 5.14 can be deduced from eq. 5.6 by a model of the structure factor based on the hypothesis that the photon-scattering aggregates are *fractal* objects.[17] This hypothesis can be represented in terms of the mathematical form of the fraction $p(r)$ of colloids making up an aggregate to be found within a distance r from a colloid placed at the center-of-mass. If the aggregate is fractal:

$$p(r) \underset{r \gg r_0}{\sim} r^D \qquad (1 < D < 3) \tag{5.15}$$

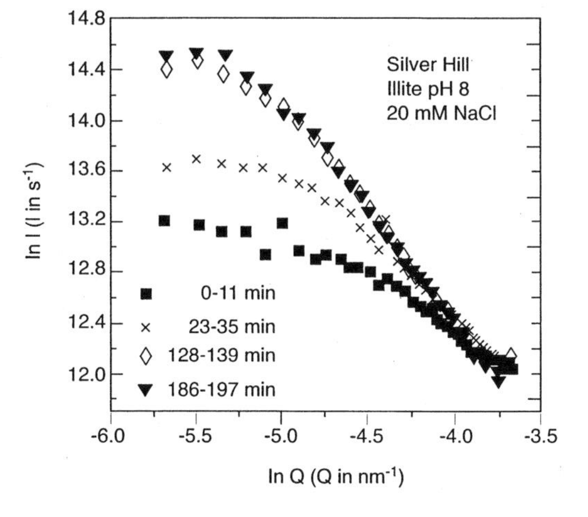

Fig. 5.4 Log-log plot of the photon scattering intensity for a suspension of illite versus the scattering wavenumber, with time as a variable parameter (L. Derrendinger and G. Sposito[15]). At fixed Q, the scattering intensity increases with time in the Guinier regime (low Q) because of particle growth. At large Q values, the scattering intensity exhibits a power-law dependence on Q (linear log-log plot).

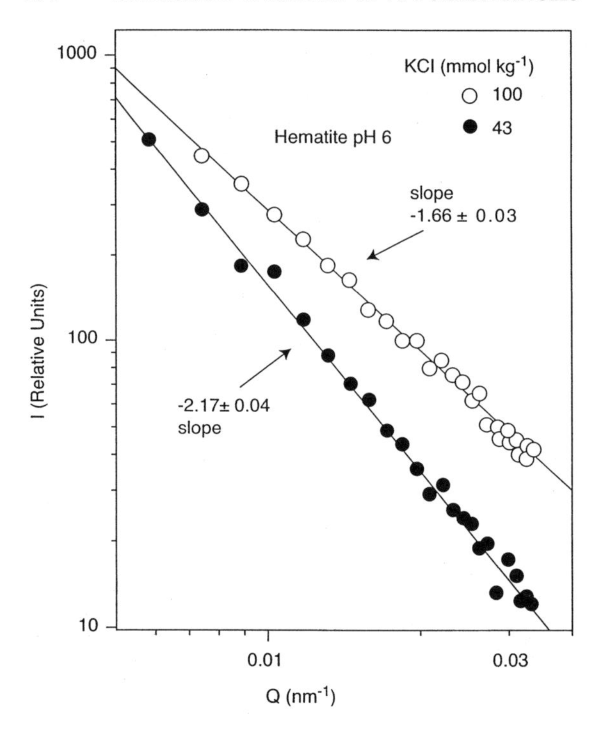

Fig. 5.5 Same type of plot as in fig. 5.4 for hematite particles suspended in KCl solution, showing that the slope of the linear portion decreases as the KCl concentration increases (J. Chorover et al.[16]).

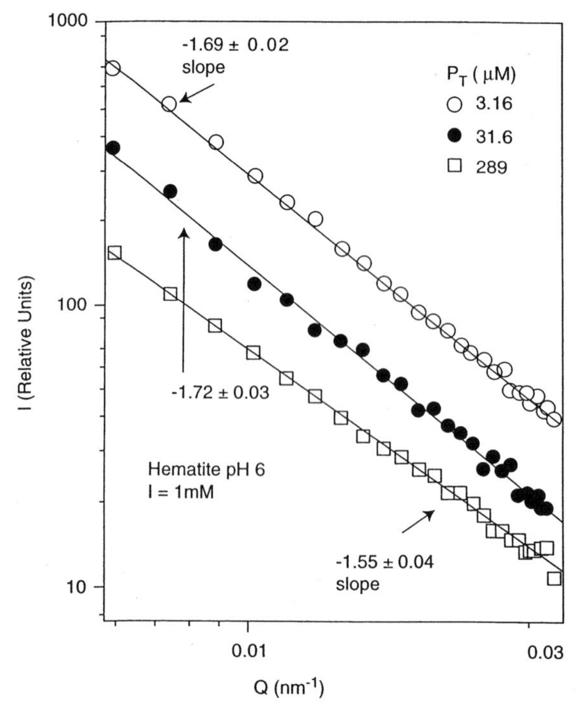

Fig. 5.6 Same type of plot as in fig. 5.4 for hematite particles suspended in KH_2PO_4 solution, showing unimodal behavior of the slope of the linear portion as the KH_2PO_4 concentration increases (J. Chorover et al.[16])

where D is a positive exponent termed the (mass or cluster) *fractal dimension* of the aggregate.[17] *The parameter D provides a measure of how completely the colloids in an aggregate fill three-dimensional space.* Thus, a value of D near 3 implies a dense, close-packed arrangement of colloids tantamount to a solid sphere, whereas a D-value near the lower limit of 1 implies an essentially linear (if convoluted) arrangement of colloids, with significant pore space left within the dendritic ball of strands that constitute the aggregate. It is likely that the first-mentioned type of spatial arrangement would not provide for much Q-dependence in photon scattering, since the distance separating the colloidal constituents would necessarily be very small relative to typical Q^{-1} values accessed in a photon scattering experiment. The aggregate with low fractal dimension, on the other hand, would exhibit significant internal spatial variability and, therefore, lead to significant Q dependence of the photon scattering intensity.

The function $p(r)$ is related to the number density $\rho_q(r)$ in eq. 5.7 by the defining equation:

$$dp \equiv 4\pi\rho_q(r)r^2 dr \tag{5.16a}$$

or, in terms of the number density, $n(\mathbf{r})$, of colloids at the point $\mathbf{r}$:[8]

$$p(R) \equiv 4\pi \int_0^R \int n(\|\mathbf{r}-\mathbf{r}'\|)n(\mathbf{r}')d^3r' r^2 dr \tag{5.16b}$$

Thus $p(R)$ can be interpreted as the fraction of all colloids located within the distance R of any other colloid in the aggregate, averaged over all possible positions of the other colloid, with $\mathbf{r}$ and $\mathbf{r}'$ referenced to the center-of-mass of the aggregate. For a fractal aggregate, eqs. 5.15 and 5.16a imply

$$\rho_q^F(r) \underset{r \gg r_0}{\sim} r^{D-3} \qquad (1 < D < e) \tag{5.17}$$

as seen by forming the differential of $p(r)$ and comparing it to eq. 5.16a. Recalling that $\rho_q(r)$ is a conditional probability density for finding a colloid at r,[8] one sees that this probability density is uniform for a nonfractal aggregate ($D = 3$), whereas it decreases as a negative power of increasing r for a fractal aggregate, thus reflecting the porous nature of the latter.

To be consistent with the asymptotic character of fractal aggregates, eq. 5.7 should now be considered in the limit $q \gg 1$:

$$S_q^F(Q) \underset{q \gg 1}{\sim} \pi \int_0^\infty \rho_q^F(r) \frac{\sin(Qr)}{Qr} r^2 dr \tag{5.18}$$

To be consistent with the normalization condition in eq. 5.8, the right side of eq. 5.17 should be supplemented to yield an equality:

$$\rho_q^F(r) = A_q r^{D-3} h(r/\xi_q) \tag{5.19}$$

where A_q is a constant to be determined by eq. 5.8 and $h(r/\xi_q)$ is a "cutoff function" with the properties:

$$\lim_{x \downarrow 0} h(x) = 1, \qquad \lim_{x \uparrow \infty} h(x) = 0 \tag{5.20}$$

Thus, $h(r/\xi_q)$ allows $\rho_q^F(r)$ to have a power-law form at distances r that are small when compared to the overall aggregate size, but it stifles the power law at large distances so as to respect the normalization condition.[18] It follows that

$$A_q = \left[4\pi\xi_q^D \int_0^\infty x^{D-1} h(x)dx\right]^{-1} \tag{5.21}$$

after substitution of eq. 5.19 into eq. 5.8. The cutoff function $h(x)$ is assumed to decay to zero fast enough to ensure the convergence of the integral in eq. 5.21.

Substitution of eq. 5.19 into eq. 5.18 yields:

$$\begin{aligned} S_q^F(Q) &= 4\pi A_q \int_0^\infty r^{D-1} h(r/\xi_q) \frac{\sin(Qr)}{Qr} dr \\ &= \frac{4\pi A_q \xi_q^D}{Q\xi_q} \int_0^\infty x^{D-2} h(x) \sin(Q\xi_q x)dx \equiv S_q^F(Q\xi_q) \end{aligned} \tag{5.22}$$

which demonstrates that the structure factor depends only on the dimensionless product, $Q\xi_q$. In keeping with eq. 5.14, we wish to evaluate the integral in eq. 5.22 under the condition, $Q\xi_q \gg 1$. This can be accomplished by making an asymptotic expansion of the integral in inverse powers of $Q\xi_q$, with the result:[19]

$$\int_0^\infty x^{D-2} h(x) \sin(Q\xi_q x)dx \underset{Q\xi_q\uparrow\infty}{\sim} \frac{\Gamma(D-1)\sin[(D-1)\pi/2]}{(Q\xi_q)^{D-1}} \tag{5.23}$$

where $\Gamma(x)$ is the gamma function.[20] Therefore,

$$S_q^F(Q\xi_q) \underset{Q\xi_q\uparrow\infty}{\sim} 4\pi A_q \xi_q^D \Gamma(D-1)\sin[(D-1)\pi/2](Q\xi_q)^{-D} \tag{5.24}$$

where A_q is given by eq. 5.21. It then follows from eq. 5.6 that

$$I(Q,t) \underset{Q\xi_q\uparrow\infty}{\sim} \frac{\Gamma(D-1)\sin[(D-1)\pi/2]}{\int_0^\infty x^{D-1} h(x)dx} \sum\nolimits_q q^2 n(q,t)(Q\xi_Z)^{-D} \tag{5.25}$$

where, analogously to eq. 5.13,

$$\xi_Z^{-D} \equiv \frac{\sum_q q^2 n(q,t)\xi_q^{-D}}{\sum_q q^2 n(q,t)} \tag{5.26}$$

is a "z-average" of ξ^{-D}. Equation 5.25 implies that a log-log plot of $I(Q, t)$ versus Q under the conditions in eq. 5.14 will be a straight line whose slope is minus the fractal dimension (figs. 5.4 to 5.6). Note that $D = 3$ in eq. 5.25 yields no photon scattering (the sine factor vanishes).

How is the aggregate length scale ξ_q to be interpreted physically? Insight can be gained by returning to eq. 5.22 and examining it in the Guinier approximation (eq. 5.9):

$$S_q^F(Q\xi_q) \approx \left[\int_0^\infty x^{D-1} h(x)dx\right]^{-1} \left\{\int_0^\infty x^{D-1} h(x)dx\right.$$

$$-\frac{Q^2}{6}\int_0^\infty x^{D+1}h(x)dx\Bigg\} \equiv 1-\frac{1}{3}\left(QR_{Gq}\right)^2 \tag{5.27}$$

which means that

$$R_{Gq}^2 = \frac{1}{2}\frac{\int_0^\infty x^{D+1}h(x)dx}{\int_0^\infty x^{D-1}h(x)dx}\xi_q^2 \tag{5.28}$$

Equation 5.28 is identical with eq. 5.11 after substitution of eq. 5.19 into the latter expression while noting eq. 5.21. Thus, ξ_q is proportional to the radius of gyration, the precise relationship being dependent on the mathematical form of the cutoff function $h(r/\xi_q)$.[18] For example, if $h(r/\xi_q) = \exp(-r/\xi_q)$, then[20]

$$R_{Gq}^2 = \frac{1}{2}\frac{\Gamma(D+2)}{\Gamma(D)}\xi_q^2 = \frac{1}{2}D(D+1)\xi_q^2 \tag{5.29}$$

For the values of D apparent in fig. 5.4 to 5.6, ξ_q is between one-half and three-quarters of the value of R_{Gq}. Other choices of cutoff formation yield ξ_q that are somewhat larger than R_{Gq}.[18] Thus, the upper bound on Q^{-1} in eq. 5.14 can be replaced by R_{GZ} for practical purposes, and the same kind of substitution can be done on both sides of eq. 5.26, since the integrals in eq. 5.28 do not depend on q. (Note that this latter substitution together with eq. 5.15 applied to q and R_{Gq} will yield eq. 5.57 in section 5.2.) Overall then, in eqs. 5.22, 5.24, and 5.25, photon scattering by fractal q-mers depends only on the dimensionless parameter, QR_{Gq}.

5.2 Particle Formation Kinetics

Natural particles whose size falls into the middle of the colloidal range (i.e., from approximately 100 nm to 1 μm) are observed to remain suspended in surface or subsurface waters for extended periods of time.[21] Colloids whose size is below this range appear to coalesce rather quickly to form larger particles, whereas colloids whose size is above the midrange appear to settle rather quickly under the influence of gravity, at least in quiescent suspending fluids. Because of these general observations, the experimental study of particle formation kinetics has tended to focus on the coalescence behavior of the midrange particles, including in its purview the influence of surface chemistry, with the goal of understanding the conditions that either ensure continued suspension or promote further particle growth with subsequent settling.[22]

Similarly to common practice in the broader discipline of chemical kinetics, the investigation of particle coalescence processes has benefited from the application of *initial-rate methods*. The value of this approach is that the concentrations of reactants are known with precision, and the complications attendant to product accumulation or transformation are avoided. Moreover, photon scattering is once again a useful experimental technique because of the sensitive dependence of the scattering intensity on particle size, not to mention simplicity in view of the inherent tedium of alternative methods, such as microscopy or direct particle counting.[23] Particle formation kinetics are exemplified implicitly in fig. 5.3, which shows the scattering intensity increasing with time at fixed Q while illite particles in aqueous suspension collide and coalesce.

The application of photon scattering to measure an initial rate of particle coalescence can be understood in terms of eq. 5.6, which models the photon scattering intensity as a function of wavenumber and time. For a colloidal suspension that comprises only monomers initially, there is need to consider just the first two terms in the sum on the right side:

$$I(Q, t) = n(1, t)S_1(Q) + 4n(2, t)S_2(Q) \tag{5.30}$$

The initial-rate method then involves consideration of the relative rate at which $I(Q, t)$ increases at "time zero":

$$\begin{aligned}\frac{1}{I(Q,0)}\left(\frac{dI}{dt}\right)_{t=0} &= \frac{1}{n(1,0)}\left(\frac{dn(1,t}{dt}\right)_{t=0} \\ &+ \frac{1}{n(1,0)}\left(\frac{dn(z,t)}{dt}\right)_{t=0}\frac{4S_2(Q)}{S_1(Q)}\end{aligned} \tag{5.31}$$

where the initial condition, $n(2, 0) \equiv 0$, has been noted. The left side of eq. 5.31 can be determined experimentally by calculating the initial slope of a graph of normalized scattering intensity (at a given Q-value) versus time.[24] The right side of eq. 5.31 requires a model of the two time-derivatives before it becomes accessible to a kinetics interpretation. Perhaps the simplest model is that based on the dimer formation reaction:

$$2 \text{ monomer} \xrightarrow{k} \text{dimer} \tag{5.32}$$

with a second-order rate coefficient k. The rate of this reaction is (see section 3.1):

$$\frac{d(\xi/V)}{dt} = -\frac{1}{2}\frac{dn(1,t)}{dt} = \frac{dn(2,t)}{dt} = k[n(1,t)]^2 \tag{5.33}$$

where V is the volume of the suspension and a second-order rate law has been invoked to describe eq. 5.32. Combination of eqs. 5.31 and 5.33 then produces the initial-rate model:

$$\frac{1}{I(Q,0)}\left(\frac{dI}{dt}\right)_{t=0} = \left[\frac{2S_2(Q)}{S_1(Q)} - 1\right]2kn(1,0) \tag{5.34}$$

Further development of eq. 5.34 is possible by introducing a suitable model of $\rho_q(r)$ in eq. 5.7. This is done again in a simple manner by assuming that $\rho_q(r)$ has a very sharp peak corresponding to the position of a monomer or to the separation of two monomers forming a dimer, that is,

$$\rho_1(r) = \delta(r), \qquad \rho_2(r) = \delta(r - d_0) \tag{5.35}$$

where $\delta(r)$ is a delta-"function" (see eq. 4.83):

$$4\pi\int_0^\infty g(r)\delta(r-a)r^2 dr \equiv g(a) \tag{5.36}$$

and d_0 is the radius of a dimer. The resultant expressions for the structure factor are (eq. 5.7):

$$S_1(Q) = 1, \qquad S_2(Q) = \frac{1}{2}\left[1 + \frac{\sin(Qd_0)}{Qd_0}\right] \tag{5.37}$$

Equation 5.34 now can be written in the form:

$$\frac{1}{I(Q,0)}\left(\frac{dI}{dt}\right)_{t=0} = 2kn(1,0)\frac{\sin(Qd_0)}{Qd_0} \tag{5.38}$$

Equation 5.38 shows that measurements of the time dependence of $I(Q, t)$ at several Q-values can be used to compute the rate coefficient k.[24]

Figure 5.7 illustrates an application of eq. 5.38 for a suspension of polystyrene latex colloids in 0.133 M $CaCl_2$ solution at pH 5.7.[24] These model colloids have sulfate surface functional groups and an average radius equal to 101 ± 8 nm; their refractive index is 1.591. The range of Q-values in the scattering experiments was 0.005 to 0.024 nm^{-1}, selected to satisfy the interference condition in eq. 5.5, with L equal to the diameter of a monomer. Under this condition, which differs completely from that in eq. 5.14 for the probing of fractal structure in large aggregates, a coherent photon scattering experiment will reveal the presence of monomers and dimers in a suspension. The graphs in fig. 5.7 are straight lines whose slopes increase with the value of $n(1, 0)$. According to eq. 5.38, a plot of the left side versus $\sin(Qd_0)/Qd_0$ should exhibit just this behavior. The value of the second-order rate coefficient that results from dividing the slopes of the lines by $2n(1, 0)$ is $2.31 \pm 0.08 \times 10^{-18}$ m^3 s^{-1} at 23°C, the temperature at which the scattering experiments were conducted. An additional experiment with colloids whose average radius was 49 ± 5 nm produced $k = 2.37 \times 10^{-18}$ m^3 s^{-1}, which does not differ from the value found with the larger colloids.[24] The characteristic time scale (half-life) for the coalescence process is $[n(1, 0)k]^{-1}$, appropriate to a second-order rate coefficient. Given the range of initial number of monomers per unit volume for the scattering experiments, this time scale is on the order of one hour and, therefore, is typical of a rapid coalescence process.[22] Measurements of k

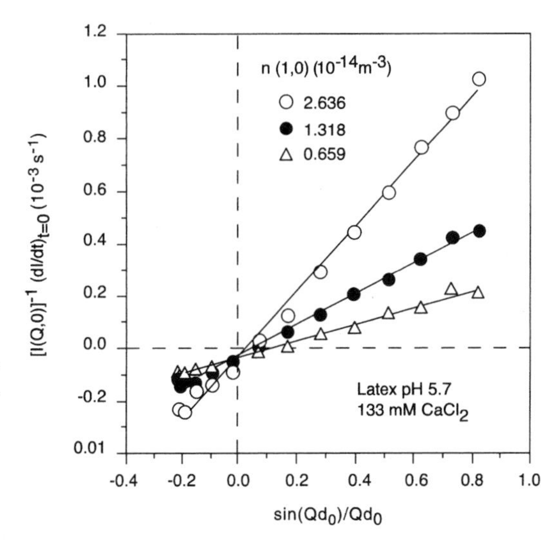

Fig. 5.7 Experimental test of eq. 5.38 for a suspension of latex colloids in $CaCl_2$ solution (J.H. van Zanten and M. Elimelech[24]). The slope of the line is equal to $2kn(1,0)$.

for rapid coalescence of polystyrene latex colloids ($r_0 = 109 \pm 5$ nm) based on eq. 5.38 also have been made with 1M $NaClO_4$ solution as the suspending electrolyte medium.[25] The value found at 25°C for $n(1, 0) = 3.8 \times 10^{14}$ m^{-3} is $k = 1.70 \pm 0.06 \times 10^{-18}$ m^3 s^{-1}, which is comparable to the values measured in $CaCl_2$.

An alternative kinetics model to eq. 5.34 can be developed from data obtained in a photon-scattering correlation experiment. This approach, termed *dynamic photon scattering*,[26] involves measurement of the time constant for the decay of correlations observed in the intensity of photon scattering by a colloid at different times during its random motion through a suspending fluid. This motion, which is diffusive in a quiescent fluid, causes the scattering intensity to be a fluctuating quantity whose time correlations persist over microsecond time scales. These correlations between the intensity at some arbitrary initial time and that at a later time die out exponentially with a time constant equal to $(Q^2\boldsymbol{D})^{-1}$, where Q is the scattering wavenumber and $\boldsymbol{D}$ is the colloid diffusion coefficient.[27] Thus, numerical analysis of correlation data from an experiment performed at fixed Q will yield a value of $\boldsymbol{D}$. For a single colloid, $\boldsymbol{D}$ can be modeled by the *Stokes-Einstein equation*:[28]

$$\boldsymbol{D}_{\mathrm{SE}} = \frac{k_B T}{6\pi\eta R_{\mathrm{H}}} \tag{5.39}$$

where $k_B(= 1.3807 \times 10^{-23}$ J K^{-1}) is the Boltzmann constant, T is absolute temperature, η is the shear viscosity of the suspending fluid, and R_H is the *hydrodynamic radius* of the colloid. At 298 K, $k_{\mathrm{B}}T/6\pi\eta = 2.451 \times 10^{-19}$ m^3 s^{-1}, which implies that colloid diffusion coefficients will range in value between 10^{-14} to 10^{-11} m^2 s^{-1} under ambient conditions. Note that eq. 5.39 predicts that colloid diffusion will be more rapid as temperature increases, the fluid viscosity decreases, or colloid size decreases.

In a suspension of diffusing colloids, a dynamic photon scattering experiment yields data on an average diffusion coefficient:[29]

$$\boldsymbol{D}(Q, t) \equiv \frac{\sum_q n(q, t) q^2 S_q(Q) \boldsymbol{D}_q}{\sum_q n(q, t) q^2 S_q(Q)} \tag{5.40}$$

which is analogous to eq. 5.6. Thus $\boldsymbol{D}(Q, t)$ is an average value of the colloid diffusion coefficient $\boldsymbol{D}_q$ weighted by the photon scattering intensity of the colloid. Given eq. 5.39 as a model for $\boldsymbol{D}_q$, it follows that eq. 5.40 can be rewritten as an expression for a "*z-average*" *hydrodynamic radius*:

$$\frac{1}{R_{\mathrm{HZ}}(Q, t)} \equiv \frac{\sum_q n(q, t) q^2 S_q(Q) R_{\mathrm{H}q}^{-1}}{\sum_q n(q, t) q^2 S_q(Q)} \tag{5.41}$$

The time derivative of both sides of eq. 5.41 can be combined with that of eq. 5.6 to derive the general relationship:

$$\frac{1}{R_{\mathrm{HZ}}} \frac{dR_{\mathrm{HZ}}}{dt} = \frac{1}{I} \frac{dI}{dt} - \frac{\sum_q \frac{dn}{dt}(q, t) q^2 S_q(Q) R_{Hq}^{-1}}{\sum_q n(q, t) q^2 S_q(Q) R_{Hq}^{-1}} \tag{5.42}$$

As in the case of eq. 5.34, interest in eq. 5.42 is focused on the initial rate of change of the "*z*-average" hydrodynamic radius:

$$\frac{1}{R_{\mathrm{HZ}}(Q,0)}\left(\frac{dR_{\mathrm{HZ}}}{dt}\right)_{t=0} = \frac{4S_2(Q)}{S_1(Q)}\left(1 - \frac{R_{H1}}{R_{H2}}\right)\mathrm{kn}(1,0) \tag{5.43}$$

upon substitution of eq. 5.34 for the first term on the right side of eq. 5.42 and noting that $n(2,0) \equiv 0$ while evaluating the second term with the help of eq. 5.33. It follows from eqs. 5.34 and 5.43 that measurements of the left sides of these two equations can be combined to determine the rate coefficient k *without* the need to model the structure factors $S_1(Q)$ and $S_2(Q)$ as in eq. 5.37:[30]

$$\frac{1}{I(Q,0)}\left(\frac{dI}{dt}\right)_{t=0} = \left(1 - \frac{R_{\mathrm{H1}}}{R_{\mathrm{H2}}}\right)^{-1}\frac{1}{R_{\mathrm{HZ}}(Q,0)}\left(\frac{dR_{\mathrm{HZ}}}{dt}\right)_{t=0} - 2\mathrm{kn}(1,0) \tag{5.44}$$

The *y*-intercept of a graph of the left side of eq. 5.34 versus the left side of eq. 5.43 will provide a value of the rate coefficient k, while the slope of the graph gives an estimate of the ratio $R_{\mathrm{H1}}/R_{\mathrm{H2}}$, the value of which is predicted by theory to lie in the interval (0.71, 0.75) for spheres.[30]

Figure 5.8 shows a test of eq. 5.44 based on measurements[31] of $I(Q, t)$ and $R_{\mathrm{HZ}}(Q, t)$ for a suspension of hematite (α-Fe_2O_3) in 100 mol m^{-3} $NaNO_3$ at pH 9.1, conditions that produced rapid coalescence. A linear plot is evident, and estimation of its *y*-intercept along with the value of $n(1,0)$ leads to $k = 0.75 \pm 0.10 \times 10^{-18}$ m^3 s^{-1}, a value substantially smaller than those cited above for latex spheres. Similar data for another sample of hematite with a different monomer size resulted in $k = 1.75 \pm 0.25 \times 10^{-18}$ m^3 s^{-1}. The average slope of all plots like the one in fig. 5.8 was equal to 3.5 ± 0.8, implying that $R_{\mathrm{Hz}}/R_{\mathrm{H2}} = 0.71 \pm 0.08$, in agreement with theory.[30,32]

Equation 5.44 also has been applied to determine k for rapid coalescence in suspensions of polystyrene latex spheres at 25°C and circumneutral pH.[25,30,32]

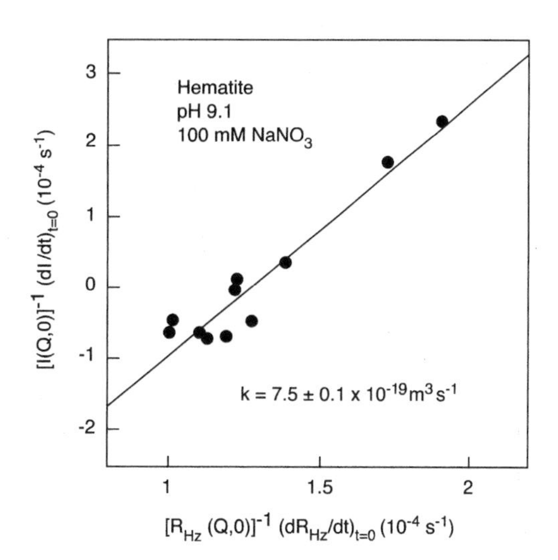

Fig. 5.8 Experimental test of eq. 5.44 for a suspension of hematite colloids in $NaNO_3$ solution (M. Schudel et al.[31]). The *y*-intercept of the line is equal to $-2kn(1,0)$.

The results represent a range of sphere radii (r_0) and initial colloid concentrations [$n(1, 0)$]:

Electrolyte	r_0(nm)	$n(1, 0)$ (10^{13} m^{-3})	k (10^{-18} m^3 s^{-1})
$NaClO_4$	109	38.0	1.68 ± 0.04
KCl	155	6.6–12.0	1.20 ± 0.20
KCl	580	4.0	3.45 ± 0.15
KCl	683	7.1	3.05 ± 0.25

The values of the second-order rate coefficient k listed above are consistent with those determined using eq. 5.34 alone for suspensions of smaller latex spheres at lower initial concentrations in a different electrolyte solution.[24] Similar values of k also have been obtained by a direct application of eq. 5.43 using the model of $S_1(Q)$ and $S_2(Q)$ in eq. 5.37:[25]

Electrolyte	k (10^{-18} m^3 s^{-1})
$NaNO_3$	1.75 ± 0.05
$NaClO_4$	1.55 ± 0.06
KCl	2.15 ± 0.10
$CaCl_2$	1.85 ± 0.10

These data were obtained for spheres of radius 109 nm in aqueous electrolyte solutions at 25°C. Overall it appears that k lies in the interval $1 - 4 \times 10^{-18}$ m^3 s^{-1} at 25°C.

Besides their use to determine the rate coefficient for colloid coalescence, measurements of the "z-average" hydrodynamic radius as a function of time can be applied to infer the fractal dimension of the aggregates formed in a coalescence process. The basis for this application can be understood through a consideration of the time dependence of the q-mer number density $n(q, t)$ when aggregates large enough to exhibit fractal characteristics are forming. Inherent to this consideration is the notion that power-law relationships, like those in eqs. 5.15, 5.17, and 5.25, reflect *the absence of intrinsic length or time scales* in the phenomena they model. (This situation is quite the opposite of what obtains for the diffuse ion swarm, for example, which is modeled by a differential equation that does exhibit an intrinsic length scale (κ^{-1}) characterizing the spatial variability of the mean electrostatic potential, as discussed in section 4.1.) The lack of intrinsic scale thereby precludes the use of a natural unit of measure to interpret spatial or temporal variability in fractal aggregate formation. Indeed, model expressions such as eq. 5.15 or 5.25 assume that the influence of the monomer radius r_0, which otherwise might serve as a natural unit of measure to assess aggregate size, has disappeared entirely from aggregate properties by the time they take on fractal characteristics. As a consequence, no absolute reference exists on which to base a determination of precisely how large an aggregate has grown to be. Accordingly, fractal aggregates will appear to have the same overall structure when examined under a microscope at varying magnification.[33]

The mathematical formulation of this kind of scaling behavior imputes to $n(q, t)$ the property of *generalized homogeneity*:[34]

$$n(\lambda q, \lambda^{\delta} t) = \lambda^{a} n(q, t) \tag{5.45}$$

where $\lambda > 0$ is a scale parameter ("the magnification") and the scaling exponents δ and a each can be any real number. Equation 5.45 states that the result of scaling (or magnifying) the number of monomers in an aggregate by λ, while scaling the time by λ^{δ}, is solely to magnify the value of $n(q, t)$ λ^{a} times. [The scale factor for time is allowed to differ from that for q; hence the exponent δ, which could even be equal to 0 (i.e., no time-scaling).] In picturesque language, if we imagine a world in which all q-mers are λ times larger than in ours, and time intervals are λ^{δ} times longer than in ours, the number density of q-mers will be λ^{a} times larger (or smaller, if $a < 0$) than in ours. If $a = 0$, $n(q, t)$ is said to be *scale-invariant*; if $a = \delta = 1$, $n(q, t)$ is said to be *homogeneous*. Generalized homogeneity allows a and δ to differ from 0 or 1.[34]

The value of the scaling exponent a can be deduced by imposing the constraint of mass conservation on $n(q, t)$. The total number of monomers in a suspension, regardless of which q-mer the monomers may inhabit at any given moment, should be a scale-invariant quantity if no monomer is destroyed by fragmentation or removed by settling out of suspension. Therefore

$$M_1(t) = \sum_q q n(q, t) \tag{5.46}$$

should not change in value under any scaling of q and t:

$$M_1(\lambda^{\delta} t) = \sum_{\lambda} q(\lambda q) n(\lambda q, \lambda^{\delta} t) = M_1(t) \tag{5.47}$$

For the large values of q associated with fractal characteristics, the sum over q in eq. 5.47 can be approximated by an integral over q:

$$\begin{aligned} M_1\left(\lambda^{\delta} t\right) &\approx \int_0^{\infty} (\lambda q)\, n\left(\lambda q, \lambda^{\delta} t\right) d(\lambda q) \\ &= \lambda^2 \int_0^{\infty} q\, \lambda^{a}\, n(q, t) dq \\ &= \lambda^{2+a} \int_0^{\infty} q\, n(q, t) dq = \lambda^{2+a} M_1(t) \end{aligned} \tag{5.48}$$

where eq. 5.45 has been applied in the second step. Given the constraint in eq. 5.47, it follows that $a = -2$ and the scaling behavior of $n(q, t)$ is:

$$n(\lambda q, \lambda^{\delta} t) = \lambda^{-2} n(q, t) \tag{5.49}$$

Additional insight as to the effect of time scaling on $n(q, t)$ can be had by applying a simple theorem concerning generalized homogeneous functions.[34] This theorem states that, if $n(q, t)$ has the scaling behavior in eq. 5.49, it can be rewritten in the equivalent form:

$$n(q, t) = t^{-\frac{2}{\delta}} g(q/t^{\frac{1}{\delta}}) \tag{5.50}$$

where $g(\ )$ is a function of the *scale-invariant* ratio, $q/t^{\frac{1}{\delta}}$. Equation 5.50 is derived from eq. 5.49 by selecting the value $t^{\frac{1}{\delta}}$ for the scale factor λ and noting that $g(u) \equiv n(u, 1)$. It is evident that the ratio $q/t^{\frac{1}{\delta}}$ does not change in value after scaling q by λ and t by λ^{δ}. Therefore, it is an invariant of the scaling operation expressed in eq. 5.45. Substitution of eq. 5.50 into eq. 5.46 yields the result:

$$M_1(t) \approx \int_0^{\infty} q t^{-\frac{2}{\delta}} g(q/t^{\frac{2}{\delta}}) dq = \int_0^{\infty} u\, g(u) du \tag{5.51}$$

which suggests normalization of $g(q/t^{\frac{1}{\delta}})$ such that the integral over u in eq. 5.51 has unit value. Then eq. 5.50 can be rewritten in the form:

$$n(q, t) = M_1 t^{-\frac{2}{\delta}} f(q/t^{\frac{1}{\delta}}) \tag{5.52}$$

where $f(u) \equiv g(u)/M_1$ and $\int_0^{\infty} uf(u)du \equiv 1$. Equation 5.52 is a model expression for $n(q, t)$ that reflects the scaling behavior in eq. 5.45 while respecting mass conservation. It can be transformed further after consideration of the time dependence of the *mass-weighted average* value of q in a suspension:

$$\langle q \rangle_t \equiv \frac{\sum_q q^2 n(q, t)}{\sum_q qn(q, t)} \equiv \frac{M_2(t)}{M_1(t)} \tag{5.53}$$

where the definition

$$M_{\alpha}(t) \equiv \sum_q q^{\alpha} n(q, t) \qquad (\alpha = 1, 2) \tag{5.54}$$

has been invoked to simplify notation. The second q-moment $M_2(t)$ weights each value of q by $qn(q, t)$, which is proportional to the mass of a q-mer per unit volume. This q-moment appears implicitly as the first term in the Guinier approximation for the photon scattering intensity (eq. 5.12) and, therefore, is accessible to measurement by observations of $I(Q, t)$ under the condition $Q^{-1} \gg R_{GZ}$.[4,11] The introduction of eq. 5.52 into eq. 5.53 together with approximation of the sum over q by an integral leads to the model expression:

$$\langle q \rangle_t \sim \int_0^{\infty} q^2 t^{-\frac{2}{\delta}} f\left(q/t^{\frac{1}{\delta}}\right) dq = t^{\frac{1}{\delta}} \int_0^{\infty} u^2 f(u) du \tag{5.55}$$

where, as in eq. 5.51, $u \equiv q/t^{\frac{1}{\delta}}$. Equation 5.55 predicts that the mass-weighted average value of q will vary with time as $t^{\frac{1}{\delta}}$ under the conditions where fractal aggregates are forming and eq. 5.52 becomes an accurate model of $n(q, t)$. Thus the value of the scaling exponent δ can be determined from measurements of $\langle q \rangle_t$ as a function of time.

Equation 5.45 can be tested experimentally in two distinct ways. First, eqs. 5.52 and 5.55 can be combined to derive a useful equation for $n(q, t)$:

$$n(q, t) = M_1 \langle q \rangle_t^{-2} \psi(q/\langle q \rangle_t) \tag{5.56}$$

where the integral factor in eq. 5.55 has been combined with $f(q/\langle q \rangle_t)$ to define the function $\psi(q/\langle q \rangle_t)$. Equation 5.56, which exposes clearly the lack of an intrinsic time scale for $n(q, t)$, implies that graphs of $\langle q \rangle_t^2 n(q, t)$ versus $q/\langle q \rangle_t$ corresponding to different values of the time variable will superpose. This prediction has been verified for fractal aggregates formed by gold and polystyrene latex spheres using both transmission electron microscopy and photon-scattering methods.[35] Second, eq. 5.55 can be tested directly by plotting log $\langle q \rangle_t$ versus log t and examining the slope of the resulting curve. This kind of test has been performed for fractal aggregation by a variety of colloids, the overall result being that $\delta \approx 1$:[36]

Colloid	Scaling Exponent δ
aerosol	0.9 ± 0.5
gold	1.01
illite	0.8 ± 0.2
latex	0.95, 0.97, 1.0
silica	1.00

Equation 5.55 provides the basis for measuring the fractal dimension of aggregates forming in aqueous suspension under conditions that are consistent with the model expression for $n(q, t)$ in eq. 5.56. If the aggregates formed are mass fractals, then, by analogy with eq. 5.15,

$$\langle q \rangle_t \underset{R_{GZ} \gg r_0}{\sim} R_{GZ}^D \tag{5.57}$$

where R_{GZ} is the "z-average" radius of gyration of the aggregates, defined in eq. 5.13. It follows from combining Eqs. 5.55 and 5.57 that

$$R_{GZ} \underset{R_{GZ} \gg r_0}{\sim} t^{\frac{1}{\delta D}} \tag{5.58}$$

for aggregate formation at sufficiently long times t. This power-law behavior of the radius of gyration is a direct result of the scaling behavior imputed to $n(q, t)$ in eq. 5.45.

Finally, a connection to dynamic photon scattering is made by assuming that R_{HZ}, defined in eq. 5.41, can be substituted for R_{GZ} in eq. 5.58 to obtain the relation:

$$R_{HZ} \underset{R_{HZ} \gg r_0}{\sim} t^{\frac{1}{\delta D}} \tag{5.59}$$

The assumption that R_{GZ} and R_{HZ} are proportional has been examined experimentally through measurements of both parameters made on the same suspension. These measurements, involving colloids of gold, hematite, latex, and silica, indicate that the ratio $R_{GZ}/R_{HZ} \approx 1$, independently of all quantities on the right side of eqs. 5.58 and 5.59.[37] Therefore, given $\delta = 1$, a log-log plot of the "z-average" hydrodynamic radius versus time can be used to infer the fractal dimension.

Figure 5.9 shows a test of eq. 5.59 for a suspension of hematite colloids undergoing rapid coalescence in KCl solution at pH 3 and 25°C.[38] The slope of this linear log-log plot is 0.534 ± 0.02, implying that $D = 1.87 \pm 0.01$, in general agreement with values of D that can be inferred from figs. 5.5 and 5.6. A variety of published data based on measurements of $R_{HZ}(t)$ and their interpretation using

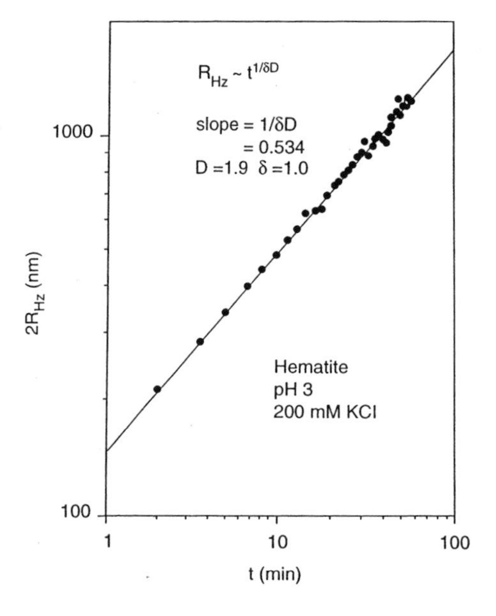

Fig. 5.9 Experimental test of eq. 5.59 for a suspension of hematite colloids in KCl solution (J. Zhang and J. Buffle[38]). The slope of the line indicates rapid coalescence of the colloids.

eq. 5.59 (with $\delta \equiv 1$) give similar results for the fractal aggregates formed by rapid coalescence:[39]

Colloid	Range of D
gold	1.7–1.9
hematite	1.7–1.9
illite	1.2–1.9
latex	1.7–1.8
silica	1.7–1.9

These experimental values of D are quite comparable to 1.7 – 1.8, the range of fractal dimension calculated with eq. 5.57 for aggregates formed in a computer simulation in which colloids are permitted to diffuse randomly with a Stokes-Einstein diffusion coefficient (eq. 5.39) until they collide, after which they coalesce instantly to form an aggregate.[40] The results in the table above also are comparable to the range of D-values inferred for aerosols, 1.7–1.8, on the basis of photon scattering experiments interpreted with eq. 5.25.[39] Thus the available data and calculations indicate that rapid coalescence processes lead to aggregates whose size is a power-law function of the time and whose fractal dimension lies in a narrow interval around 1.8.

5.3 The Stability Ratio

Equations 5.34, 5.43, and 5.44 provide a basis for measuring the second-order rate coefficient k that describes dimer formation from monomers in colloidal suspensions (eqs. 5.32 and 5.33). It is expected that, like the rate coefficients discussed in chapter 3, k will be an implicit function of the chemical composition and pH of the suspending aqueous solution, along with physical variables such as particle charge, shape, and size; temperature; and pressure. For example, the value of k, reported for the coalescence of 109 nm latex spheres suspended in 1 M $NaClO_4$ solution at 25°C as 1.55×10^{-18} m^3 s^{-1} (listed in the second table in section 5.2), is observed to *decrease* significantly as the electrolyte concentration is decreased, irrespective of the method of measuring k:[25]

	$k/k([NaClO_4] = 1M)$	
Method	250 mol m^{-3}	125 mol m^{-3}
Eq. 5.34	0.38 ± 0.07	0.079 ± 0.07
Eq. 5.43	0.37 ± 0.06	0.073 ± 0.07
Eq. 5.44	0.34 ± 0.05	0.069 ± 0.05

As the concentration of $NaClO_4$ decreases well below 1M, the value of k drops almost proportionally, with the result that the characteristic time scale for the coalescence process ($[n(1, 0)k]^{-1}$) increases from about 28 min, characteristic of rapid coalescence, to about 6 h [$n(1, 0) = 3.8 \times 10^{14}$ m^{-3}]. This trend has been observed for a wide variety of colloidal systems involving particles of either positive or negative charge suspended in 1:1, 1:2, 2:1, or 2:2 electrolyte solutions.[22]

Figure 5.10 shows the change in the ratio $k([KCl] = 80$ mmol $kg^{-1})/k$ provoked by an increase in suspending electrolyte concentration for hematite particles in KCl solution at pH 6 and 25°C.[16] This ratio of rate coefficients is the inverse of that in the table above and, therefore, decreases as the concentration of KCl increases. It was measured by dynamic photon scattering using eq. 5.43 under the assumption that all variables except the time derivative remain constant as [KCl] changes [$R_{HZ}(Q, 0) \approx 100$ nm, $Q = 0.013$ nm^{-1}, $n(1, 0) = 2.4 \times 10^{16}$ m^{-3}].[16] No change in this time-derivative was observed at KCl concentrations above 80 mmol kg^{-1}; hence all data were normalized to this concentration. The ratio

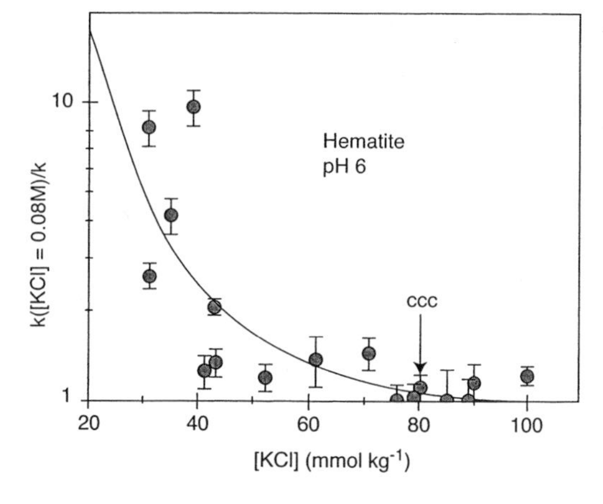

Fig. 5.10 Dependence of the coalescence rate coefficient k (eq. 5.33) on electrolyte concentration for hematite colloids suspended in KCl solution (Chorover et al.[16]). Rapid coalescence begins at a concentration of 80 mM (denoted "ccc").

$k([\text{KCl}] = 80 \text{ mmol kg}^{-1})/k$ displays a gradual decline toward unit value as [KCl] is increased over an order of magnitude. To the extent that KCl behaves as an indifferent electrolyte (section 1.4) in respect to adsorption by hematite, this decline can be interpreted solely as an effect of background electrolyte concentration on the diffuse ion swarm (section 4.1), particularly as regards Cl^-, which should be the screening ion, given the positive surface charge of hematite expected at pH 6 (i.e.p. $\approx$ 9.2). Corresponding to the range of concentration in fig. 5.10 there is a decrease by a factor of 3 in the intrinsic length scale κ^{-1} (eq. 4.7), the range of the electrostatic potential in the diffuse ion swarm. This diminishment of the influence of surface potential on the region near a hematite colloid evidently is sufficient to provoke a drop in the characteristic time scale for particle coalescence by an order of magnitude.

The quantity plotted against concentration in fig. 5.10 is termed the *stability ratio*:[41]

$$W \equiv \frac{\text{initial rate of rapid coalescence}}{\text{initial rate of coalescence observed}} \tag{5.60}$$

Given the accuracy of the model of coalescence represented in eq. 5.32, W is equal to the ratio of characteristic time scales for coalescence observed under prescribed conditions to that observed under conditions deemed to produce rapid coalescence. Usually these latter conditions are associated simply with the value of a variable parameter that corresponds to a maximal rate of particle coalescence. If the varied parameter is the concentration of an indifferent electrolyte, the *smallest* value that produces *rapid* coalescence is termed the *critical coagulation concentration* (ccc):[22,41]

$$\lim_{c \uparrow \text{ccc}} W = 1 \tag{5.61}$$

where c is the concentration of the ion in an indifferent electrolyte which is the *counterion* in a diffuse ion swarm adsorbed by the coalescing particles (section 4.1). Remarkably, an abundant published literature indicates that ccc values for monovalent counterions generally lie in the relatively narrow range 5 to 100 mol m^{-3}, whereas those for bivalent counterions typically lie in the range 0.1 to 2.0 mol m^{-3}, if the coalescing particles are inorganic, or largely so.[22,41,42] If the particles contain significant quantities of humus, however, the midpoint of the range of ccc values for either monovalent or bivalent counterions increases by nearly an order of magnitude.[42] Conversely, the values of W observed for aquatic colloids suspended in surface freshwaters fall generally into the interval (5, 100), increasing as the humus content of the colloids increases.[21] These two trends illustrate the greater efficacy of bivalent counterions, relative to monovalent counterions, in promoting particle coalescence, as well as the inhibitory effect of humus on the same process.[3,21,22]

The relationship between ccc and counterion valence can be understood and interpreted quantitatively on the basis of a simple consideration of length scales operative in the diffuse ion swarm. Adsorption of counterions in the diffuse swarm acts to screen particle charge, in that the range of the mean electrostatic potential created by a charged particle surface is diminished by adsorption (section 4.1). It is reasonable to assign to each counterion of charge Ze ($e = 1.6022 \times 10^{-19}$ C) a surface patch of equal and opposite charge that is screened by the counterion. If the screening is to be effective, the coulomb potential energy restricting the counterion to the vicinity of the charged surface patch should be of the same order of magnitude in absolute value as the thermal kinetic energy of the counterion, $\frac{1}{2}k_BT$, where k_B is the Boltzmann constant and T is absolute temperature (cf. eq. 5.39):

$$\frac{(Ze)^2}{4\pi\varepsilon_0 DL} \approx \frac{1}{2}k_BT \tag{5.62}$$

where ε_0 is the permittivity of vacuum and D is the dielectric constant of water (cf. eq. 4.1). The parameter L characterizes a "screening length" whose magnitude depends only on physical variables. Equation 5.62 can be rewritten as a more compact expression for L in terms of the diffuse-swarm parameter β introduced in eq. 4.6:

$$L \approx Z^2(\beta/4\pi N_A) \tag{5.63}$$

where N_A is the Avogadro constant. At 298.15 K, $\beta/4\pi N_A = 1.43$ nm.

Effective charge-screening by a counterion in the diffuse swarm also means that the intrinsic length scale for the mean electrostatic potential created by the charged surface (eq. 4.1) must be comparable to L in eq. 5.63. Otherwise, the influence of surface charge would "leak out" beyond the region of bound counterions. The intrinsic length scale for the mean electrostatic potential is, of course, κ^{-1}, where $\kappa = Z(\beta c_0)^{\frac{1}{2}}$ and c_0 is the bulk concentration of the screening counterion (eq. 4.19). Therefore, effective screening requires the constraint,

$$\kappa L \approx 1 \tag{5.64}$$

from which an equation for the ccc can be derived by introducing eq. 5.63 and the definition of κ:

$$\text{ccc} \approx (16\pi^2 N_A^2/\beta^3)Z^{-6} \tag{5.65}$$

The prefactor in eq. 5.65 is equal to 45 mol m^{-3} at 298.15 K. Given the typical range of ccc values for monovalent counterions, 5 to 100 mol m^{-3}, the estimate of ccc provided by eq. 5.65 for $Z = 1$ is reasonable. For bivalent counterions, the estimate of ccc is 0.7 mol m^{-3}, which also lies within the range of typical values, 0.1 to 2 mol m^{-3}.

Equation 5.65 is a quantitative expression of the *Schulze-Hardy rule*,[22] which states that the critical coagulation concentration applies to a diffuse-swarm *counterion* and is dependent on an *inverse power of the counterion valence*. The present derivation of eq. 5.65 [43] reveals that this broad generalization is a manifestation of *effective charge-screening*, a condition induced by decreasing the range of the diffuse-swarm potential generated by a charged surface until it is small enough to make the resultant coulomb attraction of a counterion toward the surface strong enough to quench the thermal kinetic energy that would otherwise send the counterion wandering off into the bulk solution.

Concomitant with the decline in the stability ratio as the concentration of an indifferent electrolyte is increased is a decline in the fractal dimension of the particles formed by coalescence.[44] Figure 5.11 illustrates this phenomenon for the hematite particles whose stability ratio is plotted in fig. 5.9 against [KCl].[16] The values of D, measured by photon scattering using eq. 5.25 (fig. 5.5) are seen to drop from about 2.1 to near 1.7 as the stability ratio declines by an order of magnitude, until it matches the condition under which eq. 5.61 holds. This decrease in the fractal dimension implies the concurrent development of colloidal aggregates whose porosity increases as they form more rapidly (eqs. 5.17 and 5.50). Denser aggregates that are formed more slowly evidently permit enough time for the participating colloids to seek out pathways of coalescence leading to more compact structures, as is indeed demonstrated in electron micrographs.[16] Similarly to the colloidal aggregates formed by rapid coalescence, those formed by slow coalescence are characterized by a relatively narrow range of fractal dimension, irrespective of the chemical composition of the participating colloids:[39,44,45]

Colloid	Range of D
goethite	1.9–2.1
gold	2.0–2.1
hematite	2.0–2.2
illite	2.2–2.3
latex	2.0–2.2
silica	2.0–2.2

These values, measured by photon scattering using eq. 5.25, are in agreement with computer simulations of colloid coalescence in which fractal aggregates with $D \approx$ 2.0–2.1 are formed by assigning a very small probability to coalescence after the collision of two particle clusters.[46]

Effective charge screening to induce rapid coalescence of two colloids is the result of weak interactions between a swarm of adsorptive counterions and a charged particle surface. Attractive van der Waals interactions, which are always present,[22,47] then can act to promote coalescence of the colloids. The mechanism of rapid coalescence is not unique, however, because the repulsive coulomb interaction between two colloids having the same surface charge sign can be vitiated by

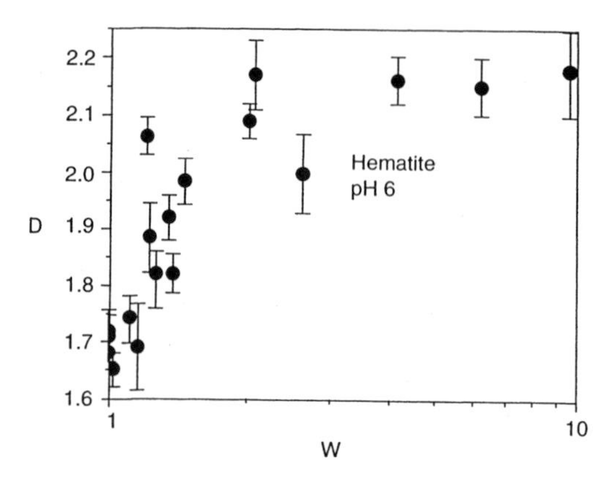

Fig. 5.11 Dependence on the stability ratio for the fractal dimension of the aggregates formed by hematite colloids in the system whose coalescence rate coefficients are graphed in fig. 5.10 (Chorover et al.[16]). As the stability ratio increases, so does the fractal dimension, indicating a less-porous aggregate structure formed under slow coalescence.

charge neutralization as well as charge screening. This mode of inducing rapid coalescence is the result of *strong* interactions between adsorptive counterions and a charged particle surface, those typically associated with specific adsorption processes (see chapter 1). These include protonation reactions and inner-sphere surface complexation of metal cations or inorganic/organic anions, as discussed in section 1.5.

Figure 5.12 illustrates the effect of pH on the rate coefficient k for dimer formation (eqs. 5.32 and 5.33) by hematite colloids [$r_0 \approx 60$ nm] suspended in $NaNO_3$ solution at 25°C.[31] Rapid coalescence was observed at any pH value for [NaCl] = 100 mol m^{-3}, the value of k [determined by photon scattering using eq. 5.44 (fig. 5.8)] then being $1.75 \pm 0.25 \times 10^{-18}$ m^3 s^{-1}, which is equal to that reported for 109 nm latex spheres suspended in the same electrolyte solution at the same temperature.[25] At lower [NaCl], an effect of pH is apparent, with the value of k decreasing by up to three orders of magnitude as pH is varied below or above approximately 9.2. The graph in fig. 5.12 is essentially a plot of −log W versus pH, since $\log k = -\log W + \log k$ ([$NaNO_3$] = 0.1 M) according to the definition of W in eq. 5.60. The value of pH required to produce a maximal value of k and, therefore, $W = 1.0$ is termed the *point of zero* charge (p.z.c.):

$$\lim_{\text{pH}\to\text{p.z.c.}} \log W = 0 \tag{5.66}$$

In the present example, direct measurement indicated that i.e.p. = 9.2 ± 0.1 and p.z.s.e. ≈ 9.2 (section 1.4). To the extent that the condition $\sigma_p = 0$ is represented by eq. 5.66 and these latter measurements, the terminology introduced here and that introduced in section 1.5 are mutually consistent. Given the truth of this consistency, note that there is *no* charge screening by the diffuse ion swarm when pH = p.z.c. (eq. 1.43).

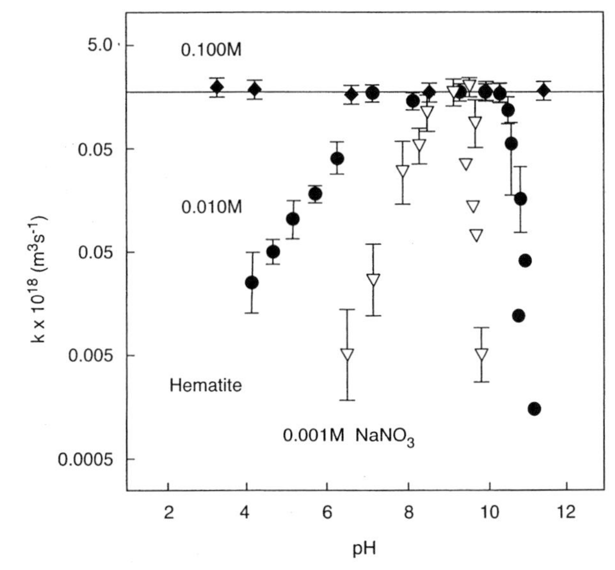

Fig. 5.12 Effect of pH on the coalescence rate coefficient k (eq. 5.33), with electrolyte concentration as a variable parameter, for hematite colloids suspended in $NaNO_3$ solution (M. Schudel et al.[31]). Note the absence of pH dependence at the ccc (100 mM $NaNO_3$). The p.z.c. is approximately 9.2 (inverted open triangle on the horizontal line).

The interplay between the definitions in eqs. 5.61 and 5.66 adds complexity to the concept of the stability ratio. Unlike the monotonic behavior of W in response to increases in the concentration of an indifferent electrolyte (fig. 5.10), W exhibits two branches quite typically as pH is increased from below to above the p.z.c.[48] The "hairpin" shape of the $-\log W$ versus pH plot in fig. 5.12 broadens considerably as the concentration of the suspending electrolyte solution is increased. This "straightening out" effect can be understood as a result of synergism between pH and [$NaNO_3$], both taken as controlling variables. The "hairpin" is bent into a straight, horizontal line as [$NaNO_3$] approaches the c.c.c. because charge screening is contributing more and more to the production of rapid coalescence conditions for *any* pH value. On the other hand, the asymmetry of the "hairpin" about the p.z.c. signals the existence of factors controlling the stability ratio other than electrolyte concentration and pH, e.g., both particle surface and morphological heterogeneities.[31] The fact that there is a "hairpin" shape at all, of course, stems from the change in sign of σ_p from positive to negative as pH is increased through the p.z.c. Interparticle coulomb repulsion prevents rapid coalescence irrespective of a given sign of σ_p. Note accordingly that anions become the counterions for pH $<$ p.z.c., whereas cations are the counterions for pH $<$ p.z.c., whereas cations are the counterions for pH $>$ p.z.c.

Figure 5.13 illustrates the effect of a strongly adsorbing inorganic counterion, $H_2PO_4^-$, on the stability ratio for hematite colloids suspended in 1 mol m^{-3} KCl solution at pH 6.[16] The plot of log W versus log[$H_2PO_4^-$] displays a characteristic "inverted hairpin" shape that signals surface charge reversal when log[$H_2PO_4^-$] -4.5, under the experimental conditions selected [$r_0 = 26 \pm 2$ nm, $n(1,0)$ 2.6×10^{16} m^{-3}, 25°C]. By analogy with eq. 5.66, this value of log[$H_2PO_4^-$] is termed *the p.z.c. with respect to* $H_2PO_4^-$:[48]

$$\lim_{\log[\,]\to \text{p.z.c.}} \log W = 0 \qquad (5.67)$$

Note that p.z.c. with respect to a *strongly adsorbing ion* is a *negative* quantity, whereas p.z.c. with respect to *protons* is a *positive* quantity. In the present exam-

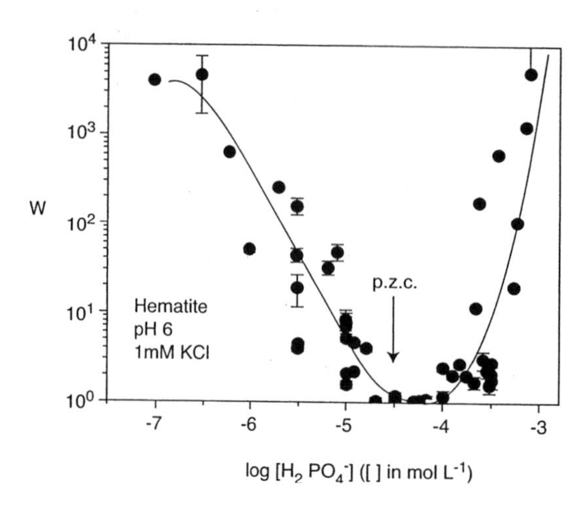

Fig. 5.13 Effect of a strongly adsorbing ion ($H_2PO_4^-$) on the stability ratio of hematite colloids suspended in 1 mM KCl at pH 6 (Chorover et al.[16]). The p.z.c. is approximately -4.5.

ple, p.z.c. with respect to $H_2PO_4^-$ is −4.5. Direct measurement[16] indicates that the i.e.p. (point of zero electrophoretic mobility) of hematite with respect to $H_2PO_4^-$ also is equal to −4.5 under the experimental conditions attendant to the data in fig. 5.13. Thus the definition of p.z.c. in eq. 5.67, like that in eq. 5.66, is consistent with the definition of p.z.c. given in section 1.4 ($\sigma_p = 0$).

It is apparent from figs. 5.12 and 5.13 that the conditions implied by eqs. 5.60, 5.66, and 5.67 all lead to rapid coalescence of colloids in an aqueous suspension. Moreover, it is known that rapid coalescence induced at the ccc is associated with the formation of aggregates having a fractal dimension between 1.7 and 1.9 (section 5.2), indicative of a rather open, porous structure. Figure 5.14 shows that this process also occurs in the case of hematite colloids for rapid coalescence induced at the p.z.c. with respect to $H_2PO_4^-$.[16] According to eq. 5.59, a log-log plot of the "z-average" hydrodynamic radius versus time, as determined by dynamic photon scattering, should be linear with a slope equal to the inverse of the fractal dimension, given that the scaling exponent $\delta \approx 1$ (eq. 5.55). Figure 5.14 demonstrates the superposability of this kind of log-log plot for both KCl and $H_2PO_4^-$ under conditions leading to rapid coalescence (see also fig. 5.5 with its log-log plots of eq. 5.25 based on "static" photon scattering measurements). However, unlike the situation that obtains for slow coalescence induced by charge screening, no change of the fractal dimension with changes of $[H_2PO_4^-]$ was observed for hematite aggregates within experimental precision.[16] For example, the three values of D given in fig. 5.5, all of which cluster near 1.6–1.7, correspond to values of W ranging over nearly two orders of magnitude.

Figure 5.15 illustrates the effect of a strongly adsorbing organic counterion, the aliphatic anion $CH_3(CH_2)_nCOO^-$, on the stability ratio for hematite colloids suspended in 50 mol m^{-3} NaCl solution at pH 5.2.[49] The log-log plots indicate p.z.c. values in the range −5 to −3, and these values are very close to the i.e.p.

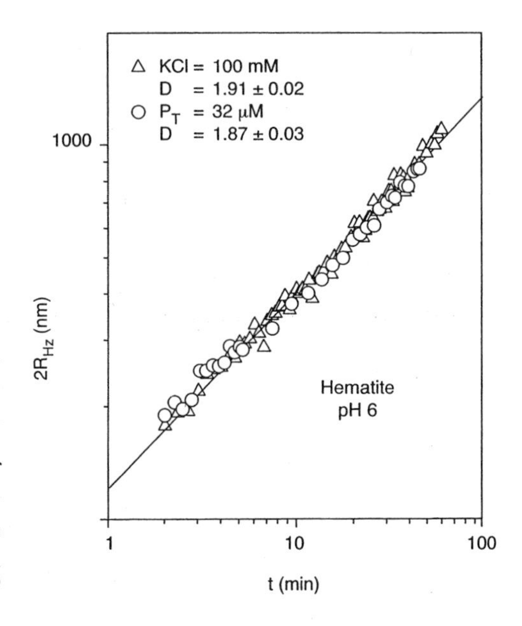

Fig. 5.14 Plot of the z-average diameter of hematite aggregates in a rapidly flocculating suspension versus time, showing that aggregate growth and, therefore, fractal dimension (eq. 5.59) is the same irrespective of whether coalescence is caused by charge screening (KCl) or charge neutralization (KH_2PO_4). Data from Chorover et al.[16]

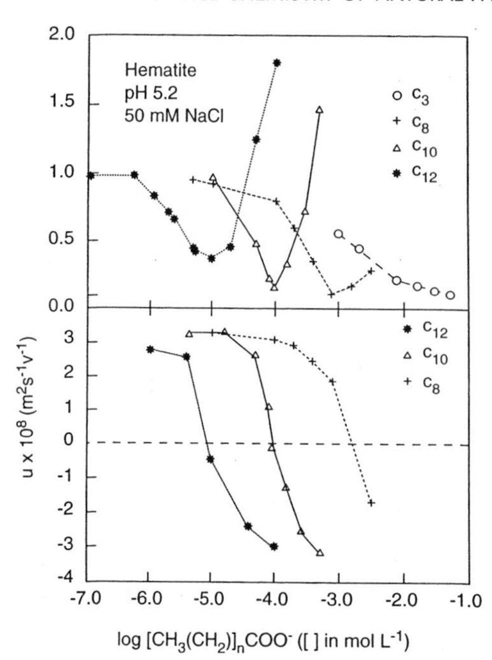

Fig. 5.15 Stability ratios of hematite suspensions in NaCl solution at pH 5.2 as influenced by varying concentrations of aliphatic acid anions that adsorb strongly (propionic, $n = 2$; caprylic, $n = 7$; capric, $n = 9$; lauric, $n = 11$). The lower set of graphs show that the electrophoretic mobility of the hematite particles goes through a zero (i.e.p.) at about the same aliphatic anion concentration as corresponds to the p.z.c. in the upper set of graphs (L. Liang and J. J. Morgan[49]).

values for hematite in the presence of the three anions investigated ($n = 6$, 8, and 10), as shown in the lower part of the figure. The "inverted hairpin" shape of the log-log plots exhibits an asymmetry about the p.z.c. similar to what is apparent in fig. 5.12. Moreover, the p.z.c. shows a decrease as the number of carbon atoms in the anion increases, suggesting that hydrophobic interactions play a role in promoting rapid coalescence.[48,49] This result implies that humic substances, which also are organic anions with hydrophobic moieties, should be effective at promoting rapid coalescence of hematite colloids, as is indeed observed experimentally.[48–50] Evidently a polymeric organic anion provokes rapid coalescence of positively charged colloids when adsorbed at low concentrations, but mitigates coalescence once it is adsorbed in substantial amounts, as indicated by the sharp increase in log W above the p.z.c. in fig. 5.14, and by the observed trend toward higher values of W for suspensions of colloids in natural waters as their content of dissolved humus increases.[21]

The kinetics of rapid coalescence induced by the polymeric anion, polyacrylate (a polymer of acrylate, $CH_2 = CHCOO^-$), is illustrated and compared with that induced by NaCl for hematite colloids in fig. 5.16.[51] The power-law shape of the time dependence of R_{HZ} is apparent, as is the resultant similarity of the fractal dimension calculated with the kinetics data, $D = 1.88 \pm 0.02$, irrespective of the coagulating anion. Measurement of the fractal dimension for smaller concentrations of the polymer than required for rapid coalescence (polymer/hematite 0.022–0.026, which also produces an i.e.p. for the hematite colloids[51]) gave $D \approx 1.9$–2.1, which is indicative of slow coalescence.[52] The correspondingly dense structure of the aggregates formed was confirmed by electron micrographs. Thus, for the polymeric anion, a transition of the fractal dimension from values near 2.1 to

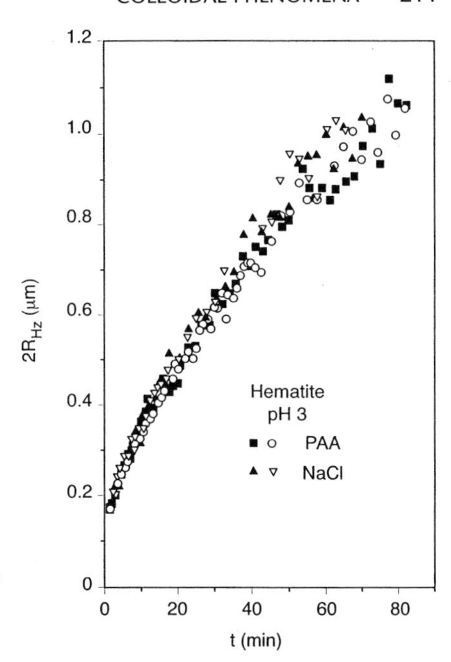

Fig. 5.16 A plot analogous to that in fig. 5.14, showing that rapid coagulation induced by charge screening (NaCl) or by charge neutralization (strong adsorption of polyacrylate, PAA) is essentially the same in suspensions of hematite (R. Ferretti et al.[51]).

those near 1.8 occurs as its concentration increases, quite in parallel with the trend shown in fig. 5.11.

5.4 The von Smoluchowski Rate Law

The second-order rate law describing the coalescence of two monomers to form a dimer (eq. 5.33) subsumes all mechanistic information about the coalescence process into the rate coefficient k, the values of which for *rapid* coalescence are on the order of 10^{-18} m^3 s^{-1}. This latter phenomenon is associated with *transport control,* as opposed to *slow* coalescence, which is associated with *surface-reaction control,* when considered from the perspective of processes that determine the observed value of the rate coefficient.[22] It is perspicacious to note in this respect that the scale factor, $k_BT/6\pi\eta$, in the Stokes-Einstein model of a colloid diffusion coefficient (eq. 5.39) has the very same units and is of the same order of magnitude as the second-order rate coefficient for rapid coalescence.

Three models of the transport-control mechanism for rapid coalescence are commonly applied to interpret data on the kinetics of particle formation in suspensions (fig. 5.17).[53] The best known of these models is *Brownian motion* (or "perikinetic aggregation"), which applies to quiescent suspensions of diffusing particles, each of whose size lies in the lower to middle portion of the colloidal range ($\lesssim 1\mu$m). Alternatively, coalescence caused by stirring a colloidal suspension can be described as *shear-induced* (or "orthokinetic aggregation"), whereas that caused by the settling of particles under gravitational or centrifugal force can be described by a *differential sedimentation* model. In all three models, the result-

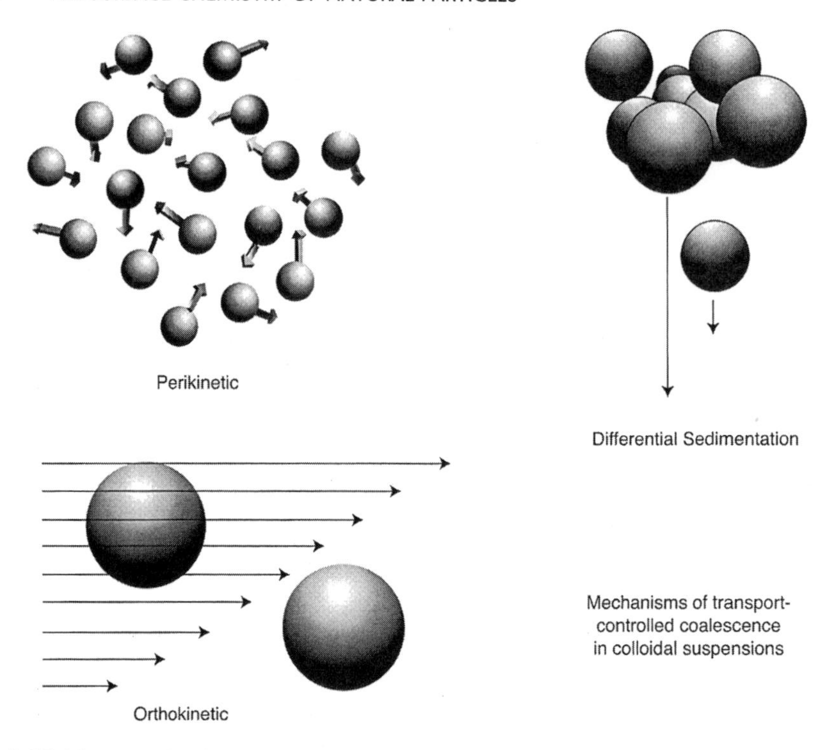

Fig. 5.17 Three mechanisms of rapid coalescence, each of which scales differently with colloid size (after K. H. Gardner[53]).

ing second-order rate coefficient is equal to the product of an effective cross-section for two-particle collisions (a geometric factor) times an effective two-particle relative velocity (a kinematic factor), thus leading to dimensions of volume per unit time. Large rate coefficients for rapid coalescence are accordingly produced by optimal combinations of particle size (geometry) and opposing interparticle velocities (kinematics).

Perikinetic aggregation entails the random thermal motion of particles that, once they collide by chance, are able to coalesce instantaneously to form a dimer. The second-order rate coefficient describing this formation process (eq. 5.33) is:[54]

$$k_p = 2\pi R_{11} \boldsymbol{D}_{11} \tag{5.68}$$

where R_{11} is the radius of the dimer and $\boldsymbol{D}_{11}$ is the diffusion coefficient of one of the colliding monomers relative to that of the other, as depicted from their center of mass taken as a reference point. In a first approximation, R_{11} is just twice the monomer radius, and $\boldsymbol{D}_{11}$ is just twice the monomer diffusion coefficient as modeled by eq. 5.39. With these simplifications, the rate coefficient for dimer formation in eq. 5.68 becomes

$$k_p^{\mathrm{SE}} = 8\pi R \boldsymbol{D}_{\mathrm{SE}} \equiv K_{\mathrm{SE}} \tag{5.69}$$

where

$$K_{SE} = 4k_BT/3\eta \tag{5.70}$$

is a constant parameter equal to 6.16×10^{-18} m^3 s^{-1} at 298 K, if water is the suspending fluid. Under this additivity approximation for dimer size and the relative diffusion coefficient, the use of the Stokes-Einstein model leads to an exact cancellation of the two opposing effects of the monomer radius R on the effective collision cross-section ($2\pi R_{11}$) and relative velocity factor ($\boldsymbol{D}_{11}$) in the perikinetic rate coefficient.

Equation 5.69 predicts a rate coefficient that is somewhat larger than any of the measured values of k for rapid coalescence that are cited in section 5.2. This discrepancy is often rationalized by including a "hydrodynamic stability ratio" in the definition of k appearing in eq. 5.68:

$$k_p = K_{SE}/W_{11}^H \tag{5.71}$$

where

$$W_{11}^H \equiv 4R\boldsymbol{D}_{SE}/R_{11}\boldsymbol{D}_{11} \tag{5.72}$$

to be consistent with the original expression for the perikinetic model rate coefficient. The stability ratio W_{11}^H should range in value from about 2 to 5 to account for the differences between the measured values of k and K_{SE}. This order of magnitude has in fact been estimated also by calculations based in theoretical fluid mechanics.[55] They indicate that the hydrodynamic radius of a dimer is only 30–40% larger than the monomer radius, not twice as large, and that the slowed motion of fluid trapped between two colliding spheres reduces $\boldsymbol{D}_{11}$ below the value $2\boldsymbol{D}_{SE}$ that obtains when the spheres are far apart.

Orthokinetic aggregation involves the capture of a monomer in the streamlines around another monomer while the former attempts to pass the latter.[53] This phenomenon requires the presence of a fluid velocity gradient (or shear rate), G, that permits one monomer to overtake the other while they both are being convected by the fluid. The rate coefficient for orthokinetic aggregation is the product of the cross-sectional area of a dimer times the relative velocity of the overtaking monomer, this product then being integrated over all possible points of contact of the overtaking monomer on the monomer with which it collides.[54] The result of the integration is a "capture sphere" that encloses the overtaken monomer and whose radius is that of the dimer formed, with the orthokinetic model rate coefficient then being:[53,54]

$$k_o = \frac{2}{3}R_{11}^3G \tag{5.73}$$

Usually, R_{11} is approximated once again by twice the monomer radius, such that

$$k_o = \frac{16}{3}GR^3 \tag{5.74}$$

serves as a model rate coefficient for shear-induced coalescence. In this case, the geometric factor increases strongly with particle size as the third power. Typical values for G are in the range 1 to 10 s^{-1} for surface waters,[53] but values up to two orders of magnitude larger can be observed in vigorously stirred fluids.

It is evident in eq. 5.74 that the importance of orthokinetic aggregation increases with particle radius for any nonzero value of G. This point can be appreciated in a simple fashion by forming the dimensionless ratio of the right sides of eqs. 5.69 and 5.74:

$$\frac{\text{perikinetic rate}}{\text{orthokinetic rate}} = \frac{1.16}{GR^3} \tag{5.75}$$

where R is in units of micrometers if G is in units of s^{-1}. It follows that orthokinetic aggregation rates exceed those of perikinetic aggregation whenever $R > G^{-1/3}$ numerically; that is, for monomers in the mid to upper range of colloidal size, given the typical values of G cited.

Transport control of rapid coalescence by differential sedimentation in the gravitational field of the earth is modeled typically by applying the well-known equation for the terminal speed of a particle settling in a viscous fluid under Stokes friction[53] to a pair of monomers with different radii, then multiplying the resulting relative velocity of the particles by the dimer cross-sectional area:

$$k_{\text{DS}} = \frac{g}{9\eta}(\rho_s - \rho_f)\pi R_{12}^2 |R_1^2 - R_2^2| \tag{5.76}$$

where g is gravitational acceleration, ρ_s and ρ_f are mass densities of the monomers and the fluid in which they are settling, respectively, and η is the coefficient of viscosity of the fluid. In this case, one monomer ("1") overtakes the other ("2") because it is larger, and, therefore, has a larger terminal speed. The dimer radius R_{12} is again usually approximated by the sum of the monomer radii. Given the fourth-power dependence on monomer size and the typical magnitude of the constant prefactor in eq. 5.76 (about 6×10^6 m^{-1}), it is apparent that differential sedimentation becomes important for particles larger than about 1 μm, as is also borne out in illustrative examples.[53]

Under the assumption that binary encounters between particles are the principal cause of particle coalescence and flocculation, eqs. (5.32) and (5.33) can be generalized to apply to a suspension containing many different particle sizes as a result of the coalescence process. Equation 5.32 is generalized to the two-step reaction sequence (cf. eq. 3.39):

$$m + n \overset{k(m,n)}{\rightarrow} q \qquad (q = m + n)$$

$$q + p \overset{k(q,p)}{\rightarrow} r \qquad (r = q + p) \tag{5.77}$$

where m, n, q, p, and r refer to aggregates that contain m, n, q, p, or r monomers, and $k(m, n) = k(n, m)$ is a second-order rate coefficient for the formation of a q-mer from the coalescence of an m-mer and an n-mer. The reaction sequence in eq. 5.77 thus describes q-mer formation from binary encounters of smaller particles and q-mer destruction from subsequent coalescence with a p-mer. It does not, however, describe q-mer destruction by fragmentation (the reverse of the first reaction in the sequence).[56]

The generalization of eq. 5.33 corresponding to the reaction sequence in eq. 5.77 is:

$$\frac{dn}{dt}(q, t) = k(m, n)n(m, t)n(n, t) - k(q, p)n(q, t)n(p, t) \tag{5.78}$$

Note that both rate coefficients in eqs. 5.77 and 5.78 refer to the same coalescence mechanism and both are symmetric under exchange of their arguments. In a suspension with a broad range of particle size, eq. 5.78 must be generalized yet further to permit myriad independent and parallel reactions of the type in eq. 5.77, with the proviso that $q = m + n$ for all m, n:

$$\frac{dn}{dt}(q, t) = \frac{1}{2}\sum_{m}\sum_{\substack{n \\ (q=m+n)}} k(m, n)n(m, t)n(n, t) - n(q, t)\sum_{p} k(q, p)n(p, t) \tag{5.79}$$

A factor $\frac{1}{2}$ must be introduced on the first summation in eq. 5.79 because of the symmetry of the rate coefficient $k(m, n)$; otherwise, the combinations $m + n$ and $n + m$, which yield the same q-mer, would be counted as distinct. Equation 5.79 is known as the *von Smoluchowski rate law* for particle coalescence through binary encounters.[57]

Without yet having to specify the dependence of the second-order rate coefficients in eq. 5.79 on their two arguments, one can derive from it a rate law for the *q-moments* $M_\alpha(t)$ that were introduced in eq. 5.54:

$$\begin{aligned}\frac{dM_\alpha}{dt} &= \frac{1}{2}\sum_{q}\sum_{m}\sum_{\substack{n \\ (q=m+n)}} q^\alpha k(m, n)n(m, t)n(n, t) - \sum_{q} q^\alpha n(q, t)\sum_{p} k(q, p)n(p, t) \\ &= \frac{1}{2}\sum_{m}\sum_{n} (m+n)^\alpha k(m, n)n(m, t)n(n, t) - \frac{1}{2}\sum_{q}\sum_{p} (q^\alpha + p^\alpha)\, k(q, p)n(q, t)n(p, t) \\ &= \frac{1}{2}\sum_{m}\sum_{n} [(m+n)^\alpha - m^\alpha - n^\alpha]\, k(m, n)n(m, t)n(n, t)\end{aligned} \tag{5.80}$$

where the second step obviates the need for summing over q in the first term on the right side by imposing the constraint $q = m + n$ directly then symmetrizes the second term on the right side by repeating it and dividing the sum by 2. Equation 5.80 is a model rate law for the q-moments based on the von Smoluchowski rate law. The quantity within the square brackets on the right side is subject to the exact mathematical inequality:[57]

$$[(m+n)^\alpha - m^\alpha - n^\alpha] \geq 0 \text{ if } \alpha \geq 1 \tag{5.81a}$$

for any m, n. It follows that *$M_\alpha(t)$ is a nondecreasing* function of time for any $\alpha \geq 1$. For example, this result implies that the two q-moments $M_1(t)$ and $M_2(t)$, introduced in eqs. 5.46 and 5.53, will never be observed to decrease in a coagulating suspension. In particular, eq. 5.80 yields the rate equations:

$$\frac{dM_1}{dt} = 0 \tag{5.82}$$

which affirms the constancy of $M_1(t)$ that was assumed in deriving the scaling relationship in eq. 5.47, and

$$\frac{dM_2}{dt} = \sum_m \sum_n mn\, k(m, n)n(m, t)n(n, t) \tag{5.83}$$

which demonstrates the bilinear dependence of $M_2(t)$ on the number of monomers in the two aggregates which coalesce to form a q-mer. This q-moment can be determined by measuring the coherent photon scattering intensity in the Guinier limit (eq. 5.12). It is proportional to the mass-weighted average value of q in a suspension ($\langle q \rangle_t$ in eq. 5.53), given the assumed constancy of $M_1(t)$.

The only q-moment that does not increase with time is $M_0(t)$:

$$M_0(t) = \sum_q n(q, t) \tag{5.84}$$

which is the total number of q-mers of any size per unit volume of suspension.[4] This q-moment is closely related to the colligative properties of aqueous suspensions, such as osmotic pressure,[4] through the *number-weighted average* value of q,

$$\bar{q}_t \equiv \frac{\sum_q q\, n(q, t)}{\sum_q n(q, t)} = \frac{M_1}{M_0(t)} \tag{5.85}$$

which can be combined with the mass-weighted average $\langle q \rangle_t$ (eq. 5.53) to determine the *polydispersity* of a suspension:[4]

$$\begin{aligned} P_t &\equiv \frac{1}{\bar{q}_t}\left[\frac{\sum_q (q - \bar{q}_t)^2 n(q, t)}{\sum_q n(q, t)}\right]^{\frac{1}{2}} \\ &= \frac{1}{\bar{q}_t}\left[\frac{\sum_q q^2\, n(q, t)}{\sum_q n(q, t)} - \bar{q}_t^2\right]^{\frac{1}{2}} \\ &= \left(\frac{\langle q \rangle_t}{\bar{q}_t} - 1\right)^{\frac{1}{2}} \end{aligned} \tag{5.86}$$

Equation 5.86 shows that the deviation of $\langle q \rangle_t / \bar{q}_t$ from unit value gives a measure of the coefficient of variation (standard deviation divided by the mean) about the number-weighted average of q in a colloidal suspension. The rate at which $\bar{q}_t$ increases with time is determined by that at which $M_0(t)$ decreases with time according to eq. 5.80 applied to the case $\alpha = 0$:

$$\frac{dM_0}{dt} = -\frac{1}{2}\sum_m \sum_n k(m, n)n(m, t)n(n, t) \tag{5.87}$$

This rate of decrease is necessarily smaller in absolute value than the rate of increase of $M_2(t)$, given in eq. 5.83, implying therefore that polydispersity increases with time, as expected from coalescence processes that create a range of aggregate sizes.

If the von Smoluchowski rate law is to be consistent with the formation of aggregates having a fractal structure, it must admit scaling of its solution, $n(q, t)$, depicted in eq. 5.45, as a symmetry. This means that scaling of the q and t variables in the rate law must leave the rate law *invariant* in mathematical form, such that the scaled rate law has exactly the same appearance as the original unscaled rate law. Physically speaking, scaling invariance of the von Smoluchowski rate law is tantamount to scale invariance of the coalescence reactions in eq. 5.77 as time passes and aggregates with fractal structure are formed. However, it is not likely that an *arbitrary* dependence of the rate coefficient $k(m, n)$ on its two arguments will be consistent with scale invariance of the rate law. Instead, scaling symmetry should act as a constraint that selects from all possible models of the rate coefficient only those that are consistent with scale invariance of the rate law. An immediate first consequence of this approach is the stipulation that the rate coefficient be a generalized homogeneous function, similar to $n(q, t)$ in eq. 5.45:

$$k(\lambda m, \lambda n) = \lambda^{\theta} k(m, n) \tag{5.88}$$

where θ is a scaling exponent analogous to δ and a in eq. 5.45.

With eq. 5.88 in hand, the scaling invariance of eq. 5.79 can be evaluated by introducing it into the three transformations of q, t, and $n(q, t)$ indicated in eq. 5.49 and equating both sides of the rate law to derive a condition on the scaling exponents δ and θ, thus following the method used to investigate the scale invariance of $M_1(t)$ in eq. 5.48.[58] The initial step in this process involves rewriting eq. 5.79 in the form:

$$\begin{aligned}\frac{dn}{dt}(q, t) \approx \frac{1}{2}\int_0^{\infty} k(m, q-m)n(m, t)n(q-m, t)dm \\ - n(q, t)\int_0^{\infty} k(q, p)n(p, t)dp\end{aligned} \tag{5.89a}$$

where explicit account of the constraint on m and n in the double summation on the right side of eq. 5.79 has been taken. The introduction of eqs. 5.49 and 5.88 then yields the relation:

$$\begin{aligned}\lambda^{-2-\delta}\frac{dn}{dt}(q, t) = \lambda^{\theta-3}\Bigg[\frac{1}{2}\int_0^{\infty} k(m, q-m)n(m, t)n(q-m, t)dm \\ -n(q, t)\int_0^{\infty} k(q, p)n(p, t)dp\Bigg]\end{aligned} \tag{5.89b}$$

for the scaled version of eq. 5.89a. Scale invariance requires cancellation of all factors in the scale parameter λ, which yields the condition:

$$\delta + \theta = 1 \tag{5.90}$$

Equation 5.90 is *the constraint that selects acceptable model rate coefficients for a scale-invariant von Smoluchowski rate law.*

As an application of eq. 5.90, the value of δ obtained from experimental studies of rapid coalescence under conditions leading to fractal aggregates (eqs. 5.55 and 5.56) can be introduced: $\delta \approx 1$ (section 5.2). It follows that a model rate coefficient consistent with this result and a scale-invariant von Smoluchowski rate law must have $\theta = 0$. Therefore, *the model rate coefficient for rapid coalescence must be scale-invariant.* This stipulation is met only by the rate coefficient for perikinetic aggregation in eq. 5.68. Any physical model of this rate coefficient that leaves the product $R_{11}D_{11}$ independent of the radius of the dimer formed [for example, any model leading to eq. 5.69, $k_p^{\mathrm{SE}}(m, n) = K_{\mathrm{SE}}$] will be an acceptable model of fractal aggregate formation by rapid coalescence. Neither orthokinetic nor differential sedimentation can meet this stipulation because of the strong dependence of the associated model rate coefficients on dimer size (eqs. 5.73 and 5.76). We may conclude that *rapid coalescence and fractal structure are consistent only for perikinetic aggregation.*

The constraint in eq. 5.90 appears in an especially cogent manner when eqs. 5.53 and 5.83 are combined to produce an ordinary differential equation for the growth of the mass-weighted average value of q in a flocculating colloidal suspension:

$$\frac{d}{dt}\langle q\rangle_t \approx M_1^{-1} \int_0^\infty \int_0^\infty mn\, k(m, n) n(m, t) n(n, t) dm dn \qquad (5.91)$$

again approximating the sums by integrals, as in eq. 5.55, and noting eq. 5.82. The rate coefficient in eq. 5.91 is assumed to have the property of generalized homogeneity (eq. 5.88). If the scale factor in eq. 5.88 is set equal to the value of $1/\langle q\rangle_t$, the rate coefficient can be expressed in a mathematical form analogous to that for $n(q, t)$ in eq. 5.56:

$$k(m/\langle q\rangle_t, n/\langle q\rangle_t) = \langle q\rangle_t^{-\theta} k(m, n)$$

or

$$k(m, n) = \langle q\rangle_t^{\theta} \kappa(m/\langle q\rangle_t, n/\langle q\rangle_t) \qquad (5.92)$$

where κ must be a *scale-invariant* rate coefficient analogous to the function $\psi\,(q/\langle q\rangle_t)$ in eq. 5.56. It is evident that introduction of the scale factor λ as a multiplier of m, n, and q on both sides of eq. 5.92 will reproduce the relationship in eq. 5.88. On combining eqs. 5.56, 5.92, and 5.91, one recasts eq. 5.91 into the scaling form:

$$\frac{d}{dt}\langle q\rangle_t \approx M_1 \langle q\rangle_t^{\theta} \int_0^\infty \int_0^\infty xy\kappa(x, y)\psi(x)\psi(y) dx dy \qquad (5.93)$$

where $x \equiv m/\langle q\rangle_t$ and $y \equiv n/\langle q\rangle_t$ are scale-invariant variables. The integral on the right side of eq. 5.93 can be interpreted as a mass-weighted average rate coefficient. Its value is independent of time precisely because of the absence of an intrinsic time scale for $n(q, t)$ when it is considered as a generalized homogeneous function (eq. 5.49). Thus, only the factor $\langle q\rangle_t^{\theta}$ is time-dependent on the right side of eq. 5.93. If $\theta = 0$, as deduced in the case of rapid coalescence discussed above, it follows from eq. 5.93 that $\langle q\rangle_t$ must be a linear function of time, in agreement with eq. 5.55 and the experimental result, $\delta = 1$. Conversely, a scale-invariant rate coefficient always must yield a mass-weighted average value of q in a flocculating suspension that varies linearly with time, which is an alternative way to see the implications of eq. 5.90. Exactly analogous results can be derived for the number-weighted average value of q, defined in eq. 5.85, by consideration of eq. 5.87.

5.5 Solving the von Smoluchowski Equation

Consideration of its scaling properties shows that the von Smoluchowski rate law describing perikinetic aggregation that produces floccules having a fractal structure should be the special case of eq. 5.79 in which $k(m, n)$ is independent of m or n; say, $k(m, n) = 2\,k_p$:[59]

$$\frac{dn}{dt}(q,t) = 2k_p\left[\frac{1}{2}\sum\nolimits_m \sum\nolimits_n \underset{(q\,=\,m+n)}{n(m,t)n(n,t)} - n(q,t)\sum\nolimits_p n(p,t)\right] \tag{5.94}$$

The solution of this integro-differential equation can be obtained analytically by several methods.[60] Perhaps the most straightforward one is through the use of a *generating function* whose mathematical form is designed to produce solutions of eq. 5.94 as partial derivatives:

$$F(u,t) = \sum\nolimits_q (u^q - 1)n(q,t) \tag{5.95}$$

The ℓth partial derivative of $F(u, t)$ with respect to the variable u follows from eq. 5.95 as:

$$\left(\frac{\partial^\ell F}{\partial u^\ell}\right)_t = \sum_{q=\ell} q(q-1)\cdots(q-\ell+1)u^{q-\ell}n(q,t) \qquad (\ell \geq 1) \tag{5.96}$$

The term in eq. 5.96 for which $q = \ell$ is $\Gamma(q+1)n(q,t)$,[20] and it is only this term that will contribute to a MacLaurin expansion of $F(u, t)$ in powers of u:

$$F(u,t) = \sum_{\ell=0}[\Gamma(\ell+1)]^{-1}\left(\frac{\partial^\ell F}{\partial u^\ell}\right)_{t,u\,=\,0} u^\ell = \sum\nolimits_\ell n(\ell,t)u^\ell - M_0(t) \tag{5.97}$$

since $F(0, t) = -M_0(t)$ by eqs. 5.84 and 5.95. Thus $n(q, t)$ can be found as $[\Gamma(q+1)]^{-1}(\partial^q F/\partial u^q)_t$ where it is evaluated at $u = 0$, if the mathematical form of $F(u, t)$ that is consistent with eq. 5.95 can be deduced.

Combination of eq. 5.95 and eq. 5.94 yields a partial differential equation for $F(u, t)$:

$$\frac{\partial F}{\partial t}+\frac{dM_0}{dt}=2k_p\left[\frac{1}{2}\sum_q u^q\sum_m\sum_n n(m,t)\,n(n,t)\right.$$
$$\left.-\sum_q u^q\,n(q,t)\sum_p n(p,t)\right]$$
$$=2k_p\left[\frac{1}{2}\sum_m u^m\,n(m,t)\sum_n u^n n(n,t)\right.$$
$$\left.-\sum_q u^q\,n(q,t)M_0(t)\right]$$
$$=k_p\{[F(u,t)+M_0(t)]^2$$
$$-2[F(u,t)+M_0(t)]M_0(t)\}$$
$$=k_p[F(u,t)]^2-k_p[M_0(t)]^2 \tag{5.98}$$

After reference to eq. 5.87 in the special case, $\alpha = 0, k(m, n) = 2k_p$, eq. 5.98 reduces to the nonlinear partial differential equation:

$$\frac{\partial F}{\partial t}=k_p[F(u,t)]^2 \tag{5.99}$$

whose solution is:

$$F(u,t)=\frac{F(u,0)}{1-F(u,0)k_pt} \tag{5.100}$$

If the initial condition is imposed that the suspension consists entirely of monomers, then $n(\ell,0)=n(1,0)\delta_{\ell 1}$ and $M_0(0) = M_1$, where M_1 is the (constant) monomer number density (Eqs. 5.46 and 5.82). It follows from Eqs. 5.97 and 5.100 that, under this initial condition,

$$F(u,t)=\frac{(u-1)M_1}{1-(u-1)M_1k_pt} \tag{5.101}$$

the qth partial derivative of the right side of eq. 5.101 with respect to u is ($q \geq 1$):

$$\left(\frac{\partial^q F}{\partial u^q}\right)_t=\Gamma(q+1)M_1\frac{(M_1k_pt)^{q-1}}{[1-(u-1)M_1k_pt]^{q+1}} \tag{5.102}$$

and, therefore,

$$n(q,t)=M_1\frac{(M_1k_pt)^{q-1}}{[1+M_1k_pt]^{q+1}} \tag{5.103}$$

is the time dependent number density of q-mers that solves eq. 5.94.

It is straightforward also to obtain analytical expressions for the q-moments, $M_0(t)$ and $M_2(t)$, from eq. 5.80 and the condition, $k(m, n) = 2k_p$:[60]

$$\frac{dM_0}{dt}=-k_p[M_0(t)]^2\Rightarrow M_0(t)=\frac{M_0(0)}{1+M_0(0)k_pt} \tag{5.104a}$$

$$\frac{dM_2}{dt} = 2k_p M_1^2 \Rightarrow M_2(t) = M_2(0) + 2M_1^2 k_p t \tag{5.104b}$$

under an arbitrary initial condition. If $M_0(0) = M_1$ and $M_2(0) = 0$, then

$$\langle q_t \rangle = \frac{M_2(t)}{M_1} = 1 + 2M_1 k_p t \tag{5.105a}$$

$$\bar{q}_t = \frac{M_1}{M_0(t)} = 1 + M_1 k_p t \tag{5.105b}$$

and

$$P_t = \left(\frac{M_1 k_p t}{1 + M_1 k_p t}\right)^{\frac{1}{2}} \tag{5.105c}$$

according to eqs. 5.53, 5.85, and 5.86, respectively. The expected linearity of $\langle q_t \rangle$ and $\bar{q}_t$ as functions of time is thereby confirmed, and the polydispersity is found to grow asymptotically to unit value. Note that the difference between twice $\bar{q}_t$ and $\langle q_t \rangle$ is constant and equal to 1.0, a property uniquely related to a constant $k(m, n)$.[60]

Figure 5.18 illustrates an experimental confirmation of eq. 5.103 for latex spheres ($r_0 = 584 \pm 22$ nm) suspended in 500 mmol L^{-1} KCl solution at pH 6

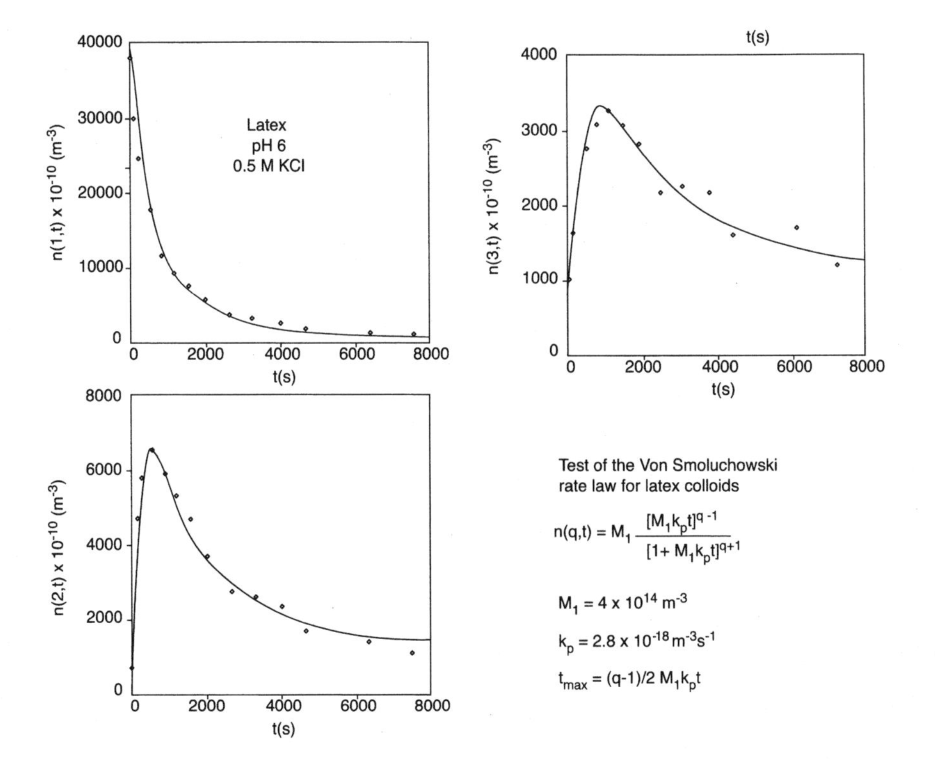

Fig. 5.18 Experimental test of eq. 5.103 for suspensions of latex colloids (A. Fernández-Barbero et al.[35]). The characteristic time scale is 892 s ($1/M_1 k_p$).

and 293 K.[61] Coherent light scattering by individual floccules was measured under the condition $Qr_0 \ll 1^{10}$ applied to eq. 5.6, such that the intensity of photons scattered is proportional to q^2 and the number of scattering events for a given q^2 is proportional to $n(q, t)$.[61] The curves through the data points in fig. 5.18 represent eq. 5.103 for $M_1 = 4 \times 10^{14}$ m^{-3} and $k_p = 2.8 \pm 0.3 \times 10^{-18}$ m^3 s^{-1}. At 293 K, eq. 5.70 yields $K_{SE} = 6.83 \times 10^{-18}$ m^3 s^{-1}, which indicates that the correction in eq. 5.71 is applicable, with $W_{11} \approx 2.4 \pm 0.3$ in agreement with typical measurements of the rate coefficient k_p for perikinetic aggregation. The function in eq. 5.103 exhibits a single maximum value at $t_{\max} = \frac{1}{2}\ (q-1)/M_1 k_p = (q-1)$ 446 s, a property that is reproduced well in the experimental data. Evaluation of W_{11} using the value of r_0 and the theoretical approach alluded to in section 5.4[55] gives the estimate, $W_{11} = 1.97$,[61] which is close to the experimental value.

Equation 5.94 can be generalized to include steady-state q-mer loss by gravitational settling or by any other physical process whose rate is proportional to the q-mer mass:[62]

$$\frac{dn}{dt}(q, t) = 2k_p\left[\frac{1}{2}\sum_m^q \sum_n^q n(m, t)n(n, t) - n(p, t)M_0(t)\right] - \frac{q}{\tau}n(q, t) \tag{5.106}$$

where more explicit note has been taken of the mass constraint on the floccule creation term (and of eq. 5.84), while τ is a characteristic time scale for the steady-state loss process (i.e., the residence time of a q-mer in a suspension undergoing the loss process). On noting the mathematical identity,

$$\underset{(q=m+n)}{\sum_m^q \sum_n^q} n(m, t)n(n, t) = \sum_{\ell=1}^{q-1} n\,(\ell, t)n(q-\ell, t) \tag{5.107}$$

which can be verified by explicit calculation for any $q \geq 2$, one can rewrite eq. 5.106 in the form:

$$\frac{dn}{dt}(q, t) = k_p \sum_\ell^{q-1} n(\ell, t)n(q-\ell, t) - 2k_p n(q, t)M_0(t) - \frac{q}{\tau}n(q, t) \tag{5.108}$$

With the definition, $C(q, t) \equiv \exp(qt/\tau)n(q, t)$, eq. 5.108 is transformed to the differential equation,

$$\frac{dC}{dt}(q, t) = k_p \sum_\ell^{q-1} C(\ell, t)C(q-\ell, t) - 2k_p C(q, t)M_0(t) \tag{5.109}$$

after taking advantage of the relationship:

$$\exp(qt/\tau)n(\ell, t)n(q-\ell, t) \equiv \exp(\ell t/\tau)n(\ell t) \times \exp[(q-\ell)t/\tau]n(q-\ell, t) = C(\ell, t)C(q-\ell, t) \tag{5.110}$$

Equation 5.109 has the same mathematical form as eq. 5.94 after incorporation of eq. 5.84; that is, the rate law for $C(q, t)$ is formally the same as that for $n(q, t)$.

Note, however, that solution of the equation now requires consideration of $M_0(t)$ simultaneously with $C(q, t)$.[63]

According to the premises of generalized homogeneity as epitomized in eqs. 5.45 and 5.88, the solution of eq. 5.94 presented in eq. 5.103 should take on the mathematical form given in eq. 5.56 when the independent variables q and t are sufficiently large. [Note that the experimental value of the exponent $\delta = 1$ (eq. 5.55) for time-scaling in rapid coalescence processes, implying that q and t scale congruently.] To see that this expectation is fulfilled, eq. 5.103 can be rewritten in the equivalent form:

$$n(q, t) = M_1 \langle q_t \rangle^{-2} 4 \frac{\left(1 - \langle q_t \rangle^{-1}\right)^{q-1}}{\left(1 + \langle q_t \rangle^{-1}\right)^{q+1}} \tag{5.111}$$

where eq. 5.105a has been introduced to eliminate the explicit time dependence. As q and $\langle q_t \rangle$ grow very large,

$$\begin{aligned} n(q, t) &\underset{q \gg 1}{\sim} M_1 \langle q_t \rangle^{-2} 4 \left[\frac{1 - \langle q_t \rangle^{-1}}{1 + \langle q_t \rangle^{-1}}\right]^q \\ &\underset{\langle q_t \rangle \gg 1}{\sim} M_1 \langle q_t \rangle^{-2} 4 \left(1 - 2\langle q_t \rangle^{-1}\right)^q \\ &= M_1 \langle q_t \rangle^{-2} 4 \left[1 - 2\frac{(q/\langle q_t \rangle)}{q}\right]^q \\ &\underset{q \uparrow \infty}{\sim} M_1 \langle q_t \rangle^{-2} 4 \exp[-2(q/\langle q_t \rangle)] \end{aligned} \tag{5.112}$$

where the last step makes use of the formal definition of the exponential function, $\exp(x)$.[64] Equation 5.112 indeed has the form of eq. 5.56 if the identification

$$\psi(q/\langle q_t \rangle) \equiv 4 \exp[-2(q/\langle q_t \rangle)] \tag{5.113}$$

is made. Therefore, as q and $\langle q_t \rangle$ grow larger in such a way as to keep their ratio finite, $n(q, t)$ for rapid coalescence described by a scale-invariant rate coefficient should decline *exponentially* with q, a prediction that also has been confirmed by experiment.[35]

Slow coalescence processes ($W \gg 1$) produce floccules with a larger fractal dimension than do rapid coalescence processes (section 5.3). It is also observed experimentally that $\langle q_t \rangle$ grows *exponentially* with time during the slow coalescence of a suspension.[65] Given this experimental result, it follows from eq. 5.93 that $\theta = 1$ and, therefore, that the scaling exponent $\delta = 0$ (eq. 5.90). We conclude from the value of θ that, for slow coalescence processes, the model rate coefficient $k(m, n)$ must be a *homogeneous* function:

$$k(\lambda m, \lambda n) = \lambda k(m, n) \tag{5.114}$$

Moreover, because $\delta = 0$, the scaling of the von Smoluchowski rate law does not involve the time variable, but instead q and the rate coefficient now scale congruently. These special characteristics and the larger fractal dimension serve to distinguish slow coalescence from rapid coalescence processes.[66]

A model rate coefficient that satisfies eq. 5.114 and has the necessary symmetry property $[k(m, n) = k(n, m)]$, while still yielding to analytical solution of eq. 5.79 is:[67]

$$k(m, n) = k_s(m + n) \tag{5.115}$$

where k_s is a parameter that incorporates pH, ionic strength, and temperature dependence. This model rate coefficient reflects the simple premise that van der Waals attractive forces, which increase proportionally with the mass of a particle,[47] become more effective at overcoming electrostatic repulsion to promote slow coalescence processes as the coalescing particles become larger.

The corresponding von Smoluchowski rate law is:

$$\begin{aligned}\frac{dn}{dt}(q, t) &= \frac{1}{2}k_s \sum_m \sum_n (m+n)n(m, t)n(n, t) \\ &\quad\;\; {\scriptstyle (q=m+n)} \\ &\quad - k_s \sum_p (q+p)n(q, t)n(p, t) \\ &= k_s\left\{\frac{1}{2}\sum_m \sum_n (m+n)n(m, t)n(n, t)\right. \\ &\quad\;\; {\scriptstyle (q=m+n)} \\ &\quad \left. - n(q, t)[qM_0(t) + M_1]\right\}\end{aligned} \tag{5.116}$$

where the second step involves eqs. 5.46 and 5.84. This integro-differential equation also can be solved by use of the generating function in eq. 5.95.[60] The associated partial differential equation is:

$$\begin{aligned}\frac{\partial F}{\partial t} + \frac{dM_0}{dt} &= k_s\left\{\sum_m mu^m n(m, t)\sum_n u^n n(n, t)\right. \\ &\quad - M_0(t)\sum_q qu^q n(q, t) \\ &\quad \left. - M_1 \sum_q u^q n(q, t)\right\} \\ &= k_s\left\{\frac{\partial F}{\partial u}u[F(u, t) + M_0(t)]\right. \\ &\quad \left. - M_0(t)\frac{\partial F}{\partial u}u - M_1[F(u, t) + M_0(t)]\right\} \\ &= k_s\left\{F(u, t)\left[\frac{\partial F}{\partial u}u - M_1\right] - M_1M_0(t)\right\}\end{aligned} \tag{5.117}$$

Reference to eq. 5.87 in the special case given by eq. 5.115 shows that the zeroth q-moment satisfies the ordinary differential equation:

$$\frac{dM_0}{dt} = -k_sM_1M_0(t) \tag{5.118}$$

and, therefore, $F(u, t)$ satisfies the partial differential equation:

$$\frac{\partial F}{\partial t} = \left(\frac{\partial F}{\partial u}u - M_1\right)k_sF(u, t) \tag{5.119}$$

The resulting solution of eq. 5.116 is ($q \geq 1$):[60]

$$n(q, t) = \frac{M_0(t)}{\Gamma(q+1)}\left\{q\frac{[1 - M_0(t)]}{M_1}\right\}^{q-1} \exp\left\{-q\frac{[1 - M_0(t)]}{M_1}\right\} \tag{5.120}$$

where

$$M_0(t) = M_1 \exp(M_1 k_s t) \tag{5.121}$$

is the solution of eq. 5.118 subject to the initial condition, $M_0(0) = M_1$, which also was used in conjunction with eq. 5.104a. The qualitative behavior of $n(q, t)$ as a function of time is similar to that of $n(q, t)$ in eq. 5.103. If $q = 1$, the number density of q-mers declines as time passes, whereas it exhibits a single maximum at $t_{\max} = \frac{1}{2} \ln q/M_1 k_s$ if $q > 1$.

The ordinary differential equation for $M_2(t)$ analogous to eq. 5.118 follows from combining Eqs. 5.83 and 5.115:

$$\begin{aligned} \frac{dM_2}{dt} &= k_s \sum_m \sum_n mn(m+n)\, n(m, t)n(n, t) \\ &= 2k_s M_1 M_2(t) \end{aligned} \tag{5.122}$$

The solution of this equation subject to the initial condition, $M_2(0) = M_1$, is then:

$$M_2(t) = M_1 \exp(2M_1 k_s t) \tag{5.123}$$

which exhibits the expected exponential growth with time. The resultant model equations for $\langle q_t \rangle$, $\bar{q}_t$, *and* P_t then follow from eqs. 5.53, 5.85, and 5.86:

$$\langle q_t \rangle = \exp(2M_1 k_s t) \tag{5.124a}$$

$$\bar{q}_t = \exp(M_1 k_s t) \tag{5.124b}$$

$$P_t = [\exp(M_1 k_s t) - 1]^{\frac{1}{2}} \tag{5.124c}$$

Equation 5.124a predicts the exponential growth in floccule size that characterizes slow coalescence.[65]

As with $n(q, t)$ in eq. 5.103, a scaling form of $n(q, t)$ in eq. 5.120 that mimics eq. 5.56 is expected as q grows larger, given that the rate coefficient in eq. 5.115 has the property of generalized homogeneity. The scaling form of $n(q, t)$ becomes apparent after rewriting eq. 5.120 in terms of $\langle q_t \rangle$ based on eqs. 5.121 and 5.124a:

$$\begin{aligned} n(q, t) = M_1 \langle q_t \rangle^{-2} \left[\frac{q}{\langle q_t \rangle}\right]^{-\frac{3}{2}} &\left\{\frac{q^{q+\frac{1}{2}}}{\Gamma(q+1)}\left(1 - \langle q_t \rangle^{-\frac{1}{2}}\right)^{q-1}\right. \\ &\left.\times \exp\left[-q\left(1 - \langle q_t \rangle^{-\frac{1}{2}}\right)\right]\right\} \end{aligned} \tag{5.125}$$

The gamma function in eq. 5.125 has the asymptotic representation ("Stirling approximation"):[20]

$$\Gamma(q+1) \underset{q \gg 1}{\sim} (2\pi)^{\frac{1}{2}} q^{q+\frac{1}{2}} \exp(-q) \tag{5.126}$$

which is accurate to within 10 % or less for $q > 2$. Therefore, eq. 5.125 takes the approximate form:

$$
\begin{aligned}
n(q,t) \underset{q>2}{\sim} & M_1 \langle q_t \rangle^{-2} (2\pi)^{-\frac{1}{2}} \left[\frac{q}{\langle q_t \rangle} \right]^{-\frac{3}{2}} \\
& \times \left(1 - \langle q_t \rangle^{-\frac{1}{2}} \right)^{q-1} \exp\left(q / \langle q_t \rangle^{\frac{1}{2}} \right) \\
= & M_1 \langle q_t \rangle^{-\frac{1}{2}} (2\pi)^{-\frac{1}{2}} \left[\frac{q}{\langle q_t \rangle} \right]^{-\frac{3}{2}} \\
& \times \left[1 - \frac{\left(q / \langle q_t \rangle^{\frac{1}{2}} \right)}{q} \right\}^{q-1} \exp\left(q / \langle q_t \rangle^{\frac{1}{2}} \right) \\
& \underset{q>2}{\sim} M_1 \langle q_t \rangle^{-2} (2\pi)^{-\frac{1}{2}} \left[\frac{q}{\langle q_t \rangle} \right]^{-\frac{3}{2}}
\end{aligned}
\tag{5.127}
$$

where, as in eq. 5.112, the last step makes use of the formal definition of the exponential function.[64] Equation 5.127 has the expected asymptotic form of eq. 5.56, with

$$\psi(q/\langle q_t \rangle) \equiv (2\pi)^{-\frac{1}{2}} [q/\langle q_t \rangle]^{-\frac{3}{2}} \tag{5.128}$$

Therefore, as q grows larger, $n(q, t)$ for slow coalescence described by a homogeneous rate coefficient should decline as the $-3/2$ power of q, a prediction that has been confirmed both experimentally and theoretically.[68]

These asymptotic features of rapid and slow coalescence processes described by the von Smoluchowski rate law permit a concise summary of the two types of flocculation:[66]

Rapid Coalescence

- mass fractal dimension < 2
- scale-invariant rate coefficient
- linear time dependence of $\langle q_t \rangle$
- exponential decline of $n(q, t)$ with $q/\langle q_t \rangle$

Slow Coalescence

- mass fractal dimension > 2
- homogeneous rate coefficient
- exponential time dependence of $\langle q_t \rangle$
- power-law decline of $n(q, t)$ with $q/\langle q_t \rangle$

Notes

1. For an introduction to the properties of aquatic colloids, see J. Buffle and G.G. Leppard, *Environ. Sci. Technol.* **29**:2169, 2176 (1995). Marine particles are described by G.A. Jackson and A.B. Burd, *Environ. Sci. Technol.* **32**:2805 (1998). Soil particles are characterized by A. Golchin, J.A. Baldock, and J.M. Oades, pp. 245–266 in *Soil Processes and the Carbon Cycle*, R. Lal, J.M. Kimble, R.F. Gollett, and B.A. Stewart (Eds.), CRC Press, Boca Raton, FL, 1998. The critical role played by humus in natural particle structure is emphasized in

detailed studies by J.M. Oades, pp. 463–483 in *Soil Colloids and Their Associations in Aggregates*, M.F. DeBoodt, M.H.B. Hayes, and A. Herbillon (Eds.), Plenum Press, New York, 1990; D. Heil and G. Sposito, *Soil Sci. Soc. Am. J.* **59**:266 (1995); and D. Perret, J.-F. Gaillard, J. Dominik, and O. Atteia, *Environ. Sci. Technol.* **34**:3540 (2000).

2. See, e.g., R. Kretzschmar, M. Borkovec, D. Grolimund, and M. Elimelech, *Advan. Agron.* **66**:121 (1999) for a review of colloid-facilitated transport in subsurface zones.
3. The discussion to be presented is taken from J. Buffle, K.J. Wilkinson, S. Stoll, M. Filella, and J. Zhang, *Environ. Sci. Technol.* **32**:2887 (1998). See also E.M. Murphy and J.M. Zachara, *Geoderma* **67**:103 (1995); K.J. Wilkinson, J.-C. Nègre, and J. Buffle, *J. Contamin. Hydrol.* **26**:229 (1997).
4. Introductory discussions of photon scattering by small particles in suspension are presented in chapter 5 of P.C. Hiemenz and R. Rajagopalan, *Principles of Colloid and Surface Chemistry*, Marcel Dekker, New York, 1997; chapter 11 of E. Kissa, *Dispersions*, Marcel Dekker, New York, 1999; and sections 3.3 and 5.7 of R.J. Hunter, *Foundations of Colloid Science*, Oxford University Press, New York, 2001.
5. The photon momentum, $h/\lambda = \hbar k$, is the analog of the photoelectron momentum defined in section 2.2. For an introductory discussion, see section 2.4 in G. Sposito, *An Introduction to Quantum Physics*, John Wiley, New York, 1970.
6. Constructive scattered photon interference will occur if $(\overline{AB} + \overline{BC})$ in fig. 5.2 is an integral multiple of the wavelength of the radiation, assuming that the scattering process is elastic (no change in photon momentum). Thus, for example, constructive interference occurs if $\left(\hat{\boldsymbol{k}} - \hat{\boldsymbol{k}}'\right) \cdot (\boldsymbol{r}_2 - \boldsymbol{r}_1) = \lambda$. This equation is in fact a form of the Bragg Law for photon diffraction by two parallel planes containing scattering centers. The relationship becomes apparent after noting eq. 5.3 and the fact that the angle between $\hat{\boldsymbol{k}}$ or $\hat{\boldsymbol{k}}'$ and either of the planes is equal to half of the scattering angle θ. See, e.g., chapter 2 in P.M. Chaikin and T.C. Lubensky, *Principles of Condensed Matter Physics*, Cambridge University Press, New York, 1995.
7. Conditions (1) and (2) apply to *Rayleigh-Gans* scattering phenomena, careful discussions of which are presented in sections 8.25 and 8.26 of D.S. Jones, *The Theory of Electromagnetism*, Macmillan, New York, 1963; chapter 7 of H.C. van de Hulst, *Light Scattering by Small Particles*, Dover, New York, 1981; and chapter 10 of A. Guinier, *X-ray Diffraction in Crystals, Imperfect Crystals, and Amorphous Bodies*, Dover, New York, 1994. Discussions of photon scattering with conditions (1) and (2) relaxed (*Mie* scattering processes) are given in the references cited in note 4 and in the book by van de Hulst.
8. The radius of gyration of a nonuniform particle is defined by the equation:

$$R_G^2 \equiv \int \|\boldsymbol{r} - \boldsymbol{R}\|^2 n(\boldsymbol{r})\, d^3r$$

where the integral is over the particle volume and $n(\boldsymbol{r})[= n(-\boldsymbol{r})]$ is the density of particle constituents at a point $\boldsymbol{r}$, normalized such that

$$\int n(\boldsymbol{r})\, d^3r = 1$$

The vector $\boldsymbol{R}$ locates the center-of-mass of the particle; thus R_G^2 is the second central moment of $n(\boldsymbol{r})$ considered as a probability density function.

An alternate definition of R_G can be given in terms of the number density $\rho_q(r)$ which appears in eq. 5.7:

$$\rho(\boldsymbol{r}) = \int n\left(\boldsymbol{r} - \boldsymbol{r}'\right) n\left(\boldsymbol{r}'\right) d^3r' \qquad (q \text{ suppressed})$$

with the same unit normalization as $n(\boldsymbol{r})$ (eq. 5.8). This quantity is the *conditional* probability density function for finding a particle constituent at the point $\boldsymbol{r}$ in the vicinity of another point $\boldsymbol{r}'$ (i.e., it is given that a particle constituent already exists at $\boldsymbol{r}'$), averaged over all possible $\boldsymbol{r}'$. Thus $\rho(\boldsymbol{r})$ gives a measure of *spatial correlations* among the constituents within a particle. Upon expressing $\|\boldsymbol{r}-\boldsymbol{r}'\|^2$ relative to the center-of-mass vector

$$\boldsymbol{R} \equiv \int \boldsymbol{r} n(\boldsymbol{r}) d^3 r$$

in the integral

$$\int\int \|\boldsymbol{r}-\boldsymbol{r}'\|^2 n(\boldsymbol{r}) n(\boldsymbol{r}') d^3 r d^3 r' = \int s^2 \rho(\boldsymbol{s}) d^3 s$$

and applying the definition of R_G^2, one can demonstrate (see note 10):

$$R_G^2 = \frac{1}{2}\int s^2 \rho(\boldsymbol{s}) d^3 s$$

The factor $\frac{1}{2}$ arises because the integral of $\rho(\boldsymbol{r})$ above includes contributions from both $\|\boldsymbol{r}-\boldsymbol{R}\|^2$ and $\|\boldsymbol{r}'-\boldsymbol{R}\|^2$.

9. Momentum balance in coherent photon scattering by a crystalline solid particle requires that $\boldsymbol{Q}$ be equal to a reciprocal lattice vector for the solid, which occurs if $\boldsymbol{Q}\cdot\boldsymbol{a}_i = 2\pi h_i$ $(i = 1, 2, 3)$ where $\boldsymbol{a}_i$ is one of three vectors defining the unit cell of the crystalline solid and h_i is an integer. This latter condition, similar to eq. 5.5, indicates that Q^{-1} defines a length scale in the solid structure. For a discussion of these concepts, see chapter 2 in C. Kittel, *Introduction to Solid State Physics*, John Wiley, New York, 1996.
10. The measured intensity of photons scattered at any wavenumber Q is a linear function of $I(Q, t)$ in eq. 5.6. The two coefficients in this relation, termed calibration parameters, also depend on Q and represent the effects of absorption, multiple scattering by the colloids within a particle, and unwanted scattering by the apparatus, the individual colloids in a particle, and the individual particles in the suspension under investigation. For a discussion of multiple scattering effects on $I(Q, t)$, see M.V. Berry and I.C. Percival, *Optica Acta* **33**:577 (1986).

    In general, the q^2 term in eq. 5.6 is replaced by a *form factor* $P_q(Q)$ that accounts for coherent photon scattering by the individual colloids that make up a scattering particle. If these colloids are modeled as identical hard spheres of radius r_0, then, aside from optical parameters (cf. note 7),

$$P_q(Q) = \frac{9q^2}{(Qr_0)^4}\left[\frac{\sin(Qr_0)}{Qr_0} - \cos(Qr_0)\right]^2$$

which reduces to q^2 in the limit $(Qr_0)^4 \ll 1$. Thus, eq. 5.6 applies to Q-values that meet this limiting condition, which means that the length scales probed by photon scattering are very much larger than r_0. Equations for $P_q(Q)$ in the case of non-spherical particles are presented by A. Guinier, *X-ray Diffraction in Crystals, Imperfect Crystals, and Amorphous Bodies*, Dover, New York, 1994, and by P. Schurtenberger and M.E. Newman, pp. 37–115 in *Environmental Particles*, vol. 2, J. Buffle and H.P. van Leeuwen (Eds.), CRC Press, Boca Raton, FL, 1993.

Equation 5.7 is a special case of the definition,

$$
\begin{aligned}
S_N(\boldsymbol{Q}) &\equiv N^{-2}\sum_{i=1}^{N}\sum_{j=1}^{N}\langle\langle \exp[i\boldsymbol{Q}\cdot(\boldsymbol{r}_i-\boldsymbol{r}_j)]\rangle\rangle \\
&= N^{-1}+(1-N^{-1})\langle\langle \exp[i\boldsymbol{Q}\cdot(\boldsymbol{r}_1-\boldsymbol{r}_2)]\rangle\rangle \\
&= N^{-1}+(1-N^{-1})\int\int n(\boldsymbol{r})n(\boldsymbol{r}')\exp[i\boldsymbol{Q}\cdot(\boldsymbol{r}-\boldsymbol{r}')]d^3r\,d^3r' \\
&= N^{-1}+(1-N^{-1})\int\int n(\boldsymbol{r})n(\boldsymbol{r}-\boldsymbol{s})\exp(i\boldsymbol{Q}\cdot\boldsymbol{s})d^3r\,d^3s \\
&= N^{-1}+(1-N^{-1})\int \rho_N(\boldsymbol{s})\exp(i\boldsymbol{Q}\cdot\boldsymbol{s})d^3s
\end{aligned}
$$

where the final step comes from note 8. This definition applies to a system containing N scattering centers, where $\ll \gg$ is an ensemble average in the sense of statistical thermodynamics, the quantity averaged being proportional to the elastic scattering intensity for an assembly of identical particles. The first term in $S_N(\boldsymbol{Q})$ results from setting $i = j$ in the double summation (N such terms), while the second term collects the $N(N-1)$ identical terms in the summation for which $i \neq j$. The integral representation of this latter term employs the interpretation of ensemble averages as probabilistic quantities, with the density $n(\boldsymbol{r})$ playing the role of a probability density function as in note 8. It reduces to the form on the right side of eq. 5.7 when $\rho_N(\boldsymbol{r})$ depends only on the magnitude of the vector $\boldsymbol{r}$. For a thorough discussion of these concepts, see chapter 3 in J.K.G. Dhont, *An Introduction to Dynamics of Colloids*, Elsevier, Amsterdam, 1996, and C.M. Sorensen, *Aerosol Sci. Technol.* **35**:648 (2001). Note that $S_N(\boldsymbol{Q})$ defined by Dhont (p. 128) is equal to $NS_N(\boldsymbol{Q})$ as defined in the present chapter.

11. Introductory discussions of the Guinier approximation are given by A. Guinier, *X-ray Diffraction in Crystals, Imperfect Crystals, and Amorphous Bodies,* Dover, New York, 1994, and C.M. Sorensen, *Aerosol Sci. Technol.* **35**:648 (2001).
12. S. Gangopadhyay, I. Elminyawi, and C.M. Sorensen, *Appl. Optics* **30**:4859 (1991). An *aerosol* is a suspension of a solid (or a liquid) in a gas (e.g., smoke or fog). The data analysis leading to fig. 5.3 is discussed in detail by C.M. Sorensen, *Aerosol Sci. Technol.* **35**:648 (2001).
13. If the condition in eq. 5.1 is met reasonably well, the refractive index of an aggregate comprising solid colloids and the fluid in which they aggregate is suspended can be approximated by the equation:

$$n_{agg} - n_f(\rho_{agg}/\rho_s)(n_s - n_f)$$

where ρ_{agg} is the dry bulk density of the aggregate, ρ_s is the density of the colloids, n_s is their refractive index, and n_f is the refractive index of the suspending fluid. By definition, $\rho_{agg} = \rho_s(1-\phi)$, where ϕ is the porosity of the aggregate. For aerosols, $n_f = n_{air} \approx 1.0$, whereas for aqueous suspensions, $n_f = n_w = 1.33$. Values of ρ_{agg} in the range 0.04 to 0.17×10^3 kg m^{-3} for carbonaceous soot aggregates are reported by C.M. Sorensen, C. Oh, P.W. Schmidt, and T.P. Rieker, *Phys. Rev. E* **58**:4666 (1998), with $\rho_s = 1.85 \times 10^3$ kg m^{-3}.
14. T.L. Farias, U.O. Koylu, and M.G. Carvalho, *Appl. Optics* **35**:6560 (1996). See also fig. 16 in C.M. Sorensen, *Aerosol Sci. Technol.* **35**:648 (2001). If $|m-1| <$ 0.75 and $kr_0 < 0.3$, where r_0 is the radius of a colloid contained in an aggregate,[10] $R \approx 3/2\ k|m-1|$ appears to be a useful estimate of the size constraint in Rayleigh-Gans scattering.
15. L. Derrendinger and G. Sposito, *J. Colloid Interface Sci.* **222**:1 (2000).
16. J. Chorover, J. Zhang, M.K. Amistadi, and J. Buffle, *Clays Clay Miner.* **45**:690 (1997). Electron micrographs of the aggregates formed under the conditions

indicated in figs. 5.5 and 5.6 show increasingly compact structure as the absolute value of the slope of a log-log plot increases.

17. The standard reference on fractal aggregates is T. Vicsek, *Fractal Growth Phenomena,* World Scientific, Riveredge, NJ, 1992. Useful discussions of photon scattering by fractal aggregates are given by J. Teixeira, pp. 145–162 in *On Growth and Form*, H.E. Stanley and N. Ostrowsky (Eds.), Martinus Nijhoff, Dordrecht, 1986; J.E. Martin and A.J. Hurd, *J. Appl. Cryst.* **20**:61 (1987); P.W. Schmidt, pp. 67–79 in *The Fractal Approach to Heterogeneous Chemistry*, D. Avnir (Ed.), John Wiley, New York, 1989; and C.M. Sorensen, *Aerosol Sci. Technol.* **35**:648 (2001).
18. Model equations for the cutoff function and their impact on the transition from the domain of the Guinier approximation to the fractal domain are discussed by C.M. Sorensen, J. Cai, and N. Lu, *Langmuir* **8**:2064 (1992) and T. Nicolai, D. Durand, and J.-C. Gimel, *Phys. Rev. B* **50**:16, 537 (1994). The principal issues, which are well-summarized by C.M. Sorensen, *Aerosol Sci. Technol.* **35**:648 (2001), are whether the resultant $S_q^F(Q)$ follows experimental data closely over a broad range of QR_{GZ} values (R_{Gq} is proportional to ξ_q) and precisely how large QR_{GZ} must be in order to estimate the fractal dimension accurately from a log-log plot of photon-scattering data.
19. Equation 5.23 is an application of the symptotic expansion,

$$\int_\alpha^\beta \exp(ixt)(t-\alpha)^{\lambda-1}\Phi(t)dt \underset{x\uparrow\infty}{\sim} \frac{\Gamma(\lambda)}{x^\lambda}\exp\left[i\left(\frac{\pi\lambda}{2}+x\alpha\right)\right]\Phi(\alpha)$$

which is proved in sections 2.8 and 2.19 of A. Erdélyi, *Asymptotic Expansions*, Dover, New York, 1956. The conditions on the above result are $\lambda > 0$ and $\Phi(\beta) \equiv 0$. Note that, by definition, $\sin xt \equiv (2i)^{-1}[\exp(ixt) - exp(-ixt)]$, $i = \sqrt{-1}$.

20. Properties and values of the gamma function,

$$\Gamma(x) \equiv \int_0^\infty t^{x-1}\exp(-t)dt$$
$$\equiv (x-1)! \qquad \text{(positive integer } x\text{)}$$

are described in chapter 6 of M. Abramowitz and I. Stegun, *Handbook of Mathematical Functions*, Dover, New York, 1972.

21. M. Filella and J. Buffle, *Colloids Surf. A* **73**:255 (1993). Observations also suggest that aerosols formed over urban regions remain suspended for long periods if their size is in the midrange, 0.1 to 1 μm (see, e.g., chapter 7 in J.H. Seinfeld, *Atmospheric Chemistry and Physics of Air Pollution*, John Wiley, New York, 1986).
22. Basic concepts of particle coalescence in suspension are discussed in chapter 10 of G. Sposito, *The Chemistry of Soils*, Oxford University Press, New York, 1989, and in chapter 9 of R.J. Hunter, *Introduction to Modern Colloid Science,* Oxford University Press, New York, 1993.
23. An overview of particle-size methodology is given in chapter 2 of E. Kissa, *Dispersions*, Marcel Dekker, New York, 1999.
24. D. Giles and A. Lips, *J. Chem. Soc. Faraday Trans 1* **74**:733 (1978); J.H. van Zanten and M. Elimelech, *J. Colloid Interface Sci.* **154**:1 (1992). Strictly speaking, the factor q^2 in eq. 5.6 should be replaced by the form factor $P_q(Q)$ (see note 10) in this application. However, the Q-dependent part of the form factor cancels from $I(Q, t)$ after division by $I(Q, 0)$, as in eq. 5.31.
25. H. Holthoff, S.U., Egelhaaf, M. Borkavec, P. Schurtenberger, and H. Sticher, *Langmuir* **12**:5541 (1996). Note that the rate coefficient k_{11} in this article is equal to $2k$.
26. Dynamic photon scattering theory is discussed in chapter 3 of J.K.G. Dhont, *An Introduction to Dynamics of Colloids*, Elsevier, Amsterdam, 1996, whereas instrumentation is described in chapter 11 of E. Kissa, *Dispersions*, Marcel Dekker,

New York, 1999. See also P. Schurtenberger and M.E. Newman, *Environmental Particles*, J. Buffle and H.P. van Leewen (Eds.), for an applications-oriented review.

27. The role of the colloid diffusion coefficient in dynamic photon scattering can be appreciated by consideration of the *dynamic structure factor*, a time-dependent generalization of $S_N(\boldsymbol{Q})$ (cf notes 10 and 26):

$$S_N^D(\boldsymbol{Q}, t) \equiv N^{-1} \sum_{i=1}^{N} \sum_{j=1}^{N} \langle\langle \exp\left[i\boldsymbol{Q} \cdot \left(\boldsymbol{r}_i(0) - \boldsymbol{r}_j(t)\right)\right]\rangle\rangle$$

[Note that $S_N^D(\boldsymbol{Q}, t)$ is normalized by N^{-1}, not N^{-2} as is $S_N(\boldsymbol{Q})$.] Thus $S_N^D(\boldsymbol{Q}, t)$ represents Q-dependence in the *fluctuating* photon scattering intensity caused by the relative motions of the N scattering particles.

However, the terms in the double summation for which $i \neq j$ are negligible in dilute suspensions because particle motions are uncorrelated and the $\langle\langle\ \rangle\rangle$ average factorizes:

$$\begin{aligned} \langle\langle \exp\left[i\boldsymbol{Q} \cdot \left(\boldsymbol{r}_i(0) - \boldsymbol{r}_j(t)\right)\right]\rangle\rangle &= \langle\langle \exp[i\boldsymbol{Q} \cdot \boldsymbol{r}_i(0)]\rangle\rangle \\ \times \langle\langle \exp\left[-i\boldsymbol{Q} \cdot \boldsymbol{r}_j(t)\right]\rangle\rangle &= \int n(\boldsymbol{r}, 0) \exp(i\boldsymbol{Q} \cdot \boldsymbol{r}) d^3r \\ \times \int n(\boldsymbol{r}', t) \exp(-i\boldsymbol{Q} \cdot \boldsymbol{r}') d^3r' &\qquad (i \neq j) \end{aligned}$$

For each of the integrals, the spatial scale of r (*inter*particle) distance) is much larger than that of Q^{-1} (*intra*particle distance), such that $\boldsymbol{Q} \cdot \boldsymbol{r} \gg 1$ and the integral is reduced to a negligibly small value by the rapid oscillations of the exponential factor in the integrand. Therefore

$$\begin{aligned} S_N^D(Q, t) &\underset{N\uparrow\infty}{\sim} N^{-1} \sum_{i-1}^{N} \langle\langle \exp[i\boldsymbol{Q} \cdot (\boldsymbol{r}_i(0) - \boldsymbol{r}_i(t))]\rangle\rangle \\ &= \int n(\boldsymbol{r}, t) \exp(i\boldsymbol{Q} \cdot \boldsymbol{r}) d^3r \end{aligned}$$

provides an accurate model of the dynamic structure factor, where r is now the distance traveled by a single scattering particle as it engages Brownian motion (random motion induced by the fluctuating thermal energy of the water molecules in the suspension).

A simple model of Brownian motion (see, e.g., chapter 3 in W.B Russel, D.A. Saville, and W.R. Schowalter, *Colloidal Dispersions*, Cambridge University Press, New York, 1989) yields a connection to colloid diffusion that can be readily epitomized in the normalized number density,

$$n(r, t) = (4\pi \boldsymbol{D}t)^{-\frac{3}{2}} \exp\left(-r^2/4\boldsymbol{D}t\right)$$

which resembles a gaussian probability density function for an ensemble with standard deviation $2\boldsymbol{D}t$, where $\boldsymbol{D}$ is the diffusion coefficient of a single colloid. Then

$$\begin{aligned} S^D(Q, t) &= \left[2\sqrt{\pi}(\boldsymbol{D}t)^{\frac{3}{2}}\right]^{-1} \int_0^\infty \exp\left(-r^2/4\boldsymbol{D}t\right) \frac{\sin Qr}{Qr} r^2 dr \\ &= \exp\left(-Q^2 \boldsymbol{D}t\right) \end{aligned}$$

upon evaluating the gaussian integral. Numerical analysis of $\ln S^D(Q, t)$ as obtained in a dynamic photon scattering experiment (see note 26) then can be applied to determine the colloid diffusion coefficient, $\boldsymbol{D}$.

28. Equation 5.39 emerges from a derivation of the partial differential equation ("diffusion equation") satisfied by $n(r, t)$:

$$\frac{\partial n}{\partial t} = \left(\frac{k_\beta T}{6\pi\eta R_H}\right)\nabla^2 n(r, t)$$

where ∇^2 is the Laplacian operator. The fundamental solution of this equation is given in note 27, if eq. 5.39 is invoked. For an introductory discussion, see chapter 3 in W.B. Russel et al., *Colloidal Dispersions*, Cambridge University Press, New York, 1989. A detailed statistical thermodynamic derivation of eq. 5.39 is presented in section 1.2 of H.J.V. Tyrrell and K.R. Harris, *Diffusion in Liquids*, Butterworths, London, 1984. Section 5.11 in this latter book describes carefully the use of dynamic photon scattering to measure diffusion coefficients of macromolecules.

29. Equation 5.40 is a short-time approximation that can be derived from the equation,

$$S^D(Q, t) = \frac{\sum_q n(q, t)q^2 S_q(Q)\exp\left(-Q^2 \boldsymbol{D}_q t\right)}{\sum_q n(q, t)q^2 S_q(Q)}$$

for the average dynamic structure factor of colloids in suspension. Given the definition of $S_N^D(\boldsymbol{Q}, t)$ in note 27, it is also apparent that *rotational* motions of colloids can affect the dynamic structure factor, but this influence is assumed to be negligible when modeling $\boldsymbol{D}_q$ with eq. 5.39. These and other important details are discussed in tutorial fashion in a series of articles by M.Y. Lin et al. which appear in *Proc. R. Soc. Lond. A* **423**:71 (1989), *Phys. Rev. A* **41**:2005 (1990), *J. Phys. Condens. Matter* **2**:3093 (1990). See also H.M. Lindsay, M.Y. Lin, D.A. Weitz, P. Sheng, Z. Chen, R. Klein, and P. Meakin, *Faraday Discuss. Chem. Soc.* **83**:153 (1987) for a detailed discussion of modeling the rotational contribution to $S^D(Q, t)$.

30. H. Holthoff, A. Schmitt, A. Fernández-Barbero, M. Borkovec, M.A. Cabrerízo-Vílchez, P. Schurtenberger, and R. Hidalgo-Álvarez, *J. Colloid Interface Sci.* **192**:463 (1997). Note that, as in eq. 5.6 (note 24), the factor q^2 in eq. 5.40 should be replaced by the form factor $P_q(Q)$ (see note 10), but that its Q-dependent part cancels in eq. 5.43 and otherwise does not alter the derivation of eq. 5.44.

31. M. Schudel, S.H. Behrens, H. Holthoff, R. Kretzschmar, and M. Borkovec, *J. Colloid Interface Sci.* **196**:241 (1997).

32. H. Holthoff, M. Borkovec, and P. Schurtenberger, *Phys. Rev. E* **56**:6945 (1997). This article also gives a useful review of the theoretical calculations of R_{H1}/R_{H2}.

33. Electron micrographs of aggregates formed by the rapid coalescence of gold colloids show similar structure over an order-of-magnitude increase in magnification. See D.A. Weitz and J.S. Huang, *Phys. Rev. Lett.* **52**:1433 (1984). The images themselves are reproduced in chapter 6 of G. Sposito, *Chemical Equilibria and Kinetics in Soils*, Oxford University Press, New York, 1994, where an introduction to scaling ideas also may be found.

34. A. Hankey and H.E. Stanley, *Phys. Rev. B* **6**:3515 (1972). This article gives a comprehensive discussion of generalized homogeneous functions in a physically-motivated context.

35. D.A. Weitz and M.Y. Lin, *Phys. Rev. Lett.* **57**:2037 (1986) (*gold*); M.L. Broide and R.J. Cohen, *Phys. Rev. Lett.* **64**:2026 (1990), A. Fernández-Barbero, A. Schmitt, M. Cabrerízo-Vílchez, and R. Martínez-García, *Physica A* **230**:53 (1996) (*latex*).

36. B.J. Olivier, C.M. Sorenson, and T.W. Taylor, *Phys. Rev. A* **45**:5614 (1992) (*aerosol*); M.Y. Lin et al. (see the first and third articles cited in note 29: *gold, latex*, and *silica*); L. Derrendinger and G. Sposito, *J. Colloid Interface Sci.* **222**:1(2000) (*illite*); D. Asnaghi, M. Carpineti, M. Griglio, and M. Sozzi, *Phys. Rev. A* **45**:1018 (1992), A. Fernández-Barbero et al., *Physica A* **230**:53 (1996)

(*latex*). In some instances, time dependence was studied in these articles using eq. 5.59, with D determined independently.

37. P. Wiltzius, *Phys. Rev. Lett.* **58**:710 (1987); M.Y. Lin et al., *Proc. R. Soc. Lond. A* **423**: 71(1989), *J. Phys. Condens. Matter* **2**:1 (2000); J. Zhang and J. Buffle, *J. Colloid Interface Sci.* **174**:500 (1995). Note that R_{GZ} is measured from the Q dependence of $I(Q, t)$, whereas R_{HZ} is measured from the Q dependence of time-correlations exhibited by the photon scattering intensity.
38. J. Zhang and J. Buffle, *Colloids Surf.* **107A**:175 (1996).
39. Fractal dimensions for aggregates of gold, hematite, latex, and silica are compiled in G. Sposito, *Colloids Surf.* **120A**:101 (1997). Values for illite, given by L. Derrendinger and G. Sposito, *J. Colloid Interface Sci.* **222**:1 (2000) are reported here as the product, δD. Aerosol fractal dimensions are summarized by C.M. Sorenson, *Aerosol Sci. Technol.* **35**:648 (2001)
40. P. Meakin, pp. 111–135, in H.E. Stanley and N. Ostrowsky, (Eds.), *On Growth and Form*, Martinus Nijhoff, Dordrecht, 1986. R.M. Ziff, E.D. McGrady, and P. Meakin, *J. Chem. Phys.* **82**:5269 (1985) have shown that $\langle q \rangle_t$ is a power-law function of the time for the diffusion-controlled coalescence process simulated by P. Meakin. See also chapter 8 in T. Vicsek, *Fractal Growth Phenomena*, World Scientific, Riveredge, NJ, 1992.
41. See, e.g., sections 1.6 and 12.8 in R.J. Hunter, *Foundations of Colloid Science*, Oxford University Press, New York, 2001, and section 6.5 in G. Sposito, *Chemical Equilibria and Kinetics in Soils*, Oxford University Press, New York, 1994, for an introductory discussion of W. Modeling of W in terms of particle interactions is discussed in chapter 8 of W.B. Russel et al., *Colloidal Dispersions*, Cambridge University Press, New York, 1989.
42. See, e.g., S. Goldberg and R.A. Glaubig, *Clays Clay Miner.* **35**:220 (1987); S. Goldberg, D.L. Suarez, and R.A. Glaubig, *Soil Sci.* **146**:317 (1988); S. Goldberg, B.S. Kapoor, and J.D. Rhoades, *Soil Sci.* **150**:588 (1990); L. Liang and J.J. Morgan, *Aquatic Sci.* **52**:32 (1990); S. Goldberg and H.S. Forster, *Soil Sci. Soc. Am. J.* **54**:714 (1990); S. Goldberg, H.S. Forster, and E.L. Heick, *Clays Clay Miner.* **39**:375 (1991); D. Heil and G. Sposito, *Soil Sci. Soc. Am. J.* **57**:1241, 1246 (1993); R. Kretzschmar, W.P. Robarge, and S.B. Weed, *Soil Sci. Soc. Am. J.* **57**:1277 (1993); R. Kretzschmar, D. Hesterberg, and H. Sticher, *Soil Sci. Soc. Am. J.* **61**:101 (1997) for data on ccc values of specimen and soil particles under a variety of experimental protocols, including the effects of humus.
43. See, e.g., section 12.6.1 in R.J. Hunter, *Foundations of Colloidal Science*, Oxford University Press, New York, 2001, which features the standard derivation of eq. 5.65 (see note 22) based on a model of particle-particle interactions requiring adjustable parameters.
44. See V.A. Hackley and M.A. Anderson, *Langmuir* **5**:191 (1989) (*goethite*); M.Y. Lin et al., *Phys. Rev. A* **41**: 2005 (1990) (*gold, latex, silica*); J. Zhang and J. Buffle, *Colloids Surf.* 107A:175 (1996), and J. Chorover et al., *Clays Clay Miner* **45**:690 (1997) (*hematite*); L. Derrendinger and G. Sposito, *J. Colloid Interface Sci.* **222**:1 (2000) (*illite*); D. Asnaghi et al., *Phys. Rev. A* **45**:1018 (1992), and J.L. Burns, Y.-d. Yan, G.J. Jameson, and S. Biggs, *Langmuir* **13**:6413 (1997) (*latex*); D. Schaefer, J.E. Martin, P. Wiltzius, and D.S. Cannell, *Phys. Rev. Lett.* **52**:2371 (1984) (*silica*).
45. The fractal dimension also has been observed to depend on $n(1,0)$ [or, equivalently, $M_1(t)$], decreasing as $n(1,0)$ increases for rapid coalescence and increasing as $n(1,0)$ increases for slow coalescence processes. See C. Aubert and D.S. Cannell, *Phys. Rev. Lett.* **56**:738 (1986); M. Carpineti, F. Ferri, M. Giglio, E. Paganini, and U. Perini, *Phys. Rev. A* **42**:7347 (1990); L. Derrendinger and G. Sposito, *J. Colloid Interface Sci.* **222**:1 (2000).
46. A.E. González, *Phys. Rev. Lett.* **71**:2248 (1993).
47. The standard reference on van der Waals interactions between colloidal particles is J. Israelachvili, *Intermolecular and Surface Forces*, Academic Press, San Diego,

CA, 1992. See also C.J. van Oss, *Interfacial Forces in Aqueous Media*, Marcel Dekker, NY, 1994, and chapter 5 in W.B. Russel et al., *Colloidal Dispersions*, Cambridge University Press, New York, 1989.
48. The chemical variables that control the stability ratio are discussed in exemplary fashion by L. Liang and J.J. Morgan, *Aquat. Sci.* **52**:32 (1990). The stability ratio *field* (i.e., W as a function of both electrolyte concentration and pH) is described graphically for the first time in J.J. Morgan, *Faraday Disc. Chem. Soc.*, No. **90**:371 (1990).
49. L. Liang and J.J. Morgan, pp. 293–308 in *Chemical Modeling of Aqueous Systems II*, D.C. Melchior and R.L. Bassett (Eds.), American Chemical Society, Washington, DC, 1990.
50. R. Amal, J.A. Raper, and T.D. Waite, *J. Colloid Interface Sci.* **151**:244 (1992).
51. R. Ferretti, J. Zhang, and J. Buffle, *Colloids Surf.* **121A**:203 (1997).
52. R. Ferretti, J. Zhang, and J. Buffle, *J. Colloid Interface Sci.* **208**:509 (1998).
53. A comprehensive discussion of the three transport control models appears in chapters 4 and 5 of T.G.M. van de Ven, *Colloidal Hydrodynamics*, Academic Press, San Diego, 1989. Brief discussions are given in chapter 8 of W.B. Russel et al., *Colloidal Dispersions*, Cambridge University Press, New York, 1989, and in M. Filella and J. Buffle, *Colloids Surf. A* **73**:255 (1993), who also compare the models as to their role in natural particle coalescence processes. Chapter 6 in M. Elimelech, J. Gregory, X. Jia, and R.A. Williams, *Particle Deposition and Aggregation*, Butterworth-Heinemann, Oxford, 1995, also surveys the three models and their applicability to the engineering design of coagulation processes. A review emphasizing simulations is given by K.H. Gardner, pp. 509–550 in *Interfacial Forces and Fields*, J.-P. Hsu (Ed.), Marcel Dekker, New York, 1999.
54. Simplified derivations of the rate coefficients for perikinetic and orthokinetic aggregation are given in section 12.8 of R.J. Hunter, *Foundations of Colloidal Science*, Oxford University Press, 2001. Note that the rate coefficient for dimer formation is one-half that for monomer consumption.
55. See, e.g., H. Holthoff et al., *Langmuir* **12**:5541 (1996); J.M. Peula, A. Fernández-Barbero, R. Hidalgo-Álvarez, and F.J. de las Nieves, *Langmuir* **13**:3938 (1997); S.H. Behrens, M. Borkovec, and P. Schurtenberger, *Langmuir* **14**:1951 (1998); and S.H. Behrens, D.I. Christl, R. Emmerzael, P. Schurtenberger, and M. Borkovec, *Langmuir* **16**:2566 (2000) for illustrative calculations of W_{11}^{H} based in fluid mechanics.
56. Fragmentation processes are discussed by E. Pefferkorn and J. Widmaier [*Colloids Surf. A* **145**:25 (1998)], who demonstrate their importance as aggregates grow larger and under conditions of decreasing ionic strength.
57. A comprehensive discussion of eq. 5.79 is given by R.L. Drake, pp. 201–376 in *Topics in Current Aerosol Research*, G.M. Hidy and R.L. Brock (Eds.), Vol. 3, Part 2, Pergamon, Oxford, 1972. An introductory discussion emphasizing physical applications is in chapter 10 of J.H. Seinfeld, *Atmospheric Chemistry and Physics of Air Pollution*, John Wiley, New York, 1986.
58. G. Sposito, *Colloids Surf. A* **120**:101 (1997).
59. The factor 2 in the relation between $k(m, n)$ and K_{SE} can be appreciated after consideration of eq. 5.79 for dimer formation ($q = 2, m = n = 1$, no dimer loss terms) and comparison of the result with eq. 5.33: $k = \frac{1}{2}k(1, 1)$.
60. See chapter 6 in G.M. Hidy and R.L. Brock (Eds.), *Topics in Current Aerosol Research*, Pergamon, Oxford, 1972, for a detailed mathematical discussion of the solution of eqs. 5.94 and 5.116 under an arbitrary initial condition on $n(q, t)$. Brief physically motivated discussions of the methods for solving eq. 5.94 are given by W.T. Scott, *J. Atmos. Sci.* **25**:54 (1968) and by R.M. Ziff, pp. 191–199 in *Kinetics of Aggregation and Gelation*, F. Family and D.P. Landau (Eds.), Elsevier, Amsterdam, 1984. The latter article is the basis for the presentation in section 5.5.

61. A. Fernández-Barbero et al., *Physica A* **250**:53 (1996). See also H. Holthoff et al., *J. Colloid Interface Sci.* **192**:463 (1997). Note that k_{11} in these articles is the same as $k(1, 1)$ in the present chapter.
62. E.M. Hendriks and R.M. Ziff [*J. Colloid Interface Sci.* **105**:247 (1985)] consider eq. 5.106 without q in the loss term and with a floccule source term. Note that eq. 5.82 is no longer a constraint on the solutions of eq. 5.106.
63. See, e.g., pp. 415f in J.H. Seinfeld, *Atmospheric Chemistry and Physics of Air Pollution*, John Wiley, New York, 1986, for a description of this method of solving eq. 5.109.
64. See, e.g., chapter 4 in Abramowitz and Stegun, *Handbook of Mathematical Functions*, Dover, New York, 1972.[20]
65. D.A. Weitz, J.S. Huang, M.Y. Lin, and J. Sung, *Phys. Rev. Lett.* **54**:1416 (1985) (*gold*); M.Y. Lin et al., *Proc. R. Soc. Lond. A* **423**:171 (1989), *J. Phys. Condens. Matter* **2**:3093 (1990), (*latex, silica,* and *gold*); R. Ferretti et al., *J. Colloid Interface Sci* **208**:509 (1996) (*hematite*); L. Derrendinger and G. Sposito, *J. Colloid Interface Sci.* **222**:1 (2000) (*illite*). Exponential growth of $\langle q_t \rangle$ also is observed in computer simulation of slow coalescence processes [A.E. González, *Phys. Rev. Lett.* **71**:2248 (1993)].
66. See M.Y. Lin et al., *Proc. R. Soc. Lond. A* **423**:71(1989) and G. Sposito, *Colloids Surf. A* **120**:101 (1997) for a discussion of the physical criteria that differentiate rapid from slow coalescence processes.
67. This model rate coefficient was postulated for slow coalescence processes by R.C. Ball, D.A. Weitz, T.A. Witten, and F. Leyvraz, *Phys. Rev. Lett.* **58**:274 (1987).
68. D.A. Weitz and M.Y. Lin, *Phys. Rev. Lett.* **57**:2037 (1986); M.Y. Lin et al., *Proc. R. Soc. Lond. A* **423**:71 (1989); M.L. Broide and R.J. Cohen, *Phys. Rev. Lett.* **64**:2026 (1990); A.E. González, *Phys. Rev. Lett.* **71**:2248(1993).

For Further Reading

J. K. G. Dhont, *An Introduction to Dynamics of Colloids,* Elsevier: Amsterdam, 1996. The first four chapters of this comprehensive textbook on the fundamental physics of colloidal suspensions provide detailed accounts of photon scattering and the kinetics of flocculation.

K. H. Gardner in *Interfacial Forces and Fields*, J.-P. Hsu, Ed., Marcel Dekker: New York, 1999, pp. 509–550. This chapter presents a useful summary of the kinetics of flocculation as investigated through the von Smoluchowski rate law and computer simulation, with emphasis on fractal floccules.

E. Kissa, *Dispersions*, Marcel Dekker: New York, 1999. Chapters on the instrumentation for investigating flocculation by wet-chemical and photon-scattering techniques are included in this comprehensive monograph.

C. M. Sorensen, *Aerosol Sci. Technology* **35**:648 (2001). This review of photon scattering by fractal aerosols provides detailed insight into the modeling and measurement of $I(\boldsymbol{Q}, t)$ in eq. 5.6.

T. Vicsek, *Fractal Growth Phenomena*, World Scientific, Singapore: 1992. This book, a standard reference on fractal floccules, reviews in detail the concept of a mass fractal and the computer models that simulate mass fractals.

Research Matters

1. A convenient mathematical model of $S_q(Q)$ in eq. 5.6 is the *Fisher-Burford* expression: $S_q^{FB}(Q) = [1 + 2(QR_{Gq})^2/3D]^{-D/2}$ where R_{Gq} is the radius of gyration of a fractal floccule (eq. 5.11) and D is its mass fractal dimension (eq. 5.17). (This model is discussed in the review article by C. M. Sorensen listed under *For Further Reading*.) M. Carpinetti et al. [*Phys. Rev. A* **42**:7347 (1990)] have applied

the model to interpret photon scattering data on suspensions of latex colloids ($r_0 = 65$ nm, $n_s = 1.58$) in 30 mol m^{-3} $MgCl_2$ solution, which leads to rapid coalescence. This process was investigated for a range of $n(1, 0) = M_1$ values, 10^{15} to 5×10^{16} m^{-3}.

(a) Show that $S_q^{FB}(Q)$ mimics the expressions for the structure factor in eqs. 5.10 and 5.24 under the appropriate conditions on QR_{Gq}.

(b) Examine how well the conditions in eqs. 5.1, 5.2b, 5.6 [$(Qr_0)^4 \ll 1$, as discussed in note 10], and 5.14 were met in the experiments reported by Carpinetti et al.

(c) Explain carefully the basis for determining the fractal dimension by: (1) a log-log plot of $I(Q, t)$ against Q at fixed t (fig. 3 in Carpinetti et al.) and (2) a log-log plot of $I(0, t)$ against $R_{GZ}(t)$ at fixed t (fig. 7 in Carpinetti et al.)

(d) Explain why a log-log plot of $R_{GZ}(t)$ against the variable $T \equiv M_1 t$ should result in a single straight line for data collected on suspensions with differing $n(1, 0)$ that are undergoing rapid coalescence (fig. 8 in Carpinetti et al.). What is the physical interpretation of the slope of this line?

2. The *turbidity* of a colloidal suspension is equal to the integrated intensity of scattered photons over all possible angles of scattering. (This property of a colloidal suspension is discussed in the references cited in note 4.) Turbidity depends on the wavelength of the incident radiation in vacuo (λ_0) because of photon scattering by the individual particles in a floccule and because of scattering by the floccule structure as epitomized in $S_q(Q)$. The wavelength dependence of the individual-particle contribution to scattering is λ_0^{-4} if the refractive index of the suspension does not vary with wavelength; otherwise it is $\lambda_0^{-4+\gamma}$, where γ is a small parameter introduced to account for dispersion in the refractive index. The wavelength dependence caused by fractal structure in turn depends on the model form of $S_q(Q)$. A convenient choice is the Fisher-Burford expression (see Problem 1), which leads to the turbidity model equation: $\tau = \tau^* k^{4-\gamma}[1 + 4k^2 R_{GZ}^2/3D]^{-D/2}$ where k is the wavenumber of the incident radiation (eq. 5.4), R_{GZ} is the z-average radius of gyration of the floccules in suspension (eq. 5.13), D is their fractal dimension (eq. 5.57), and τ^* is a parameter that depends on the floccule concentration in the suspension and other properties unrelated to wavelength. (This turbidity model is discussed in the article by C. M. Sorensen listed under *For Further Reading*.)

Inspired by a semi-empirical correlation reported by D. S. Horne [*Faraday Discuss. Chem. Soc.* **83**:259 (1987)], Senesi et al. [*Soil Sci. Soc. Am. J.* **60**:1773 (1996)] measured the wavelength dependence of the turbidity of humic acid suspensions over the pH range 3 to 7 and calculated the mass fractal dimension of the humic acid floccules using the Horne correlation equation: $D = d \ln \tau / d \ln \lambda_0 + 4.2$.

(a) Give a derivation of the Horne correlation equation based on the Fisher-Burford model of turbidity.

(b) Examine how well the conditions required for your derivation in (a) were met in the experiments reported by Senesi et al. Apply your conclusions to the values of D listed in tables 1 and 2 in their article.

3. Thermal neutrons emitted from a nuclear reactor have de Broglie wavelengths that are about two orders of magnitude smaller than those of visible photons, but their coherent scattering by floccules can be interpreted as done for photon scattering if their scattering angles are small enough to produce a similar range of scattering wavenumbers (eq. 5.4) and, therefore, a similar range of length scales probed (eq. 5.5). In practice, this condition restricts θ to be smaller than about 5° (*small-angle neutron scattering*, SANS). (Coherent neutron scattering is discussed in the article by J. Teixeira cited in note 17.)

Österberg and Mortensen [*Eur. Biophys. J* **21**:163 (1992); *Radiation Environ. Biophys.* **33**:269 (1994); *Humic Substances in the Global Environment and Implications on Human Health*, N. Senesi and T. M. Miano, Eds., Elsevier:

Amsterdam, 1994, pp. 127–132; *Naturwissenschaften* **82**:137 (1995)] have applied the Fisher-Burford model (see problem 1) to estimate the mass fractal dimension of humic acid suspensions at pH 5 and varying solid mass concentrations.

(a) Examine how well the conditions for the validity of eqs. 5.12 and 5.25 were met in the experiments reported by Österberg and Mortensen. Interpret their parameters ξ_1 and ξ_2 [see eqs. (4) and (5) in the first article cited] in light of your examination.

(b) Develop an equation for $I(Q, t)$ that generalizes eqs. 5.12 and 5.25 to incorporate the Fisher-Burford model as applied by Österberg and Mortensen. Use this equation to explain why $I(Q, t)$ data in the fractal regime taken at different solids concentrations [fig. 1A in the *Naturwissenschaften* article] superimpose to yield a single curve [fig. 1B in the *Naturwissenschaften* article] after they are divided by the ratio of the solids concentration to that in the most concentrated suspension.

4. (a) Show that the solution of eq. 5.93 is $\langle q_t \rangle = \langle q_{to} \rangle [1 + (t - t_o)/t_c]^z$ where $z = (1 - \theta)^{-1}$, $t_c = z[M_2(t_o)]^{\frac{1}{2}}/M_1^{2-\theta}K$, K is the integral in eq. 5.93, and t_o is the time after which the scaling form of $n(q, t)$ in eq. 5.56 becomes accurate. This equation is a generalization of eq. 5.105a for an arbitrary initial condition and scaling exponent $\theta < 1$. D. Asnaghi et al. [*Phys. Rev. A* **45**:1018 (1992)] have measured $\langle q_t \rangle$ by photon scattering from suspensions of latex colloids ($r_0 = 65$ nm, $M_1 = 5 \times 10^{16}$ m^{-3}) in $MgCl_2$ solutions whose concentrations varied in the range 10 to 30 mol m^{-3}. Their log-log plots of $\langle q_t \rangle$ against time were linear (their fig. 2), allowing them to infer the value of θ as the suspending solution concentration varied (their fig. 3). Examine the experimental results of Ansaghi et al. in light of the equation for $\langle q_t \rangle$ above and discuss the significance of their finding θ to increase as the solution concentration decreased.

   (b) An alternative form of eq. 5.56 has $\bar{q}_t$ (eq. 5.85) replacing $\langle q_t \rangle$ Show that the corresponding alternative form of eq. 5.93 is $d\bar{q}_t/dt \approx 1/2M_1\bar{q}_t^{\theta} \int_0^{\infty} \int_0^{\infty} \kappa(x, y) \psi(x)\psi(y)dxdy$ where the scaling variable in the integral is now $\bar{q}_t$ B. J. Olivier, C. M. Sorensen, and T. W. Taylor [*Phys. Rev. A* **45**:5614 (1992)] have derived the solution of this differential equation for arbitrary $\theta < 1$ [their eq. (8)] and applied it to measured values of $\bar{q}_t$ for an aerosol, assuming that $\theta = 0$ and $t_o = 0$ (their figs. 7 and 8). Derive an equation for the parameter t_c under the assumptions made by Olivier et al. and explain how their data on $\bar{q}_t$ then can be used to estimate the rate coefficient for flocculation.

5. M. Schudel et al. [*J. Colloid Interface Sci.* **196**:241 (1997)] have measured R_{HZ}(Q,t) (eq. 5.41) for a flocculating hematite suspension at varying pH values ($d_o = 100$ nm, I = 10 mol m^{-3} $NaNO_3$, 25°C, $Q = 0.0183$ nm^{-1}). Their data (fig. 1 in their article) show a marked increase in the growth of R_{HZ}(0.0183, t) as pH increases toward the p.z.c. (9.2). In principle, these data can be simulated using eq. 5.41 and expressions for $n(q, t)$ available from analytical solutions of the von Smoluchowski rate law (eqs. 5.94 and 5.116) for rapid and slow coalescence processes.

   (a) Calculate $R_{HZ}(0.0183, t)$ at pH 9.1, retaining only two terms in eq. 5.41. Take $M_1 = 8.2 \times 10^{14}$ m^{-3} and $R_{HZ} = 1.38\ R_{H1}$, where $R_{H1} = 70$ nm.

   (b) Calculate R_{HZ}(0.0183,t) at pH 6.1, again retaining only two terms in eq. 5.41. Take $M_1 = 2.1 \times 10^{15}$ m^{-3}.

   For both simulations, use the rate coefficient data in fig. 8a of the article by Schudel et al. Note that their rate coefficients are the same as $k(1, 1)$ in eq. 5.79.

Index

adsorbate, 3
  structure, 11
adsorbent, 3
adsorption, 3
  affinity classes, 157–170
  and anion protonation, 37, 85
  and crystal growth, 11–12, 17
  and metal ion hydrolysis, 36, 85
  complex, 3
  edge, 6, 40, 43, 153–154
  envelope, 6, 41, 153–154
  experiment, 3
  ligand-like, 8, 36, 40–41, 78
  metal-like, 8, 40
  negative, 3, 23 (*See also* coion)
  nonspecific, 33
  positive, 3, 23 (*See also* counterion)
  specific, 33–34, 85–94, 123
  temperature effects, 148, 181
adsorption isotherm, 6
  L-curve model, 9, 161, 181
  Langmuir-Freundlich, 165–169
  regularization, 160
  site affinity model, 158, 162, 166–169, 181
  temperature effects, 155
  Tóth, 10, 41
  two-site Langmuir, 9, 41, 155, 158, 164, 181
  types, 7
adsorptive, 3
affinity spectrum, 158, 162–163, 181
  Sips, 164, 181
anion exchange capacity (AEC), 23
apparent activation energy, 84, 105, 124–125
Arrhenius equation, 84, 104–106, 124

beidellite, 14
Bohr magneton, 62
  nuclear, 66
bond strength, 19
bond valence, 19–20, 41, 77

cation exchange capacity (CEC), 23, 30
chemical shielding, 66–68, 78
  mean principal, 66–67
  tensor, 66, 68
chemical shift, 66, 69–71, 78
chi-square statistic, 159–161
Chorover plot, 25–26

critical coagulation concentration (ccc), 204–205
clay minerals, 14–15
coion, 126
  exclusion, 137, 139, 166
colloid, 182
  coalescence kinetics, 193–202, 211–215
  diffusion coefficient, 196, 212–213
  size range, 193
compensation law, 105, 124
counterion, 126
  condensation, 137–138, 140

delta-"function", 158, 194
differential sedimentation model, 211–212
diffraction, neutron, 49
diffuse ion swarm, 126–140, 166 *See also* surface charge, diffuse layer
distribution coefficient, 8–9, 41

Eigen-Wilkins-Werner mechanism, 85, 91
electron spin resonance (ESR) spectroscopy, 60–65
  ENDOR, 64, 78
  ESEEM, 63, 65, 78
elementary reaction, 82, 92–93
energy level diagram, 62, 66
equilibrium species, 84
EXAFS, 50–52, 54, 77
exclusion distance, 139, 166, 180
exclusion volume, 139, 166
extent of reaction, 79, 102–104, 124, 194

ferrihydrite, 16–17
fractal dimension,191
  and turbidity, 236
  dependence on stability ratio, 206
  rapid coalescence, 202, 235–236
  slow coalescence, 206
fractal object, 189, 198

generalized homogeneity, 199, 217, 223
gibbsite, 16
goethite, 16
Guinier plot, 188

indifferent electrolytes, 32, 146
infrared spectroscopy, 55–59, 78
inner-sphere surface complex. *See* surface complex
intrinsic equilibrium constant, 142, 145, 180–181
  correlations, 147
isocoulombic reaction, 152, 156, 181
isoelectric point, 26, 29–30
isokinetic temperature, 106, 124

kaolinite, 15
kinetic species, 83–84, 86, 93, 124–125
Kirchoff equation, 149
Kurbatov plot, 6, 8, 41

Larmor frequency, 64, 66, 78
layer charge, 14, 21, 30
linear free energy relationship (LFER), 83, 99–100, 109, 115

magic angle spinning (MAS), 67, 69–71, 78
Marcus process, 94
mass-weighted average, 200, 221, 237
metal oxides, 15–16
mineral dissolution reactions, 101–115
  electron density effects, 108
  inhibition, 109–110
  ligand-promoted, 106
  mechanisms, 101–102
  oxidative, 112
  proton-promoted, 101
  reductive, 112, 125
modified Bessel function, 133
modified Gouy-Chapman theory, 127–140, 166
  charge-potential relationship, 144
mole fraction, 4
montmorillonite, 14

nuclear magnetic resonance (NMR) spectroscopy, 65–71, 78, 85
number-weighted average, 216, 221, 237

orthokinetic aggregation, 211
outer-sphere surface complex. *See* surface complex
overall reaction, 82

Pauling rules, 18
PBE theorems, 130–133
perikinetic aggregation, 211
  analytical solution, 219–222
  and fractal structure, 218

photon scattering, 183–186
  coherent, 186
  elastic, 184
  intensity, 186
  length scale, 186, 192
  momentum change, 186
  scaling exponents, 199–200
  wavenumber, 186
polydispersity, 216, 221
points of zero charge, 25–30, 40–41, 146, 180, 207–208
  relation to stability ratio, 207–208
  temperature dependence, 149–152
Poisson-Boltzmann equation, 127–128
  analytical solution, 135
  Debye-Hückel approximation, 133–134
  intrinsic scales, 129–130, 133, 136
  Monte Carlo simulations, 136–137
potential of mean force, 127–128
pseudo first-order rate coefficient, 82, 96
PZC theorems, 30–33
  and surface complexes, 20, 41
  and Triple Layer Model, 146

q-moment, 200, 215–216, 220–221

radial distribution function, 51, 186
radius, hydrodynamic, 196
  and radius of gyration, 201
  time dependence, 201–202, 209, 211, 237
  "z-average", 196, 237
radius of gyration, 187, 201, 235
  fractal object, 193
  "z-average", 187
rate coefficient, 82
  composite, 95–96, 100–101, 124
  dimer formation, 194, 198, 203, 212–214
  electron transfer, 95, 125
  homogeneous, 223
  inner-sphere surface complexation, 86
  mineral dissolution, 102
  q-mer formation, 214, 217, 237
  rapid coalescence, 211–214
  scale-invariant, 218
  slow coalescence, 224
  water exchange, 85, 90, 109
rate law, 81, 86, 123
  dimer formation, 194
  linearized, 87–88, 93, 95, 123
  mineral dissolution, 102–104, 113
  von Smoluchowski, 215
reaction order, 81
  and molecularity, 82
reaction rate, 79, 103
readily exchangeable ions, 33
regularization, 160

scaling exponent, 199
  constraint equation, 217
  number density, 199
  rate coefficient, 217, 237
  time, 201
Schindler diagram, 33–37, 41
  resonances, 7, 35–36
Schulze-Hardy rule, 205
site affinity, 10 *See also* affinity spectrum
  distribution function (SADF), 41, 157–158
small-angle neutron scattering (SANS), 236–237
smectite, 15
sorption, 11
  isotherm, 12
specific adsorption. *See* adsorption
spectroscopic methods, 43–49
  adsorbate surrogates, 45
  diffusionally averaged states, 48
  molecular time scales, 46–47
  criteria, 45
  reporter units, 45
  vibrationally averaged states, 48
stability ratio, 202–211
stoichiometry, 82
Stokes-Einstein equation, 196, 211
structure factor, 186
  Fisher-Burford model, 235–237
  fractal object, 192
  Guinier approximation, 187
surface charge, 20
  anion exchange capacity, 23
  apparent net adsorbed proton, 22, 28–29
  balance, 24, 28, 32, 144, 166
  cation exchange capacity, 23
  diffuse layer, 24, 144, 166
  intrinsic, 24
  net adsorbed ion, 20
  net adsorbed proton, 20, 26, 29, 40
  net total particle, 24
  neutralization, 207

screening, 204–205
Stern layer, 24, 32, 141
structural, 20
surface complex, 12
crystal growth and, 17
inner-sphere, 12, 17, 18, 43–44, 57, 59, 63, 65, 78, 86, 141
models, 140–148
multinuclear, 12
outer-sphere, 12–13, 43–44, 54, 57, 59, 68–69, 86, 89, 141
Pauling rules, 20, 55, 77
temperature effects, 148–157
ternary, 94, 106
surface-controlled coalescence, 211, 223–226
surface-controlled mineral dissolution, 101
surface excess, 4
isotherm, 6
surface functional groups, 18
surface Mössbauer spectroscopy, 48
surface oxidation-reduction (redox), 94–101, 124
electron density effects, 98–99, 101
generic rate law, 100, 124
surface precipitate, 12

Taube process, 94
thermodynamic stability, 31
time constant, 87–94, 103, 195
total adsorbed cation (anion) charge, 21
total net particle charge, 24
and electrophoretic mobility, 29
transport-controlled coalescence, 211–214, 219–223, 226
transport-controlled mineral dissolution, 101
Triple Layer Model, 141–148, 180
charge-potential relationships, 143
interfacial capacitance densities, 143–144
surface potentials, 143
turbidity, 236

van't Hoff equation, 148
von Smoluchowski rate law, 211–219
analytical solutions, 219–226
scaling properties, 217
with gravitational settling, 222

wavenumber vector, 183

XANES, 50, 53
X-ray absorption spectroscopy (XAS), 49–55